W0106656

Springer-Verlag Berlin Heidelberg GmbH

Yuk-Shan Wong   Nora F.Y. Tam   (Eds.)

Wastewater Treatment with Algae

 Springer

Yuk-Shan Wong

The Hong Kong University
of Science and Technology
Research Centre
Clear Water Bay, Kowloon
Hong Kong

Nora F. Y. Tam

The City University of Hong Kong
Dept. of Biology and Chemistry
Tat Chee Avenue, Kowloon
Hong Kong

Library of Congress Cataloging-in-Publication Data

Wastewater treatment with algae / [edited by] Yuk-Shan Wong, Nora F.Y. Tam.
p. cm. - (Environmental intelligence unit)
Includes bibliographical references and index.
ISBN 978-3-662-10865-9   ISBN 978-3-662-10863-5 (eBook)
DOI 10.1007/978-3-662-10863-5
1. Sewage - Purification - Phosphorus removal. 2. Chlorella. 3. Immobilized microorganisms - Industrial
applications. I. Wong, Yuk-Shan. II. Tam, Nora F.Y. III. Series.
TD758.5.P56W37 1997
628.3'5 - DC21   97-18167 CIP

This work is subject to copyright. All rights are reserved, whether the whole or part of the material
is concerned, specifically the rights of translation, reprinting, reuse of illustrations, recitation,
broadcasting, reproduction on microfilms or in any other way, and storage in datas banks. Duplication
of this publication or parts thereof is permitted only under the provisions of the German Copyright
Law of September 9, 1965, in its current version, and permission for use must always be obtained
from Springer-Verlag Berlin Heidelberg GmbH.
Violations are liable for prosecution under the German Copyright Law.

© Springer-Verlag Berlin Heidelberg 1998
Originally published by Springer-Verlag Berlin Heidelberg New York in 1998
Softcover reprint of the hardcover 1st edition 1998

The use of general descriptive names, registered names, trademarks, etc. in this publication does
not imply, even in the absence of a specific statement, that such names are exempt from the
relevant protective laws and regulations and therefore free for general use.

Product liability: The publishers cannot guarantee the accuracy of any information about dosage
and application contained in this book. In every individual case the user must check such information
by consulting the relevant literature.

Typesetting: Landes Bioscience Georgetown, TX, U.S.A.

SPIN: 10637956   31/3111 - 5 4 2 3 2 1 0 - Printed on acid-free paper

PREFACE

Untreated or improperly treated wastewater has become a major pollution problem for rivers and coastal waters. A related problem is our inability to recover its useful components (i.e., the nutrients contained in the wastewater) such that we pay expensive cost to manufacture fertilizer. Conventional approaches such as activated sludge systems involve tremendous capital investment and impose major shortcomings such as inefficient removal of phosphate, unpleasant odor and foaming problems.

Recent years have witnessed an exponential increase of knowledge in generating innovative treatment technology. The objective of publishing this book is thus to bring together scientists and engineers intimately involved in this field to discuss state-of-the-art research on the application of algae for wastewater treatment. Authors of this book describe different approaches and methodologies utilizing microalgae for removing nutrients, heavy metals and other organic pollutants in sewage and industrial effluents. This book also includes biosorption, bioaccumulation and biotransformation of metals in living and non-living algal cells; and cell immobilization technology for algae and induction of metal binding complexes such as metallothionein in metal resistant algal species. Modelling of metal ion uptake to predict the adsorption and membrane transport process of the living algal cells is also discussed.

This book will be invaluable for researchers and engineers in environmental sciences and algal biotechnology to keep abreast of the subject. Additionally, advanced research students of biology and environmental engineering will find it a useful source of information.

We would like to take this opportunity to thank all the contributing authors and the staff of Landes Bioscience for their valuable support and cooperation during the development and publication of this book. We also like to thank Miss Ellen M.Y. So for her help in preparing some of the manuscripts.

Yuk-Shan Wong
Nora F.Y. Tam
June, 1997

CONTENTS

EDITORS

Yuk-Shan Wong
Research Centre
Hong Kong University of Science and Technology
Hong Kong
Chapters 2 and 9

Nora F.Y. Tam
Department of Biology and Chemistry
City University of Hong Kong
Hong Kong
Chapters 2 and 9

CONTRIBUTORS

Zümriye Aksu
Department of Chemical
　Engineering
Hacettepe University
Beytepe, Ankara, Turkey
Chapter 3

Simon V. Avery
Department of Biology
Georgia State University
Atlanta, Georgia, U.S.A.
Chapter 4

Michael A. Borowitzka
School of Biological
　and Environmental Sciences
Murdoch University
Perth, W.A., Australia
Chapter 12

Geoffrey A. Codd
Department of Biological Sciences
University of Dundee
Dundee, Scotland, U.K.
Chapter 4

Marli F. Fiore
Molecular Biology Laboratory
Centro de Energia Nuclear
　na Agricultura
University of São Paulo
Piracicaba, SP, Brazil
Chapter 7

Geoffrey M. Gadd
Department of Biological Sciences
University of Dundee
Dundee, Scotland, U.K.
Chapter 4

P.S. Lau
Research Centre
Hong Kong University of Science
　and Technology
Clear Water Bay, Hong Kong
Chapter 9

Frank Lawson
Department of Chemical
　Engineering
Monash University
Clayton, Victoria, Australia
Chapter 8

Shigeru Maeda
Department of Applied Chemistry
　and Chemical Engineering
Kagoshima University
Korimoto, Kagoshima, Japan
Chapter 5

Nour-Eddine Mezrioui
Faculte des Sciences
　Université Cadi Ayyad
Semlalia, Bd Le Prince My Abdallah
Marrakech, Morocco
Chapter 10

David H. Moon
Molecular Biology Laboratory
Centro de Energia Nuclear
na Agricultura
University of São Paulo
Piracicaba, SP, Brazil
Chapter 7

Akira Ohki
Department of Applied Chemistry
and Chemical Engineering
Kagoshima University
Korimoto, Kagoshima, Japan
Chapter 5

Brahim Oudra
Faculté des Sciences
Université Cadi Ayyad
Semlalia, Bd Le Prince My
Abdallah
Marrakech, Morocco
Chapter 10

Ian G. Prince
Department of Chemical
Engineering
Monash University
Clayton, Victoria, Australia
Chapter 8

Gerald J. Ramelow
Department of Chemistry
McNeese State University
Lake Charles, Louisiana, U.S.A.
Chapter 6

Peter K. Robinson
Department of Applied Biology
University of Central Lancashire
Preston, U.K.
Chapter 1

Craig G. Simpson
Department of Biology
and Chemistry
City University of Hong Kong
Hong Kong
Chapter 2

Y.P. Ting
Department of Chemical
Engineering
National University of Singapore
Kent Ridge, Singapore
Chapter 8

Jack T. Trevors
Ontario Agricultural College
University of Guelph
Guelph, Ontario, Canada
Chapter 7

Xianghua Wen
Department of Environmental
Engineering
Tsinghua University
Beijing, People's Republic of China
Chapter 11

Hua Yao
Department of Chemistry
McNeese State University
Lake Charles, Louisiana, U.S.A.
Chapter 6

Wei Zhuang
Department of Chemistry
McNeese State University
Lake Charles, Louisiana, U.S.A.
Chapter 6

CHAPTER 1

Immobilized Algal Technology for Wastewater Treatment Purposes

Peter K. Robinson

General

This chapter comprises a brief review of the current literature dealing with the use of immobilized algae for wastewater treatment purposes, together with results from some of our own experimental studies on immobilized *Chlorella emersonii* for the removal of phosphate-phosphorus (PO_4-P) from waste waters. The intention of including these results is to highlight some of the practical advantages of using immobilized systems, and also to introduce some of the problems and limitations of such systems at the laboratory scale. Consideration of the scale-up of such processes is also included.

Immobilization

Immobilization may be defined as "*the accumulation of a biocatalyst (cell or enzyme) either on surfaces or within particles*" (based on ref. 1). Therefore the fundamental characteristic of an immobilized system is that the biocatalyst is fixed within a distinct phase (the biocatalyst phase) which is separate from the bulk phase, in which the substrates and ultimately the products are dissolved (Figure 1.1a). This system heterogeneity gives rise to two major advantages of immobilized systems over their non-immobilized (free) counterparts:

- A physical advantage in that the biocatalyst and product may be separated easily, enabling effective product recovery (downstream processing) and biocatalyst re-use.
- A biological advantage in that the immobilized biocatalyst is often found to be more stable than the non-immobilized form (where stability may mean an enhanced rate of catalysis, prolonged duration of catalysis, or greater operational stability to extremes of pH or temperature).

Wastewater Treatment with Algae, edited by Yuk-Shan Wong and Nora F.Y. Tam.
© Springer - Verlag and Landes Bioscience 1998.

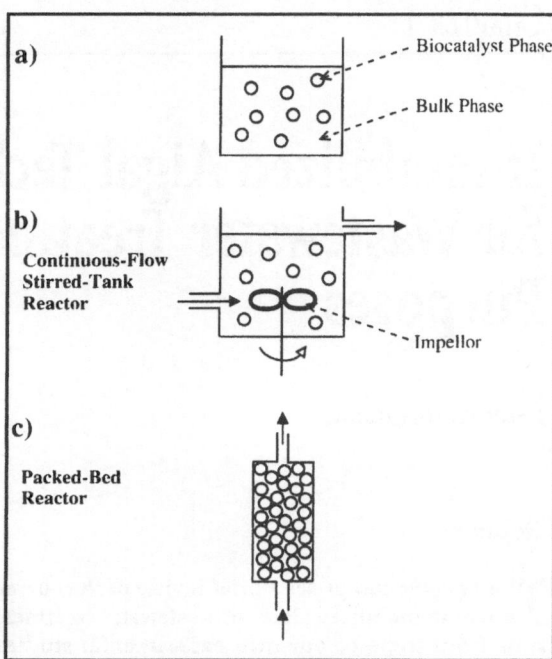

Fig. 1.1. Basic features of (a) a model immobilized system, (b) a CSTR, and (c) a PBR.

Even in cases where there is no real biological advantage conferred by immobilization, the physical advantage of ease of separation of biocatalyst from product may be sufficient to favor implementation of an immobilized system in a variety of industrial situations. Indeed, such systems are usually developed to place the biocatalyst in a continuous-flow reactor of the continuous-flow stirred-tank (CSTR) or packed-bed (PBR) type (Fig. 1.1b and 1.1c) where the catalyst is continuously being reused. In many industrial situations such enzyme reactors are considered to be "new technology" though PBRs have been employed for many years in the field of wastewater treatment in the form of percolating (trickling) filters.

Immobilized Algae

The first paper involving the study of immobilized algae was published in 1969,[2] though it was not until the early 1980s that interest in the field became widespread. When we last reviewed the relevant literature in 1986[3] there were about 50 scientific papers dealing with the immobilization of algal cells. Now, a decade later, there are in excess of 200 relevant publications, including scientific papers, book chapters and books, and of these, about 30% deal with the use of immobilized algae for wastewater treatment purposes. Indeed, since 1990 nearly 50% of all papers dealing with immobilized algae involve their use in wastewater treatment applications. A thorough review of this literature is beyond the scope of this chapter, but Table 1.1 serves to summarize much of the relevant work and reveals the great diversity of interests in this field. Studies range from those of the uptake kinetics of individual pollutants (e.g., PO_4-P [4] and Hg [5]) by axenic algal cultures in batch systems, to those involving the uptake of pollutants from complex mixtures (e.g., swine waste[6] and cattle manure[7]). It is perhaps sensible to split such studies into two main groups:

1) Studies involving viable cells with uptake often involving active mechanisms.
2) Studies involving the use of non-viable algal cells where non-active biosorption is the predominant uptake mechanism.

Most of the studies of NO_3-N, NO_2-N and PO_4-P uptake fit into the first category, while many of the studies of metal uptake involve passive biosorption by non-viable biomass. Consequently, in many of the metal uptake studies, harsh immobilization procedures may be adopted which render the cells non-viable, whereas in most studies of N- and P-uptake gentle immobilization procedures such as alginate entrapment are necessary to preserve cell viability.

One further point to make regarding the current literature is that virtually all of the studies published to date involve small laboratory-scale studies; the only exception being the work carried out by Bio-recovery Systems Inc., New Mexico, who have patented the AlgaSORB® process for metal removal and have already tested this at the pilot-plant scale with columns treating in excess of 380 l min^{-1} effluent.[8-10]

Our own work on the immobilization of algae began in the early 1980s with some fundamental studies of the physiology and biochemistry of microalgal cells immobilized in gel matrices. At that time other teams were already working on immobilized blue-green algae (cyanobacteria), including Codd's team at the University of Dundee[52,53] and Hall's team at Kings College London.[54,55] It seemed sensible, therefore to select the other major algal group—the eukaryotic green microalgae—as our organisms of interest. In addition, there are advantages to using eukaryotic microalgae rather than cyanobacteria for certain purposes, not least the fact that there are relatively few toxins produced by the eukaryotes, whereas the cyanobacteria produce numerous potent toxins which could pose a problem to their use in a variety of technological applications.

Studies on Immobilized Algal Physiology and Biochemistry

Our own studies have involved the use of the eukaryotic green microalga *Chlorella emersonii* Shihira and Kraus (CCAP 211/8a), though after more than a decade in our culture collection this strain now shows significant morphological and biochemical differences from the original strain and has been redesignated in our collection as *C. emersonii* LP8a. Though other forms of immobilization have been used, most of the work presented within this section involves the use of alginate entrapment based on the protocol of Kierstan and Bucke,[56] using 5% (w/v) sodium alginate and 0.3 M $CaCl_2$ to produce beads of 4 mm diameter. To evaluate cell growth in the immobilized state, beads were disrupted using Na_3-citrate (pH 6.5) and the released cells were counted using a Coulter Counter. During batch culture studies the growth rate of cells entrapped within alginate beads was found to be significantly less than that of free cell suspensions (Fig. 1.2). However, the alginate-entrapped systems suffered from considerable levels of cell leakage, with cells breaking out from the alginate beads and then growing rapidly within the bulk phase, making a real comparison of growth rate in the free and immobilized systems more complex. Analysis of growth rate in the immobilized state is further complicated by the fact that cells at the periphery of the alginate beads had growth rates clearly greater than those within the center, leading to zonation of the cells such that only

Table 1.1. *Summary of literature on the application of immobilized algae for wastewater treatment*

Algal Taxa	Waste Treated	Immobilization Technique	Type of Culture or Reactor	Reference
AlgaSORB*	metals	silica gel	PBR	8, 9, 10
Anabaena CH3	N	alginate	batch and semi-continuous	11
Anabaena doliolum and Chlorella vulgaris	NP and metals	agar, alginate, carrageenan and chitosan	batch and semi-continuous	12, 13, 14
Aphanocapsa pulchra	metals	alginate	PBR	15
Chlamydomonas reinhardtii	NO$_3$ NO$_2$	alginate	batch and FBR	16, 17, 18
Chlorella emersonii	P	alginate	batch and PBR	4, 19, 20, 21, 22
Chlorella emersonii	Hg	alginate, agar and agarorse	batch and PBR	23, 5, 22, 24
Chlorella homosphaera	Cd Zn Au	alginate	batch	25
Chlorella regularis	U	polyacrylamide	batch and PBR	26
Chlorella vulgaris	metals	polyacrylamide	PBR	27
Chlorella vulgaris	N and P	alginate	batch	28
Chlorella vulgaris, Chlorella kessler and Scenedesmu quadricauda	cattle manure	alginate carrageenan polystyrene polyurethane	PBR, FBR	29, 30, 7
Chlorella vulgaris and Scenedesmus bijugatus	N and P	alginate	PBR	31
cyanobacteria	Zn, Mn	mixed biofilm on glass wool	PBRs	32
Nostoc calcicola	Cu	alginate	batch	33, 34
Nostoc calcicola	methyl-Hg	alginate	batch	35
Phormidium sp.	urban effluent	chitosan	batch and semi-continuous	36, 37
Phormidium laminosum	P	polyvinyl foam	batch, PBR, FBR	38
Phormidium laminosum	NO$_3$ NO$_2$	polyvinyl and polyurethane foam	batch, PBR	39, 40
Prototheca zopfii.	kepone	agar	PBR	41
Sargassum fluitans	heavy metals	synthetic polymers	PBRs	42, 43
Scenedesmus acutu and S. obliquus	N and P	carrageenan	batch	44
Scenedesmus bicellularis	N and P	alginate and chitosan	repeated batch	45, 46, 47, 48
Scenedesmus obliquus	NO$_3$	polyurethane polyvinyl	batch and PBR	49
Spirulina maxima	swine waste	carrageenan	FBR	50, 6
Scenedesmus quadricauda	N and P	carrageenan	batch	51

N=nitrogen; P=phosphorus; PBR=packed-bed reactor; FBR=fluidized bed reactor.

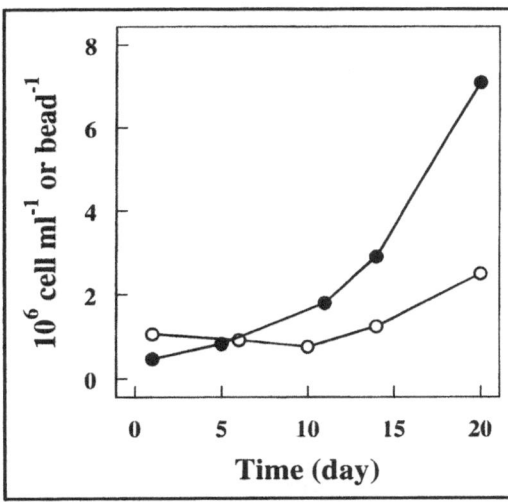

Fig. 1.2. Growth of free cell suspension (●) and alginate-entrapped cells (○) in 20-day batch culture. Reproduced with permission of the publisher from Robinson PK, Dainty AL, Goulding KH, et al. Enzyme Microb Technol. Elsevier Sciences Inc. 1985; 7:212-216.

the outer 50 µm was used by the actively growing algae (Fig. 1.3a). These two problems of peripheral growth and cell leakage pose major limitations to the use of gel-immobilized algal systems in a variety of biotechnological applications. Unfortunately, such problems are generally not reported in the literature, though our experience is that, if questioned, authors often provide verbal clarification that such problems were indeed encountered in their studies. We have therefore attempted a number of strategies to maximize the penetration of cells into gel particles and to minimize leakage from such particles.

Since algal cells are generally cultured autotrophically, or photo-heterotrophically, the necessity for illumination is clear, and light limitation is usually perceived as a major problem in immobilized algal systems. Our studies, however, have shown that alginate is a relatively transparent material which allows good light penetration, and that when immobilized and free cells were illuminated within the chamber of an oxygen electrode both types of cells gave similar responses in bicarbonate-dependent oxygen evolution when irradiance was varied.[57] In addition, we have studied the growth of algal cells in gel beads which had been formed around the end of fiber optic cables. Findings reveal that cells still grew at the periphery of the beads even when illumination was provided from the interior of the bead.[57] Clearly then light penetration was not an important limiting factor in these batch culture systems. Carbon supply was, however, of paramount importance. When the bulk phase CO_2 concentration was elevated by bubbling the culture with air enriched with 2% (v/v) CO_2, cell growth was much more uniform throughout the bead volume.[57] Similar results could be produced by culturing cells in an inorganic medium plus 0.1% (w/v) glucose (Fig. 1.3b) or even by co-immobilizing algal cells with urease and by adding urea to the bulk phase, thereby generating CO_2 within the bead itself.[58]

Work has shown that two main factors affect leakage of cells from alginate matrices. First, the physical integrity of the gel itself and second, the growth rate and location of the cells within the gel. In early studies it was shown that growth of immobilized algae in low PO_4-P medium reduced leakage since PO_4-P is known to

Fig. 1.3. SEMs of (a) a section through an alginate bead demonstrating zonation of cells on the outer periphery of the bead. Reproduced with permission of the publisher from Robinson PK, Dainty AL, Goulding KH et al. Enzyme Microb Technol, Elsevier Sciences Inc. 1985; 7:212-216. (b) a section through an alginate bead demonstrating uniform cell growth throughout, and (c) an alginate bead wrapped in a second alginate layer to minimize leakage problems.

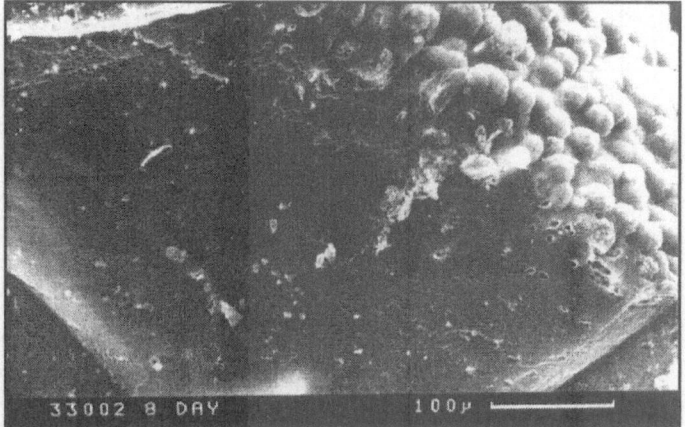

a.

b.

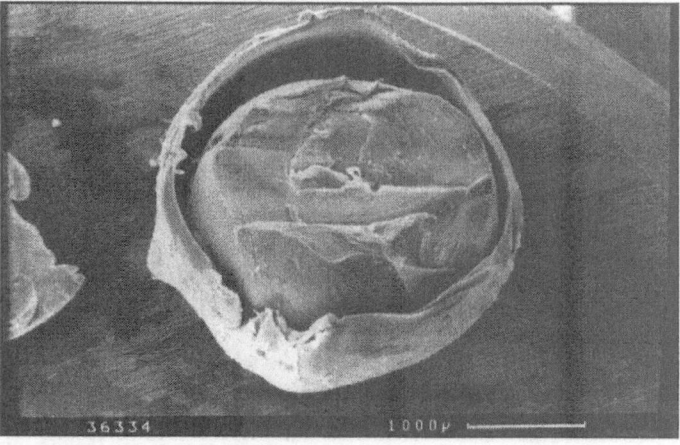

c.

chelate the calcium ions which hold the Ca-alginate matrix together.[59] Similarly, decreased bead disruption has been noted with reduced pH[60] and with alginates having a high guluronic acid : mannuronic acid ratio.[61] Not surprisingly, leakage from immobilized systems is also reduced when growth is not restricted to the periphery of the particle[57] and when the growth rate is slow.[59] Wrapping alginate beads in further layers of alginate (not stocked with cells) to prevent cell leakage has been attempted with some success (Fig. 1.3c) but seems impractical for large-scale application.

Alginate is, of course, not the only naturally occurring gel which may be utilized for entrapment purposes; agar, agarose, carrageenan and chitosan have all been used with some success (see Table 1.1). However, in the vast majority of these cases long-term bead instability and/or cell leakage are also problems. There have been few reports of the problems of cell leakage from synthetic polymers such as polyacrylamide, such matrices are mechanically rather stable and since the biomass is probably killed prior to[27] or upon[26] immobilization. Leakage from polymer foams has also been reported.[40] Small degrees of cell leakage may not pose a problem for many applications of immobilized cells,, and may, in fact, contribute significantly to reactor performance.[62] Where leakage cannot be tolerated, however, it is more likely that membrane-based confinement systems will need to be developed rather than relying on PBRs or CSTRs containing gel or foam particles—a suggestion that will be addressed later in this chapter.

PO$_4$-P Uptake Studies

The study of the removal of PO$_4$-P from the bulk phase surrounding immobilized cells has been a subject of extensive interest and has been the focus of much of our own work over the past decade. Phosphorus, in the form of phosphate, is of particular importance since it is known to be an important causal factor in the eutrophication of waterways, and it is also known that algal cells are highly effective in their uptake and accumulation of this pollutant.

Studies on PO$_4$-P removal by immobilized algae have been restricted to small laboratory-scale systems. Published studies can be divided into three main groups; batch culture studies, semi-continuous (repeated batch) systems, and studies using various types of continuous-flow photobioreactors.

Batch culture studies have generally involved measurement of the depletion in bulk phase PO$_4$-P concentration over periods of hours to days. Uptake has generally been found to be non-linear over the duration of these experiments. In our own studies, Erlenmeyer flasks containing 100 ml growth medium and 100 alginate beads (each containing 10^7 cells) or an equal number of free cells were incubated in an illuminated orbital incubator. At intervals, samples were removed from each flask and analyzed spectrophotometrically for PO$_4$-P using the molybdate-antimony method.[63] Prior to this assay, free cells were first removed in a microfuge. Typical PO$_4$-P uptake responses of free and immobilized cells are shown in Figure 1.4.

Data from such experiments may be analyzed either by incubating cells in a number of flasks containing media with a range of PO$_4$-P concentrations and calculating the initial rate of PO$_4$-P uptake (typically that which occurs over the first hour or two of experimentation, and is generally linear), or by incubating cells in one flask and calculating uptake rates throughout the duration of experimentation as the PO$_4$-P concentration within the flask falls. Data obtained using both of these

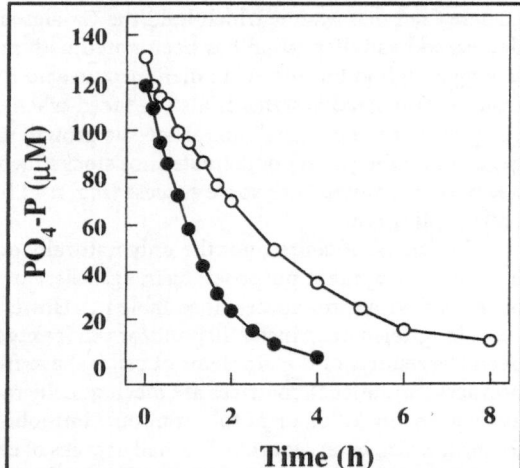

Fig. 1.4. Phosphate concentration in medium surrounding free cells (●) and alginate entrapped cells (○) incubated in medium with initial P-concentration of approximately 100 μM.

approaches fit well into a Michaelis Menten model and are incorporated in Figure 1.5, revealing that the K_M for both free and immobilized cells remains relatively unaltered at close to 100μM. The V_{max} values are, however, quite different with the free cells having a V_{max} about 2.5 times greater than the immobilized cells. Such results are consistent with the alginate gel providing a barrier to the free diffusion of PO_4-P, though one might also expect an increase in the K_M of the uptake system to be associated with such a restriction. Perhaps surprisingly there have been no other direct comparisons of PO_4-P uptake by non-immobilized (free) and immobilized cells reported in the literature. There have, however, been numerous studies of PO_4-P uptake by free cells. Cembella[64] reviewed the field and reveal that for many Chlorophyceae, V_{max} values fall within the range 1.0×10^{-16} to 4.8×10^{-14} mol cell^{-1} h^{-1}. Our results (4.7×10^{-15} mol cell^{-1} h^{-1} for immobilized cells and 1.2×10^{-14} mol cell^{-1} h^{-1} for free cells) are generally consistent with these estimates. In contrast, however, most published studies dealing with free cells reveal K_M values in the lower micromolar range[64] while in this study both free and immobilized cells demonstrate K_M values close to 100 μM. Detailed studies of the PO_4-P uptake mechanisms of *Chlorella pyrenoidosa* have, however, revealed two separate P-transport systems—an arsenate-inhibited high affinity system with a K_M of 2-4 μM determined at low PO_4-P concentrations, and a low affinity system with a K_M of 200-310 μM detectable when cells were exposed to millimolar concentrations of PO_4-P.[65, 66] Considering the PO_4-P concentrations used in this study, and indeed, in other studies of PO_4-P uptake by immobilized algae, it is probable that such studies have generally involved the low affinity P-uptake system.

Numerous other studies have similarly shown that immobilized algae may have significant potential in the removal of PO_4-P from waste waters. However, before this suggestion can really be taken seriously, it is necessary to demonstrate that such rates are sustainable and sufficient for treatment purposes. In an attempt to demonstrate the duration of PO_4-P uptake by immobilized *Chlorella*, cells were subjected to a repeated batch culture, being incubated for one hour in each flask, then rinsed and placed into fresh medium. The results (Table 1.2) demonstrate

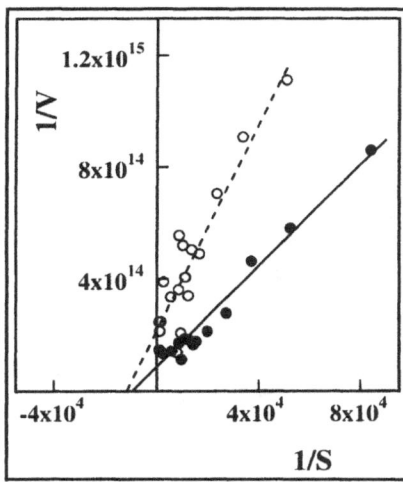

Fig. 1.5. Lineweaver-Burk plot for P-uptake by free (●) and immobilized (○) *Chlorella*. (Kinetic constants-immobilized cells K_M= 87µM, V_{max}= 4.7 x 10^{-15} mol cell^{-1} h^{-1}; free cells K_M = 106 µM, V_{max}=1.2 x 10^{-14} mol cell^{-1} h^{-1}).

that cells retain high uptake activity over a number of subcultures. When cells were subcultured at daily intervals for three cycles they could remove the 100 µM PO_4-P from the growth medium during each cycle. Similar results have been obtained by Chevalier and de la Noüe working with microalgae entrapped in K-carrageenan[51] and by Kaya and co-workers studying chitosan-entrapped cells.[48] Uptake of PO_4-P from media with higher initial PO_4-P concentrations, however, reveals that the uptake response of both free and immobilized *Chlorella* becomes saturated after they have taken up about 2 x 10^{-14} mol P cell^{-1} over a 2 or 6 hour period, respectively. Upon saturation the rate of PO_4-P uptake declines dramatically (Fig. 1.6) a point that will be further considered later in this chapter.

Continuous-flow culture studies have been carried out using small-scale packed-bed reactors containing *Chlorella* in alginate or agarose beads. Reactors were constructed from Pharmacia C10/40 chromatography columns, 40 cm in length and 1 cm in diameter, and were maintained at 25°C with Pharmacia JC10 thermostatic heater jackets connected to a Hakke G/D8 circulating water bath. Six 20 W fluorescent lights provided illumination (60 µmol^{-2} s^{-1} Photosynthetically Active Radiation at the column surface). Columns were packed with 200 alginate or agarose beads, each containing 10^7 cell bead^{-1}. Medium containing 100 µM PO_4-P was pumped into the bottom of the columns at a flow rate of 200 ml day^{-1}.

Table 1.2. Uptake of PO_4-P by alginate-entrapped Chlorella *in repeated batch culture; cells placed into fresh medium at 1 hour cycles*

PO_4-P Uptake Rate (mol cell^{-1} h^{-1})		
Cycle 1	Cycle 2	Cycle 3
3.6 x 10^{-15}	3.0 x 10^{-15}	2.2 x 10^{-15}

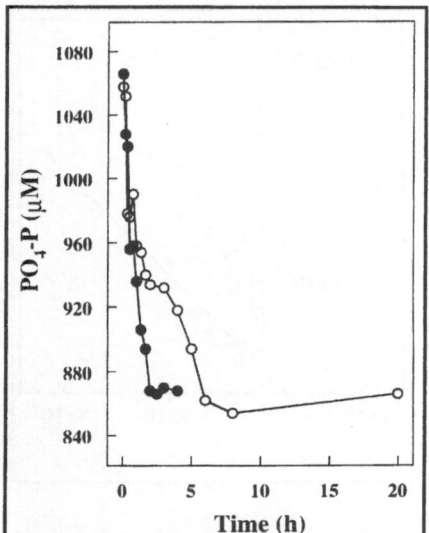

Fig. 1.6. Phosphate concentration in medium surrounding free cells (●) and alginate-entrapped cells (○) incubated in medium with initial P-concentration of approximately 1 mM. Note saturation of uptake response after 2 hour (free cells) and 6 hour (immobilized cells).

Columns containing agarose beads have been operated for up to 7 weeks duration during which time they attained a steady-state condition of almost complete removal of the influent PO_4-P, i.e., 20 µmol P day^{-1}. Columns had been originally stocked with 2 x 10^9 cells and even if cell growth in the system was ignored the cellular P-uptake activity would only approach 4.2 x 10^{-16} mol cell^{-1} h^{-1}, which is only a fraction of that attainable in short-term batch studies. This result was not particularly unexpected, however, since short-term batch culture studies had already revealed that the initial rapid P-uptake rates may decline within a matter of hours (see Fig. 1.6 with cells in 1 mM P). We can therefore consider that the uptake rates derived from batch studies represent the kinetics of short-term "luxury consumption" processes, while long-term studies involve steady state uptake where PO_4-P uptake is most likely related to the rate of PO_4-P incorporation into the biomass and therefore to the growth rate of that biomass. We have investigated the relationship of these two uptake rates by immobilizing cells of different ages which had short-term P-uptake rates in batch culture of 8.2 x 10^{-15} and 1.53 x 10^{-14} mol^{-1} cell^{-1} h^{-1}. When such cells were placed into small-scale packed-bed reactors the performance of the reactors only differed for the first five days of operation, and after that time both reactors achieved the same steady-state P-removal performance.[20] Presumably after five days within the PBRs the growth rate of the cells in each reactor would have been similar and consequently each system would be expected to show the same long-term P-uptake performance.

Given this level of P-uptake activity in continuous-flow culture, it is interesting to estimate the amount of immobilized biomass necessary to treat the effluent from a typical wastewater treatment works. Such treatment works come in a wide variety of sizes; within the U.K. a treatment works treating a village or an area of a larger town might discharge 10^6 l day^{-1} effluent, while a works treating a small town might discharge 10^7 l day^{-1} and a large city works might discharge more than 10^8 l day^{-1}. The PO_4-P content of this effluent will clearly depend both on the nature of the influent stream and the type of treatment applied, but a value of 50 μM PO_4-P may be typical for a works employing primary and secondary treatment processes. Assuming P-removal performances found in our packed-bed reactor studies, we may calculate that a 250 liter reactor containing 5 x 10^6 gel beads (165 l gel), each holding 10^9 algal cells could be used to treat the waste from a small works discharging 10^6 l day^{-1} effluent, reducing the effluent PO_4-P from 50 μM to close to zero. Such estimates, suggest that the full-scale commercial application of this type of technology is a reasonable aspiration, though biomass stocking densities of 10^9 cell bead^{-1} have, to our knowledge, not yet been attained.

One major problem with the use of gel materials, however, is the inherent instability of such structures. The employment of alginate gels for this application is particularly problematic since PO_4-P is itself known to destabilize the alginate gel by chelating the calcium ions from the Ca-alginate matrix. In our own studies with small-scale packed bed reactors we have found it difficult to maintain the integrity of alginate gels beyond a few weeks. In addition, such polysaccharide-based matrices are prone to microbial attack when placed in the natural environment. In studies on the integrity of alginate gels in open systems Vogelsang and Østgaard found microbial biodegradation of the matrix such that the beads suffered total collapse after only 3.5-6 weeks.[67] Other gels, including carrageenan and agarose may be used to form immobilized particles, and our own studies have revealed that agarose gels are certainly physically stable for periods of a few months. However, our considered opinion is that the application of this type of technology at the full-scale demands much more stable forms of immobilization, and that gel-based immobilization techniques are unsuitable for such processes. We have, therefore studied membrane entrapment as an alternative to gel entrapment, and have been particularly interested in the use of hollow fiber technology for such applications.

Hollow fiber cartridges are commercially available in a variety of sizes. In our studies we have used Diaflo cartridges (Amicon Inc.) comprising 50 cylindrical tubes of polysulfone bundled and sealed in a transparent cartridge 20 cm in length. Each fiber is of 1.1 mm internal diameter and the fiber wall is perforated by numerous pores. In our studies, we have used fibers with pore sizes as large as 0.1 μm diameter and as small as 10 kD. The surface area of each cartridge available for mass transfer is 0.03 m^2. Such cartridges may be operated with algal cells contained within the lumen of each fiber and nutrients pumped in through the shell space, or with cells suspended in the shell space and nutrients pumped through the lumen (see Fig. 1.7). In either case it would be expected that PO_4-P could diffuse through the pores in the fiber wall, yet the biomass would remain immobilized on one side of the membrane.

Preliminary studies have revealed that the PO_4-P uptake rates in such systems declined within a matter of hours, not due to any intrinsic lack of activity of the biomass, but as a result of the biomass settling within the reactor. Mixing was introduced by recycling the fluid in the shell or lumen to keep the algal cells in suspension. However, recirculation of the cell suspension at rates of up to 1.7 l h^{-1} still failed to prevent settlement of the algal cells and subsequent loss of uptake efficiency. To overcome these problems we have attempted to increase the viscosity of the liquid in the surrounding cells. Algal cells, if unmixed, quickly settle such that the turbidity of a suspension (absorbance @ 600 nm) may fall by 40-60% within an hour. Within 4 hours the solution will be clear with the cells deposited as a layer at the bottom of the container. When cells are suspended in a solution of 1% (w/v)

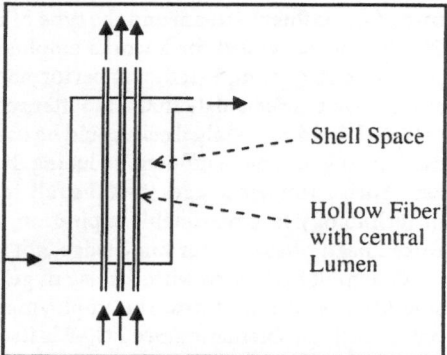

Fig. 1.7. Basic features of a hollow fiber reactor.

Na-alginate, however, settlement rates are imperceptibly slow. The molecular weight of alginate varies according to source, but is generally close to 90 kD. Thus by suspending cells in Na-alginate and partitioning them within the shell side of re-actors packed with fibers with a 10 kD "cut-off" we have been able to study P-removal performance without suffering from problems of cell settlement. While such studies are only in their preliminary stages, the results shown in Figure 1.8 reveal that such reactors quickly attain a steady-state removal rate which has been found to remain constant for several weeks. Such reactors are physically stable and can, of course, be cleaned and re-stocked with biomass as and when P-removal performance falls below acceptable standards. However, the 20% P-removal per-formance during the steady-state condition in Figure 1.8 equates to a very low level of cellular uptake activity (8.3×10^{-17} mol cell^{-1} h^{-1}) and likely results from mass transfer limitations commonly experienced within such unstirred systems. P-up-take performance will certainly be enhanced by increasing the biomass stocking density, increasing the residence time or increasing the surface area of the fibers. It is also possible to introduce stirring to the shell side of such membrane reactors; with extensive modifications to the reactor geometry this may not, in fact, en-hance performance since it may render the system more like a CSTR than a plug-flow reactor (PFR) and it is well known that CSTRs perform less well than PFRs especially, as in this study, when the substrate concentration is well below the K_M of the uptake system.[68]

The Future

Our results have therefore shown that algal cells in batch culture rapidly re-move PO_4-P from the surrounding growth medium, but that this process soon ceases and long-term uptake occurs at a much slower rate, probably correlated to the growth rate of the cells. Since the total phosphorus content of members of the Chlorophyceae is known to range from 3.3×10^{-15} mol cell^{-1} to 2.8×10^{-13} mol cell^{-1},[69] and assuming that such cells typically have doubling times of 24 hours, we can calculate the upper limit for steady-state P-removal by an algal cell to be 1.2×10^{-14} mol h^{-1}. Our results for PO_4-P uptake of 4.2×10^{-16} mol cell^{-1} h^{-1} for cells in packed-bed reactors and 8.3×10^{-17} mol cell^{-1} h^{-1} for cells in hollow fiber reactors are thus only a fraction (3.5% and 0.7%, respectively) of that theoretically attainable by a rapidly growing cell population, and likely indicate that the rate of cell growth within such reactors is low. Optimization of phosphorus uptake by encouraging

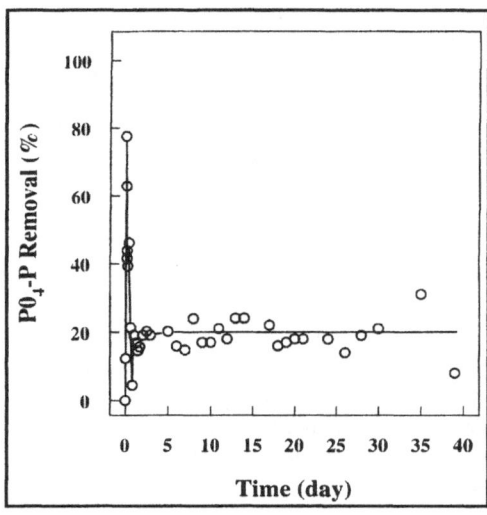

Fig. 1.8. P-uptake performance of a hollow fiber reactor with cells in shell space of unit where the cells are kept in suspension by a 5% (w/v) Na-alginate solution.

rapid cell growth within immobilized cell reactors will unfortunately lead to increased rates of cell leakage from systems adopting gel entrapment, and thereby clearly favors the development of hollow fiber technology for such applications.

Acknowledgments

PKR would like to acknowledge the support of SERC, NAB, TERF and the University of Central Lancashire for funding the work included within this chapter.

References

1. Atkinson B. Immobilized cells, their applications and potential. In: Webb C, Black GM, Atkinson B, eds. Process Engineering Aspects of Immobilized Cell Systems. Rugby: Inst Chem Eng, 1986: 3-19.
2. Hallier UW, Park RB. Photosynthetic light reactions in chemically fixed *Anacystis nidulans, Chlorella pyrenoidosa,* and *Porphyridium cruentum.* Physiol Plant 1969; 44:535-539.
3. Robinson PK, Mak AL, Trevan MD. Immobilized algae: a review. Process Biochem 1986; 21:122-127.
4. Robinson PK, Reeve JO, Goulding KH. Kinetics of phosphorus uptake by immobilized *Chlorella.* Biotechnol Lett 1988; 10:17-20.
5. Wilkinson SC, Goulding KH, Robinson PK. Mercury removal by immobilized algae in batch culture systems. J Appl Phycol 1990; 2:223-230.
6. Cañizares RO, Rivas L, Montes C et al. Aerated swine-wastewater treatment with K-carrageenan-immobilized *Spirulina maxima.* Bioresource Technol 1994; 47:89-91.
7. Còrdoba LT, Hernàndez EPS, Weiland P. Final treatment for cattle manure using immobilized microalgae. I. Study of the support media. Resources Conserv Recycling 1995; 13:167-175.
8. Walker G. Algae help to clean up contaminated water. New Scientist 1991; 130(1770):24.
9. Barkley NP. Extraction of mercury from groundwater using immobilized algae. J Air Waste Man Assoc 1991; 41: 1387-1393.

10. Darnall DW. Removal and recovery of heavy metal ions from wastewaters using a new biosorbent; AlgaSORB®. In: Freeman HM, Sferra PR, eds. Innovative Hazardous Waste Treatment Technology Series. Biological Processes. Lancaster, Pennsylvania, USA: Technomic Publishing Co. 1991; 3:65-72.

11. Lee CM, Lu CS, Lu WM et al. Removal of nitrogenous compounds from wastewaters using immobilized cyanobacteria *Anabaena* CH3. Environ Technol 1995; 16:701-713.

12. Mallick N, Rai LC. Influence of culture density, pH, organic acids and divalent cations on the removal of nutrients and metals by immobilized *Anabaena doliolum* and *Chlorella vulgaris*. World J Microbiol Biotechnol 1993; 9:196-201.

13. Rai LC, Mallick N. Removal and assessment of toxicity of Cu and Fe to *Anabaena doliolum* and *Chlorella vulgaris* using free and immobilized cells. World J Microbiol Biotechnol 1992; 8:110-114.

14. Mallick N, Rai LC. Removal of inorganic-ions from wastewaters by immobilized microalgae. World J Microbiol Biotechnol 1994; 10:439-443.

15. Subramanian VV, Sivasubramanian V, Gowrinathan KP. Uptake and recovery of heavy-metals by immobilized cells of *Aphanocapsa pulchra* (Kütz) Rabenh. J Environ Sci Health 1994; A29:1723-1733.

16. Vilchez C, Vega JM. Nitrite uptake by *Chlamydomonas reinhardtii* cells immobilized in calcium alginate. Appl Microbiol Biotechnol 1994; 41:137-141.

17. Vilchez C, Vega JM. Nitrate uptake by immobilized *Chlamydomoas reinhardtii* cells growing in airlift reactors. Enzyme Microb Technol 1995; 17:386-390.

18. Garbayo I, Braban C, Lobato MV, et al. Nitrate uptake by immobilized growing *Chlamydomonas reinhardtii* cells. In: Wijffels RH, Buitelaar RM, Bucke C et al, eds. Immobilized Cells: Basics and Applications. Amsterdam: Elsevier Science BV, 1996:410-415.

19. Robinson PK, Reeve JO, Goulding KH. The biotechnological potential of immobilized microalgal cells. In: Chang S-T, Chan K-Y, Woo NYS, eds. Recent Advances in Biotechnology and Applied Biology. Hong Kong: Chinese University Press, 1988:193-204.

20. Robinson PK. Effect of pre-immobilization conditions on phosphate uptake by immobilized *Chlorella*. Biotechnol Lett 1995; 17:659-662.

21. Robinson PK, Reeve JO, Goulding KH. Phosphorus uptake kinetics of immobilized *Chlorella* in batch and continuous-flow culture. Enzyme Microb Technol 1989; 11:590-596.

22. Robinson PK, Wilkinson SC. Removal of aqueous mercury and phosphate by gel-entrapped *Chlorella* in packed-bed reactors. Enzyme Microb Technol 1994; 16:802-807.

23. Wilkinson SC, Goulding KH, Robinson PK. Mercury accumulation and volatilization in immobilized algal cell systems. Biotechnol Lett 1989; 11:861-864.

24. Wilkinson SC. The potential of immobilized algal cell systems for the removal of mercury from aqueous solution. Univ Central Lancashire, UK: PhD thesis, 1992.

25. da Costa ACA, Leite SGF. Metals biosorption by sodium alginate immobilized *Chlorella homosphaera* cells. Biotechnol Lett 1991; 13:559-562.

26. Nakajima A, Horikoshi T, Sakaguchi T. Recovery of uranium by immobilized microorganisms. Eur J Appl Microbiol Biotechnol 1982; 16:88-91.

27. Darnall DW, Greene B, Henzl MT et al. Selective recovery of gold and other metal ions from an algal biomass. Environ Sci Technol 1986; 20:206-208.

28. Tam NFY, Lau PS, Wong YS. Wastewater inorganic N and P-removal by immobilized *Chlorella vulgaris*. Water Sci Technol 1994; 30:369-374.

29. Travieso L, Benitez F, Weiland P et al. Experiments on immobilization of microalgae for nutrient removal in wastewater treatments. Bioresource Technol 1996; 55:181-186.

30. Travieso L, Benitez F, Dupeiron R. Sewage treatment using immobilized microalgae. Bioresource Technol 1992; 40:183-187.

31. Megharaj M, Pearson HW, Venkateswarlu K. Removal of nitrogen and phosphorus by immobilized cells of *Chlorella vulgaris* and *Scenedesmus bijugatus* isolated from soil. Enzyme Microb Technol 1992; 14:656-658.

32. Bender J, Gould JP, Vatcharapijarn Y et al. Removal of zinc and manganese from contaminated water with cyanobacteria mats. Wat Environ Res 1994; 66:679-683.

33. Singh SP, Singh RK, Pandey PK et al. Factors regulating copper uptake in free and immobilized cyanobacterium. Folia Microbiol 1992; 37:315-320.

34. Singh SP, Verma SK, Singh RK et al. Copper uptake by free and immobilized cyanobacterium. FEMS Microbiol Lett 1989; 60:193-196.

35. Pant A, Srivastava SC, Singh SP. Methyl mercury uptake by free and immobilized cyanobacterium. BioMetals 1992; 5:229-234.

36. de la Noüe J, Proulx D. Tertiary treatment of urban wastewaters by chitosan-immobilized *Phormidium* sp. In: Stadler T, Mollion J, Verdus M-C et al, eds. Algal Biotechnology. London: Elsevier, 1988:159-168.

37. de la Noüe J, Proulx D. Biological tertiary treatment of urban wastewaters with chitosan-immobilized *Phormidium*. Appl Microbiol Biotechnol 1988; 29:292-297.

38. Garbisu C, Hall DO, Serra JL. Removal of phosphate by foam-immobilized *Phormidium laminosum* in batch and continuous-flow bioreactors. J Chem Tech Biotechnol 1993; 57:181-189.

39. Garbisu C, Hall DO, Serra JL. Nitrate and nitrite uptake by free-living and immobilized N-starved cells of *Phormidium laminosum*. J Appl Phycol 1992; 4:139-148.

40. Garbisu C., Gil JM, Bazin MJ et al. Removal of nitrate from water by foam-immobilized *Phormidium laminosum* in batch and continuous-flow bioreactors. J Appl Phycol 1991; 3:221-234.

41. Pore RS, Sorenson WG. Kepone removal from aqueous solution by immobilized algae. J Environ Sci Health 1981; A16:51-63.

42. Tong C, Ramelow US, Ramelow GJ. Evaluation of polymeric supports for immobilizing biomass to prepare sorbent materials for metals. Int J Environ Analyt Chem 1994; 56:175-191.

43. Ramelow US, Guidry CN, Fisk SD. A kinetic study of metal ion binding by biomass immobilized in polymers. J Hazard Materials 1996; 46:37-55.

44. Chevalier P, de la Noüe J. Wastewater nutrient removal with microalgae immobilized in carrageenan. Enzyme Microb Technol 1985; 7:621-624.

45. Kaya VM, de la Noüe J, Picard G. A comparative study of four systems for tertiary wastewater treatment by *Scenedesmus bicellularis*: new technology for immobilization. J Appl Phycol 1995; 7:85-95.

46. Kaya VM, Picard G. The viability of Scenedesmus bicellularis cells immobilized on alginate screens following nutrient starvation in air at 100% relative-humidity. Biotechnol Bioeng 1995; 46:459-464.

47. Kaya VM, Goulet J, de la Noüe J et al. Effect of intermittent CO_2 enrichment during nutrient starvation on tertiary treatment of wastewater by alginate-immobilized *Scenedesmus bicellularis*. Enzyme Microb Technol 1996; 18:550-554.

48. Kaya VM, Picard G. Stability of chitosan gel as entrapment matrix of viable *Scenedesmus bicellularis* cells immobilized on screens for tertiary treatment of wastewater. Bioresource Technol 1996; 56:147-155.

49. Urrutia I, Serra JL, Llama MJ. Nitrate removal from water by *Scenedesmus obliquus* immobilized in polymeric foams. Enzyme Microb Technol 1995; 17:200-205.

50. Cañizares RO, Domínguez AR, Rivas L et al. Free and immobilized cultures of *Spirulina maxima* for swine waste treatment. Biotechnol Lett 1993; 15:321-326.

51. Chevalier P, de la Noüe J. Efficiency of immobilized hyperconcentrated algae for ammonium and orthophosphorus removal from wastewaters. Biotechnol Lett 1985; 7:395-400.
52. Musgrave SC, Kerby NW, Codd GA et al. Sustained ammonia production by immobilized filaments of the nitrogen-fixing cyanobacterium *Anabaena* 27893. Biotechnol Lett 1982; 4:647-652.
53. Musgrave SC, Kerby NW, Codd GA et al. Structural features of calcium alginate entrapped cyanobacteria modified for ammonia production. Eur J Appl Microbiol Biotechnol 1983; 17:133-136.
54. Rao KK, Muallem A, Bruce D et al. Immobilization of chloroplasts, algae and hydrogenases in various solid supports for the photoproduction of hydrogen. Biochem Soc Trans 1982; 10:527-528.
55. Muallem A, Bruce D, Hall DO. Photoproduction of H_2 and $NADPH_2$ by polyurethane-immobilized cyanobacteria. Biotechnol Lett 1983; 5:365-368.
56. Kierstan M, Bucke C. The immobilization of microbial cells, subcellular organelles, and enzymes in calcium alginate gels. Biotechnol Bioeng 1977; 19:387-397.
57. Robinson PK, Goulding KH, Mak AL et al. Factors affecting the growth characteristics of alginate-entrapped *Chlorella*. Enzyme Microb Technol 1986; 8: 729-733.
58. Mak AL, Trevan MD. Urea as a nitrogen source for calcium-alginate immobilized *Chlorella*. Enzyme Microb Technol 1988; 10:207-213.
59. Robinson PK, Dainty AL, Goulding KH et al. Physiology of alginate-immobilized *Chlorella*. Enzyme Microb Technol 1985; 7:212-216.
60. Dainty AL, Goulding KH, Robinson PK et al. Stability of alginate-immobilized algal cells. Biotechnol Bioeng 1986; 28:210-216.
61. Smidsrod O, Skjåk-Bræk G. Alginate as immobilization matrix for cells. Trends Biotechnol 1990; 8:71-78.
62. Black GM. Characteristics and performance of immobilized cell reactors. In: Webb C, Black GM, Atkinson B, eds. Process Engineering Aspects of Immobilized Cell Systems. Rugby: The Institution of Chemical Engineers, 1986:75-86.
63. Golterman HL, Clymo RS, Ohnstad MAM. Methods for Physical and Chemical Analysis of Fresh Waters. 2nd ed. Oxford: Blackwell Scientific Publications, 1978.

CHAPTER 2

Removal of Copper by Free and Immobilized Microalga, *Chlorella vulgaris*

Nora F.Y. Tam, Yuk-Shan Wong and Craig G. Simpson

Introduction

Toxic heavy metals of industrial wastes are major pollutants which must be removed before discharge. Microalgal biomass, because of its high metal uptake capacity and high multiplication rate, has been applied as a simple and effective alternative to remove heavy metals from industrial wastewater. The capacity of adsorbing/absorbing and accumulating heavy metals in microalgal cells depend on many biotic factors, in particular, the cell density and how algal cells are pretreated before use. The effectiveness of microalgal cells to remove heavy metals can further be enhanced by immobilization which not only eliminates the necessity for separating the cells from treated wastewater, but also makes the regeneration and reuse of the immobilized microalgal cells become possible. This chapter presents and discusses the experimental findings on: 1) the effects of cell density on removal of copper by free *Chlorella vulgaris*; 2) the performance of living and dead (killed by different methods) microalgal cells on copper removal; 3) the removal of copper by alginate immobilized cells; and 4) the regeneration and reuse of immobilized algal biomass.

Heavy Metal Pollution in the Environment

Industrialization and urbanization have led to an increase in metal contamination of aquatic environments. The increased amount of heavy metals has resulted in toxicity of soil, air and water.[1] Unlike organic pollutants, which in most cases can eventually be destroyed, metallic species released into the environment tend to persist indefinitely. They circulate and eventually accumulate throughout the food chain, thus posing a series of threats to animals and man.[2] Their cumulative toxicity has promoted research in reducing the quantities of metals discharged into the environment and investigation regarding their removal from industrial wastewaters.[1,2] Among all heavy metals, copper and its derivatives have received

Wastewater Treatment with Algae, edited by Yuk-Shan Wong and Nora F.Y. Tam.
© Springer - Verlag and Landes Bioscience 1998.

significant attention in recent years due to their common existence in many industrial wastewaters. Copper presented in industrial wastes is primarily in the form of bivalent Cu(II) ion as a hydrolytic product, $CuCO_3$(aq) and/or organic complexes.[3] Several industries, for example, dyeing, paper, petroleum, copper/brass plating and copper-ammonium rayon release undesirable amounts of copper ions. In copper cleaning, plating and metal processing industries, copper ion concentrations often approach 100-120 mg L^{-1}; this value is very high in relation to water quality standards. Copper (II) concentrations of wastewaters should be reduced to a value of 1.0-1.5 mg L^{-1}.[3] In Hong Kong, due to rapid industrial development and the improper disposal of industrial effluent throughout the years, copper in wastewater is of major concern. Extremely high contents of copper (220 to 750 mg Kg^{-1} dry solids) were found to be accumulated in marine sediments, and were significantly higher than levels found in polluted harbors in the U.K.[4]

The most reasonable policy to diminish the escape of heavy metals from industrial sources is the adoption of low waste-generating technologies coupled with effective effluent purification processes.[1] Conventional methods for removing dissolved copper (II) ions include chemical precipitation, chemical oxidation and reduction, ion exchange, electrochemical reduction and evaporative recovery.[1,3] These chemical processes may be ineffective or extremely expensive when initial heavy metal concentrations are in the range of 10-100 mg L^{-1}.[5,6] The ability of microorganisms, such as microalgae to thrive in environments that are polluted by heavy metals and the mechanisms whichenable them to grow under these conditions are of interest. The potential of adsorbing/absorbing and accumulating metals in microalgal cells suggests that these algae can be utilized as an alternative to costly and often ineffective physicochemical technologies for the treatment of wastes containing high concentrations of soluble heavy metals.[7,8]

Removal of Heavy Metal by Algal Cells

Many types of biomass, including agricultural wastes, sewage sludges, and microbial cells in the whole or part, have been used to absorb heavy metals. [2,9] Lombrana et al[10] used wasted sewage sludge to remove nickel from solution, and proposed that this system could be developed to remove heavy metals from waste water. The removal efficiencies for nickel were found to be different between different bacteria species, namely, *Nostoc muscorum, Pseudomonas* sp. and *Rhodopseudomonas* sp.[11] This trend of differing efficiencies between microbial species is a common phenomenon. *Saccharomyces cerevisiae* was considered as a mediocre biosorbent, while the fungi, *Rhizopus* sp. and *Absidia* sp. were excellent biosorbents for lead, copper, cadmium, zinc and uranium.[2] Differences in copper removal efficiencies of dried biomass between *Saccharomyces cerevisiae* and *Candida* sp. have been reported,[12] indicating that this was probably due to differences in the cell wall composition.

Among all microorganisms, green algae, because of their photosynthetic activity (the ability to grow and retain biocatalytic activity without the need for exogenously supplied organic carbon) and their trophic position (the basic trophic level) in food chains have drawn specific attention. The algal genus, *Chlorella* (a spherical, unicellular, eukaryotic green alga containing chlorophylls *a* and *b*) is a frequent symbiont of many other organisms such as paramecium, hydra and sponges, and is important in fresh and marine environments, as well as in the soil.[13] Moreover, these microorganisms are capable of sequestering heavy metals and concentrating them up to several thousand times (or more) within their cellular structures.[14] The large surface area of microalgae further makes them very effective in removing and recovering metals from wastewater. For these reasons, algae, in particular, microalgae, are often the preferred cell type for removal of heavy metals from wastewater. It has been reported that dried green algae, *Chlorella vulgaris* could be practically and easily used to treat industrial wastewater containing Cr^{4+} ions up to 100-200 mg L^{-1}, and that this species was also a good adsorbing medium for Fe^{2+}, Cu^{2+}, Zn^{2+}, Pb^{2+} and Hg^{2+}.[15,16]

The adsorption of heavy metals by algae has been extensively studied in the past decades. Metal uptake by algal cells is a complex process, and, in general, two stages are thought to be involved in the kinetics of metal uptake.[7] The first is very rapid, assumed to be passive (i.e., physical and/or chemical adsorption or ion exchange at the cell surface), and occurs immediately after initial contact with the metal. This stage is often referred to as 'biosorption'. The second stage, 'bioaccumulation', is slower and possibly active, being related to some type of metabolic activity. The relative importance of the two stages may vary with algal species and metal ions.[7] From a quantitative point of view, surface sorption seems to be particularly important and may be the largest portion of the total metal uptake. There was no difference in short-term uptake kinetics of cadmium, copper and lead between whole cells of *Chlorella fusca* and different cell wall fractions from the same organism, but differences in the amount of metal accumulated have been recorded.[17] Many studies have reported that the cell walls of many green algae, for instance, *Chlorella vulgaris*, contain a complex mixture of sugars, uronic acids, glucosamine, and proteins.[14,15,18,19] These biomolecules offer potential metal-binding sites, thus the algal cells have a high sorptive capacity for a variety of metal ions.[20,12] The 'biosorption' on algal cell walls generally involves two principal mechanisms: 1) ion exchange wherein ions such as Na, Mg, and Ca become displaced by heavy metal ions, and 2) metal binds to an unoccupied site, probably by complexation between metal ions and various functional groups such as carboxyl, amino, thiol, hydroxy, phosphate, and hydroxy-carboxyl.[6,19,22-24] The relative importance of the different functional groups is dependent, at least to some degree, on the metal in question.

Crist et al[25] indicated that for the algae, *Rhizoclonium* sp. and for calcium alginate, used as a cell wall model, the sorption could best be explained by an ion exchange model. The algal surfaces of *Chlamydomonas* sp. also had a high affinity for copper and cadmium, and the processes could be considered as a fast initial pseudoadsorption equilibrium with the surface followed by a slow diffusion controlled uptake inside the cell.[28] Harris and Ramelow[26] found that *Chlorella vulgaris* and *Scenedesmus quadricauda* had similar uptake patterns for silver, copper, cadmium and zinc, with most of the adsorption occurring in the first minute. Aksu and his co-workers, in a series of studies, recorded that dead cells of *Chlorella vulgaris* and *Zooglea ramigera* removed copper,[3,6] zinc and iron,[6] lead[6,19] and chromium (VI),[6,15] from solution with high adsorption rates. Their studies also reported the optimum pH for adsorption for all metals tested, with the exception of iron (II) and chromium (VI) at a pH range of 4 - 5; the optimum pH for chromium and iron (II) adsorption was found at pH 2. Cadmium and zinc were found to bind at different sites on the cell wall of *Chlorella vulgaris*.[27] These findings indicate that different binding sites existed for different heavy metals. Therefore algal species of different cell shapes, cell sizes and cell wall compositions will have different metal binding capacities which then influence their effectiveness in removing heavy metals from wastewater. The cell densities or total algal biomass present in wastewater will also affect the heavy metal removal efficiency.

The feasibility of using algal cells to remove heavy metals depends not only on their ability to grow and accumulate heavy metal ions, but also on the ease of preparation and storage of biomass.[12] Pretreatment of algal biomass results in certain effects on their metal uptake ability, depending on on treatment methods, algal species and metal ions concerned. Results reported in the literature were conflicting and no general conclusion have been made, however. For instance, it has been found that dried algal biomass prepared from freeze-dried or oven-dried methods had similar metal uptake ability when compared with fresh biomass, but the dried one is easier to granulate and store than the fresh biomass.[12,26,28] Aksu et al[3] also reported that dead cells of *C. vulgaris* accumulate heavy metal ions to the same or greater extent as living cells. Non-viable biomass under certain conditions has also been reported to adsorb metal ions in larger quantitites than viable biomass.[29,30]

Heavy Metal Removal by Immobilized Microalgae

One drawback of using microalgae in wastewater treatment is the difficulty in recovering the planktonic algal cells (<10 μm) from the treated effluent. Immobilization technologies, developed over the past decade, offer a number of advantages for algal-wastewater treatment systems: 1) elimination of the necessity for harvesting algal cells; 2) providing a means for selective recovery of heavy metals as heavy metals accumulated on algal matrix may be stripped by pH adjustment and/or addition of specific ligands; once stripped free of metal ions, the algal cells are effectively regenerated; 3) allowing the algal biomass to be packed into columns which reduce the space required; 4) increasing the cell biomass and cells' retention time in the bioreactor; and 5) protecting the cells against the toxicity of heavy metals. The use of immobilized microorganisms to remove metals from solution has been studied by several authors. Successful removal and recovery of Cu, Zn, Au and Hg from waste streams by immobilized *Chlorella vulgaris* in polyacrylamide matrices has been reported.[20,31] Tolley et al[32] used *Citrobacter* sp., immobilized in polyacrylamide to remove lanthanum from solution. Alginate immobilized *Chlorella emersonii* were found to remove more mercury from solution than free cell systems[31,33] while alginate immobilized *C. salina* accumulated large amounts of Co, Zn and Mn.[34] *Scenedesmus quadricauda*, immobilized in a cross-linked copolymer of ethyl acrylate-ethylene glycol dimethacrylate, could adsorb copper from solution.[26] Despite all these studies, information on direct comparison between free and immobilized algal cells, and the contribution between cell biomass and immobilization agents on metal removal is still relatively scanty and deserves more in-depth study.

Objectives

A series of bench-scale experiments were designed with the aims listed below:

1) To examine the effect of cell density on removal of copper from solution by *C. vulgaris;*

2) To determine the effects of cell pretreatment on copper removal efficiency and difference between dead and living *C. vulgaris* cells;

3) To compare removal of copper from solution by free and immobilized *Chlorella* cells; and

4) To investigate the recovery of copper from immobilized algal beads.

The alga *Chlorella vulgaris* was chosen in the present study because of its easy growth in commercial culture. This unicellular green species with a cell diameter of 5 μm, has been utilized for various purposes, ranging from nutrient removal from wastewater to their use as a food source.[35] Our previous research work demonstrated that *C. vulgaris* was very effective in removing nitrogen and phosphorus simultaneously from settled domestic sewage and could be used as secondary sewage treatment process.[36-42] Its ubiquitous nature and its ability to survive in a wide range of conditions also make it a potentially useful alga for industrial wastewater treatment.

Materials and Methods

Mass Culture of Chlorella vulgaris
The unicellular green alga, *Chlorella vulgaris* from Carolina Biological Supply Company, USA was mass cultured in 17 L plastic vessels containing liquid commercial Bristol media under open batch culture conditions. Aeration and mixing was provided by filtered air using an air pump installed with stone-diffusers at the bottom of the vessel. The cultures were incubated in an environmental chamber with temperature set at 26°C, light intensity of 174 $\mu Es^{-1} m^{-2}$ and a 16 hour light/8 hour dark photoperiod. Algal cells at the end of the log phase were harvested by centrifugation (10,000 rpm for 10 minutes). The culture media were decanted and the cell pellets were resuspended in small amount of deionized water, then stored at 4°C prior to further use.

Effects of Cell Density on Copper Removal
Different quantities of the stored algal cells were resuspended in 150 mL digestion tubes, each with 50 mL copper sulphate (anhydrous) solution containing 30 mg L^{-1} Cu, to give different cell densities. A total of 11 cell densities, 2×10^7, 5×10^7, 2×10^8, 3×10^8, 4×10^8, 5×10^8, 6×10^8, 7×10^8, 8×10^8, 9×10^8 and 1×10^9 cells mL^{-1}, were prepared. Each cell density treatment was done in triplicate. The cells were kept in suspension and aerated by air-diffusers. At frequent time intervals, i.e., 0, 10, 60, 120, 180 and 240 minutes, 5 mL cell suspension from each tube was collected and centrifuged. The copper content that remained in the supernatant was determined by a flame atomic absorption spectrophotometer (Shimadzu AA-65015) using 324.7 nm wavelength and a band pass width of 1 nm. At the end of the 4-hr reaction period, 20 mL cell suspension were harvested and the cell pellets from centrifugation were determined for biomass (dry weight) and tissue copper concentration. The cell residues were filtered through prewashed and preweighed glass fiber discs. The filters and cells were then dried at 105°C to a constant weight. The filters and cells were acid digested by 10 mL concentrated nitric acid at 160°C. The digest was then filtered and made up to 50 mL with double deionized water, the copper concentration in the filtrate was determined by flame atomic absorption spectrophotometry.

The viability of the cells after use for copper removal was checked by growth in liquid culture and on agar plates. The cell pellets from centrifugation of the first and last sampling were resuspended in Bristol medium to obtain an initial cell density of 5×10^7 cells mL^{-1}. The liquid cultures were incubated in the environmental chamber for 4 days, the cultures were then stained and the uptake of neutral red observed under light microscope indicated that the cells were viable.[43] In addition to liquid culture, the cell viability was further checked by dropping 10 μL cell suspension (in duplicates) onto agar plates containing Bristol medium. The cell was considered viable if a colony was produced after 4 days incubation.

Effect of Cell Pretreatment on Copper Removal

A total of 5 sets of stored algal cells were resuspended in 50 mL deionized water, each with a cell density of 5×10^8 cells mL^{-1} (the optimum cell density as identified in the previous experiment), and treated by one of the following five methods before being used to remove copper: 1) autoclaving at 121°C for 15 minutes; 2) heating at 80°C water bath for 1 hour; 3) oven drying at 80°C for 72 hours; 4) freeze drying; and 5) no treatment, i.e., living cells, used as control. After the pretreatment, the cells were centrifuged and the pellets were resuspended in digestion tubes containing 50 mL copper solution with 30 mg L^{-1} Cu. The cells treated at 80°C in the water bath, and the control (living free cells) were easily harvested by centrifugation at 4,000 rpm for 10 minutes. The autoclaved cells required much greater centrifugation forces (10,000 rpms for 30 minutes) to spin down into pellets. The dried cells, both the oven and freeze dried, were resuspended in deionized water prior to harvesting. The freeze dried cells were easily resuspended using a vortex mixer but the oven dried cells were more difficult to resuspend and required 15 minutes blending to achieve good resuspension. These suspensions were checked microscopically to ensure proper segregation of cells, and cell morphology after pretreatment was also observed.

After placing the pretreated cells into copper solution, 1.5 mL samples were collected from each tube at 0, 2, 10, 15, 60, 120, 180 and 240 minutes. The cells were separated from copper solution by micro-centrifugation (10,000 rpms for 1 minute). The residual copper concentration in the solution, the tissue copper concentration and the cell viability were determined by the methods described in the previous experiment.

Immobilized Microalgae to Remove Copper

Immobilization of Algal Cells with Sodium Alginate

Immobilization of *C. vulgaris* cells in sodium alginate solution followed the method of Tam et al.[38] To immobilize the algae in alginate matrix, 300 mL 4% sodium alginate solution was mixed with 300 mL algal suspension to give a cell density of 2×10^7 cells mL^{-1}. To create blank alginate beads (without algae), 300 mL 4% sodium alginate was mixed with 300 mL deionized water. The 600 mL 2% alginate-algae mixture or alginate suspension was used to produce the beads for one column bioreactor (internal diameter of 10 cm and length of 50 cm) containing 1.8 L copper solution of 30 mg L^{-1} Cu. A peristaltic pump with speed set at 13 rpm was used to siphon the alginate suspension into 900 mL 2.5% $CaCl_2$ solution. The beads started to harden almost immediately when dropped into the $CaCl_2$ solution and were left in the solution overnight to fully harden. The beads were then washed in deionized water and stored at 4°C prior to use.

Removal of Copper by Alginate Immobilized Algae

Two treated sets and two controls, each in triplicate, were used in this experiment. The treated samples consisted of either blank alginate or algal-alginate beads (prepared from 600 mL alginate suspension as mentioned above). The treated reactors were aerated with diffused air to keep the beads in suspension. The controls consisted of the same copper solution but without any beads, one with aeration and the other without. The column bioreactors were incubated at room temperature ($20 \pm 2°C$) with continuous lighting throughout the experiment.

The bioreactors were sampled at regular time intervals over a period of 52 hours. At each sampling time, 50 mL solution was removed from each column and stored in plastic bottles at $4°C$ prior to analysis of the copper concentrations using the AAS. At the same time, pH of each column was recorded. Approximately 550 beads were removed at each sampling to ensure that the liquid to bead ratio remained constant. Beads collected were measured for dry weight, copper concentrations, and cell viability (after dissolving the beads in 0.1 M sodium citrate and harvesting by centrifugation at 4,000 rpm for 10 minutes) using the methods described above.

Repeated Copper Dosing, Saturation and Recovery of Copper from Immobilized Beads

The bioreactors were set up as per the previous experiment except that only three treatments were used: 1) control with aeration only; 2) alginate blank beads; and 3) immobilized algal beads. The columns were dosed 23 times with fresh 30 mg L^{-1} copper solution and then 3 times with 780 mg L^{-1} copper solution. The liquid phase of the columns was poured out and fresh copper solution added at each dosing. The retention time of the first 6 doses was kept at 24 hours, thereafter, the retention time was reduced to 3 hours as the copper uptake reached equilibrium within 3 hours retention. The beads were harvested after the columns were

Table 2.1. Copper remaining in solution after treated by free algal cells of various cell densities for 1 and 4 hours (mean and standard deviation of 3 replicates were shown, the bracket shows the dry weight of biomass)

Cell density (x 10^7 cell mL^-1)	Residual Cu conc. (mg L^-1)		Cu removal (%)	
	1 hour	4 hour	1 hour	4 hour
2 (0.163 g L⁻¹)	28.38 ± 0.54	29.06 ± 0.66	5.4	3.1
5 (0.381 g L⁻¹)	27.31 ± 1.54	29.23 ± 2.05	8.9	2.6
20 (1.494 g L⁻¹)	21.61 ± 0.17	22.15 ± 0.42	27.9	26.2
30 (3.043 g L⁻¹)	14.99 ± 3.62	15.05 ± 2.32	50.0	49.8
40 (3.866 g L⁻¹)	12.11 ± 4.51	12.95 ± 2.42	59.6	56.8
50 (5.477 g L⁻¹)	8.46 ± 0.45	9.99 ± 0.20	71.8	66.7
60 (7.217 g L⁻¹)	6.79 ± 1.79	8.74 ± 4.05	77.4	70.9
70 (8.605 g L⁻¹)	6.60 ± 2.04	7.25 ± 3.26	78.0	75.8
80 (11.12 g L⁻¹)	4.70 ± 1.27	5.40 ± 2.41	84.3	82.0
90 (12.46 g L⁻¹)	4.57 ± 0.38	3.60 ± 0.62	84.8	88.0
100 (14.38 g L⁻¹)	5.57 ± 1.53	5.27 ± 0.76	81.4	82.4

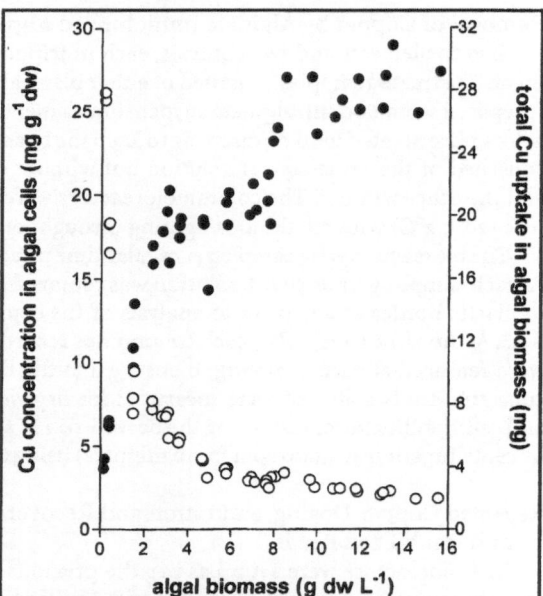

Fig. 2.1. Specific copper uptake in *C. vulgaris* cells (open circle) and total copper accumulated in cells (solid circle) at varying cell densities (Cells were exposed to 30 mg L^{-1} copper solution for 4 hours).

saturated with copper. A small amount of beads was collected to determine total Cu retained in beads (by ashing the beads and dissolving the ash in conc. HNO$_3$). The majority of beads were placed in another column containing deionized water, and HCl (0.1 M) was added gradually to acidify the pH of the column reactor. Small aliquots were collected to measure the concentration of copper desorbed from the beads at different pH.

Copper Removal Efficiency: Free Cell Versus Immobilized Viable and Killed Algae Cells

This experiment was carried out on a small scale as per the free cell experiments (150 mL digestion tube containing 50 mL copper solution), and the initial cell density was held at 5×10^8 cells mL^{-1}. There were five treatments: 1) living free cells; 2) dead free cells killed by heating the cells at 80°C water bath for one hour; 3) alginate blank beads; 4) immobilized living algal beads; and 5) immobilized dead algal beads. Approximately 550 beads were used in each reaction tube except in the first treatment. At the beginning and the end of the reaction (4 hours), cells were taken for acid digestion then analyzed for tissue copper content. The immobilized beads were digested either as whole beads or as cells from beads. To collect cells from the beads, the beads were dissolved using 0.1 M sodium citrate and the cells were separated from alginate by centrifugation. The cell fraction was then washed, weighed, oven dried and acid digested prior to copper determination. The supernatant, being the alginate fraction from the beads, was also collected and copper concentration was measured directly by the AAS.

Results and Discussion

Effects of Cell Density on Copper Removal

The residual Cu concentrations in solution decreased significantly with cell density, the higher the cell density the more reduction in copper (Table 2.1). At low cell density, very little copper was removed from the solution and the percentage of Cu reduction reached a steady level when cell density increased to 8×10^8 cells mL^{-1} (equivalent to 11 mg mL^{-1} dry biomass). This value was higher than the values reported in the literature (around 1-5 mg mL^{-1} dry biomass), probably due to differences in cell size and cell wall composition of different algal species. The decline in residual copper concentration was directly reflected in an increase in total Cu accumulated in cell biomass (Fig. 2.1), indicating that more biomass resulted in more copper uptake. On the other hand, the amounts of copper adsorbed/absorbed by algal biomass at the end of the 4-hour reaction period decreased as cell density increased (Fig. 2.1), suggesting that specific Cu uptake was lower when more biomass was used for copper removal. This phenomenon was also noted by Aksu and Kutsal[6] who concluded that it was due to aggregates of cell, leading to a reduced effective adsorption area. The lower specific Cu uptake recorded in reactors containing high cell densities might also be due to their larger biomass which diluted the amounts of copper accumulated. Such change in effective adsorption of copper with an increase in cell density has implications for both the uptake efficiency as well as possible toxicity. The low cell densities showing an increase in the amount of copper adsorbed per gram dry weight suggested an increase in toxic load per individual cell. Thus if the cells were to be kept alive it might require a higher cell density to reduce the copper load in an individual cell. In the present study, all cells harvested at the end of 4 hours copper treatment were viable with colonies found in agar plates and cells from liquid culture took up neutral red stain. This finding indicates that exposed cells, no matter under high or low cell density, to copper (30 mg L^{-1}) solution, were algastatic and not algacidal as cell recovery occurred. The low toxicity of copper recorded in this study might be explained by the fact that cells were only exposed to one dose of copper solution for a short period of time, thus the cells could recover easily. Furthermore, the amount of copper retained in cells was relatively small and might not have reached their saturation level.

The rate of copper removal by free algal cells was very quick in that most copper was taken up by algae in the first couple of minutes, and there was no difference in Cu reduction between 1 and 4 hours of treatment (Fig. 2.2). The metal uptake is defined as the concentration of metal bound to the algae, comprising all the possible locations of metals on or in the aglal cells, i.e., the cell wall, the cell membrane, the space between the cell wall and the cell membrane, and the intracellular space within the cell membrane.[44,45] The rapid adsorption recorded in the present study indicates that the copper uptake was mainly on the cell wall *via* the single phase of metabolism-independent 'biosorption' with no apparent intracellular metabolic uptake ('bioaccumulation'). Similar copper adsorption rates with the majority of adsorption happening in the first minute have also been reported.[1,26] Aksu et al[3] found that equilibrium between *C. vulgaris* and copper took around 10 to 15 minutes under their conditions but Aksu and Kutsal,[6] again using *C. vulgaris* and copper, found an equilibrium time of between 40 to 60 minutes. The reasons

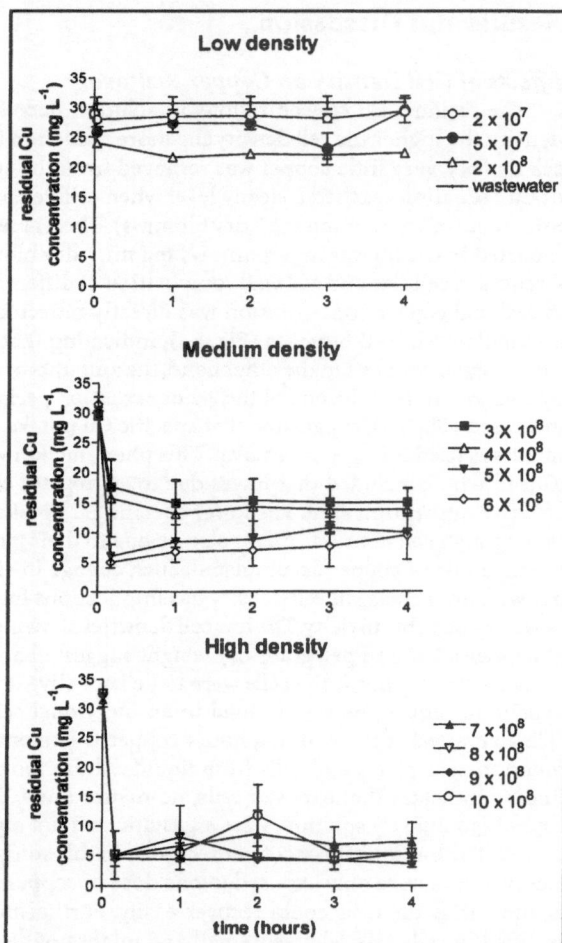

Fig. 2.2. Residual copper
concentration in reactors
containing different cell
densities (mean and stan-
dard deviation values of 3
replicates were shown).

for these differences are not immediately clear, however, differences in experimental
conditions such as pH, metal concentrations, cell density used and pretreatment
of cells may influence the time required for equilibrium.

Effect of Cell Pretreatment on Copper Removal

Figure 2.3 shows that the removal of copper by pretreated cells was very rapid
(within minutes) and the residual copper concentrations were significantly lower
than those in reactors containing living control cells. Non-viable biomass under
certain conditions has been reported to adsorb metal ions in larger quantities than
viable biomass.[12,29,30] Pretreatment of cells, in particular, autoclaving and heating
at 80°C water bath, led to an increase in the amount of copper removed from solution, when compared to untreated control cells (Table 2.2). The control cells removed only around 75% copper while the free cells subject to different heat treatments prior to use, all removed approximately 99% copper from solution. The

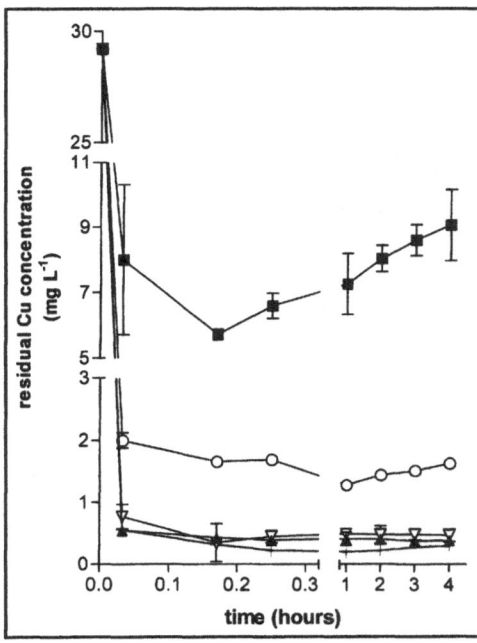

Fig. 2.3. Effects of different cell pre-treatment on removal of copper (mean and standard deviation of 3 replicates of copper concentration remained in solution were shown; ■: control living cells; ▲: heated in 80°C water bath for 1 hour; ▼: autoclaving at 121°C for 15 minutes; ✳: oven dried at 80°C; ○: freeze dried; cells with initial density of 5 x 10⁸ cells mL⁻¹ were exposed to 30 mg L⁻¹ Cu for 4 hours).

Table 2.2. *Concentrations of copper adsorbed/absorbed by pretreated and control algal cells at the end of 4 hours copper treatment (mean and standard deviation of 3 replicates were shown)*

Pretreatment of Cells	Biomass (g dw L⁻¹)	Tissue Cu content (mg g⁻¹ dw biomass)	% Cu removal after 4 hr treatment
control living cells	7.27 ± 0.35	1.56 ± 0.09	68.96 ± 3.72
80°C water bath, 1 hour	7.35 ± 0.38	1.78 ± 0.19	98.68 ± 0.12
121°C, 15 minutes	5.24 ± 0.67	2.61 ± 0.53	98.40 ± 0.35
oven dried 80°C	6.93 ± 0.27	1.77 ± 0.21	98.88 ± 0.16
freeze dried	7.38 ± 0.07	1.71 ± 0.12	94.40 ± 0.16

freeze dried cells removed around 95% copper, less effective than the heated cells. However, the difference between freeze dried and oven dried cells found in this study was not the same as that reported by Harris and Ramelow[26] who observed that cells of *C. vulgaris* and *Scenedesmus quadricauda* pretreated by either freeze dying or oven drying at 70°C or 100°C had no difference in the uptake of copper or silver. Such deviation between two studies could be due to variation in degree and efficiency of the respective freeze drying methods. Among all treated cells, the autoclaved cells appeared to have the highest copper uptake rate (Table 2.2), however, these cells had the smallest dry weights. It was observed that cells might have been lysed during autoclaving, making it difficult for cells to be separated from water resulting an overall loss of cells during this pretreatment process.

The differences in copper adsorption between living and dead cells could be due to changes in cell wall on death, making more binding sites available for cations. It has been recorded that subjecting *Chlorella vulgaris* to various physical/chemical treatments (acid- or alkaline-treated, heat treated or formaldehyde-treated cells) gave rise to changes in the specific area due to changes in the mean cell size and the size distribution of the culture, and that the specific uptake of gold by treated cells was higher than the control.[46] Drying yeast cells was also found to decrease their particle size from 300 μm to single cell diameter, but had no significant effect on the biosorption process.[12] Oven drying of cells in this study resulted in significant cell shrinkage and cell diameters much smaller than controls. In the present experiment, cells were still intact after various pretreatments. Similarly, Simmons et al[12] reported that cells from dried and ground biomass were still intact under scanning electron microscopy and light microscopy. These suggest that pretreatment of cells did not really alter the cell morphology, but cell sizes and cell weight might have changed when cells were pretreated (Table 2.3). Cells treated by oven drying and autoclaving in the present study became paler in color and were clumped. Bacterial contamination was serious in these two sets of reactors, suggesting that certain cells might have been lysed and the released cellular products then attracted more bacterial growth. The membrane permeability may have been altered due to disruption of cell membrane, and the denaturing of the cell membrane (which enabled intracellular access by the metals) ultimately lead to greater uptake by treated cells.[46] The viability tests show that only cells from the control (untreated one), after exposed to Cu, produced growth in Bristol medium. None of the pretreated cells were able to regrow in both agar plates and liquid Bristol medium.

Removal of Copper by Immobilized Microalgae

Figure 2.4 shows that both immobilized algal beads and alginate blank beads removed copper rapidly, with more than 90% Cu reduced from solution within 3 hours of reaction. There was no significant difference between alginate blank and algal beads. It has been suggested that alginate matrix, similar to agar and carrageenan, offers many anion sites for the binding of polyvalent cations.[47,48] At a high cell density (5×10^8 cells mL^{-1}), free algal cells removed more copper than the immobilized algal beads, and blank beads were less efficient than algal beads (Fig. 2.5). This was probably due to the increased surface area available in the free cell suspension, thus allowing easier exposure of the surface binding sites to the copper ions. However, previous research carried out by Granham et al[34] showed that immobilized cells removed greater amounts of metals than free cells of comparable cell densities. Such discrepancy between the present study and Granham's work might be explained by the different cell densities used for removing heavy metals. In the present study, free cells removed more copper than immobilized beads under high cell densities (5×10^8 cell mL^{-1}), but opposite results were recorded at low cell density (2×10^7 cells mL^{-1}). At this low cell density, free algal cells removed around 5% Cu while the algal bead removed more than 90% Cu. When immobilized beads were dissolved and copper content adsorbed in alginate matrix and in cell biomass were determined, it was found that cells only contributed

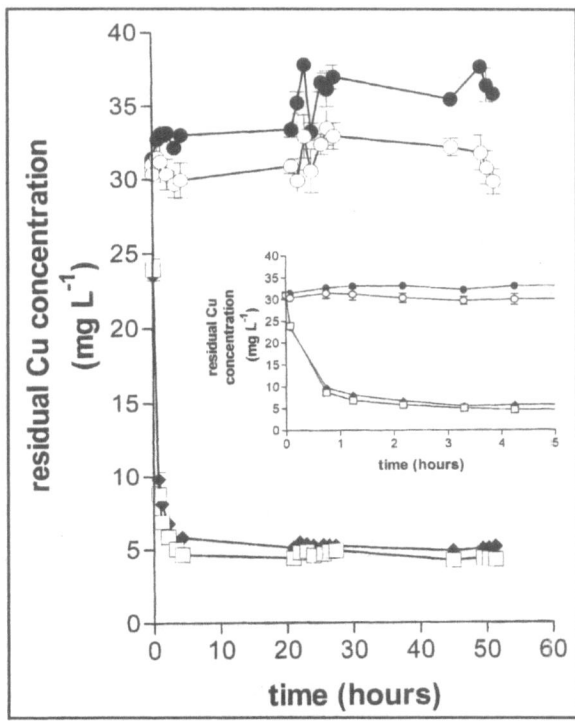

Fig. 2.4. Removal of copper by immobilized beads (mean and standard deviation of 3 replicates of copper concentration remained in solution were shown; ●: control, no aeration; ○: control, with aeration; ◆: immobilized algal beads; ❑: alginate blank beads; the insert shows the changes in residual Cu content during the initial first 5 hours of exposure to 30 mg L^{-1} Cu; initial cell density for immobilized algal beads were 2 x 10^7 cells mL^{-1}).

Table 2.3. *Microscopic appearance of cells after pretreatment and exposure to copper solution*

Cell Pretreatment	Macroscopic appearance of cells	Microscopic appearance of cell at 400x magnification	Average cell diameter (μm, n=20)
Control living cells	dark green	cells are round, clear and transparent	3.7
80 °C water bath, 1 hour	dirty brown/green	cells appeared smaller than control cells; with rough edges	3.5
121 °C, 15 min.	dirty brown/green	cells appeared smaller than control cells; some with rough edges but less obvious than cells killed in water bath; cells disintegrated	3.8
oven dried 80 °C	dirty brown/green	small cells and cells aggregated	2.7
freeze dried	dark green	cells are round, similar to control	3.4

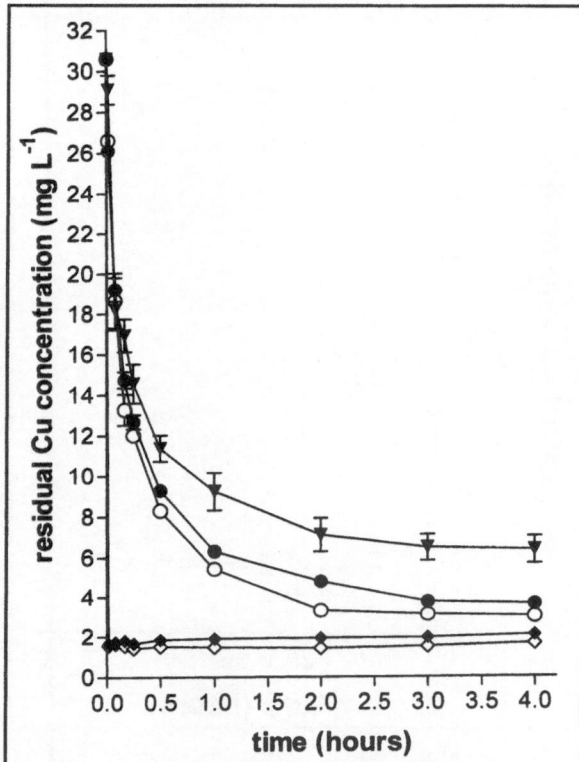

Fig. 2.5. Performance of free and immobilized algal cells (both living and dead cells) on copper removal (mean and standard deviation of 3 replicates of copper concentration remained in solution were shown; ◆: free living cells; ◇: free dead cells; ●: immobilized beads containing living cells; ○: immobilized beads containing dead cells; ▼: alginate blank beads; initial cell density for free and immobilized algal beads were 5×10^8 cells mL^{-1}).

to approximately 12-14% of the copper removal while alginate accounted for the majority of the copper uptake in the beads (Table 2.4). Crist et al[25] also found the metal sorption abilities of calcium alginate to be considerably higher than the algal biomass. Similar to free cell condition, dead cell biomass and immobilized dead algal beads removed more copper than the living cells and the algal beads formed from living cells (Table 2.4).

When the bioreactor columns containing alginate beads were repeatedly dosed (each dose of 1.8 L) with 30 mg L^{-1} copper solution, the copper removal efficiency (over 90%) was maintained in the first few doses (Fig. 2.6), then the removal efficiencies dropped gradually (Fig. 2.7). Microscopic studies showed that algal cells from alginate beads, collected from the first few doses, were still intact, transparent, clear and were light green in color, but no re-growth was found if the algal beads were exposed to more than 5 doses of copper solution. These results suggest that the immobilized algal cells might have suffered from copper toxicity and could not be recovered as more copper accumulated in the immobilized beads. The residual copper concentrations increased gradually from the fifth dose onwards, indicating that the copper adsorption ability decreased. The immobilized beads reached saturation at around 100 doses (Fig. 2.8), i.e., after retaining more than 3,000 mg Cu. These findings showed that immobilized beads had a large capacity to retain copper, and there was no difference between alginate blank and algal

Table 2.4. Copper content in cells and alginate matrix from immobilized algal beads after treated with 30 mg L⁻¹ Cu for 4 hours (initial cell density = 5 x 10⁸ cells mL⁻¹)

Component	Living beads		Dead beads	
	µg Cu per bead	% contribution	µg Cu per bead	% contribution
cell only	0.27 ± 0.04	12.21	0.38 ± 0.01	13.56
alginate matrix	1.97 ± 0.03	90.43	2.19 ± 0.12	78.07
whole beads*	2.18 ± 0.16		2.80 ± 0.16	

(*Cu content in whole bead compared with the summation from cells plus alginate matrix; the recovery was 103% in living beads and 92% in dead beads)

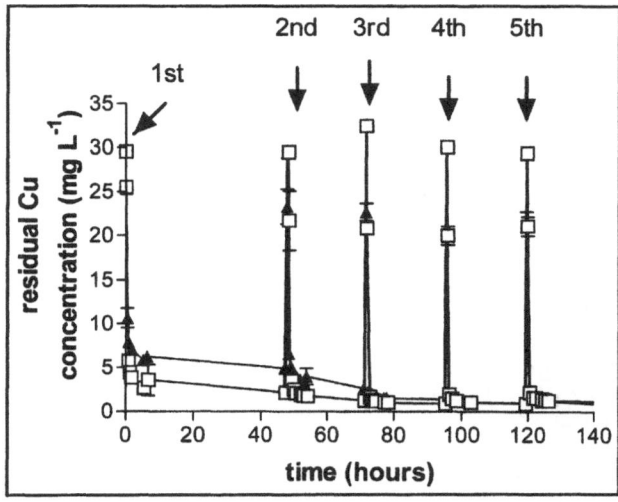

Fig. 2.6. Repeated dosing of the reactor containing immobilized beads with copper solution (copper concentration remained in solution dring the first five cycles were shown; ▲: immobilized algal beads with initial cell density of 2 x 10⁷ cells mL⁻¹; □: alginate blank beads; arrow indicates when fresh copper solution was added).

beads. It was found that the copper saturated alginate blank beads adsorbed 186.45 ± 6.77 mg Cu g⁻¹ dry weight (or 5.868 meq g⁻¹) while algal beads retained 173.29 ± 7.51 mg Cu g⁻¹ dry weight (or 5.454 meq g⁻¹). These adsorption capacities compare favorably with those of ion exchange resins. For instance, the adsorption of most Sigma cation exchange resin was in the range of 0.7-10 meq g⁻¹.[49]

Copper retained in alginate beads was easily desorbed by lowering pH of the reactor to less than 3. The optimum pH for desorption of copper was around 1.5 to 2.0, and at this pH, all copper retained were leached out (Fig. 2.9). It is obvious that the eluting solution should be non-toxic and achieve maximum recovery with minimum quantities at the lowest possible concentrations. In the present study, the immobilized beads were not damaged by 0.1M HCl after the desorption process. This suggests that the alginate beads could be repeatedly used for copper removal

Fig. 2.7. Copper removal efficiency of immobilized beads continuously dosed with 30 mg L^{-1} copper solution (mean and standard deviation of 3 replicates were shown; ●: immobilized algal beads ○: alginate blank beads).

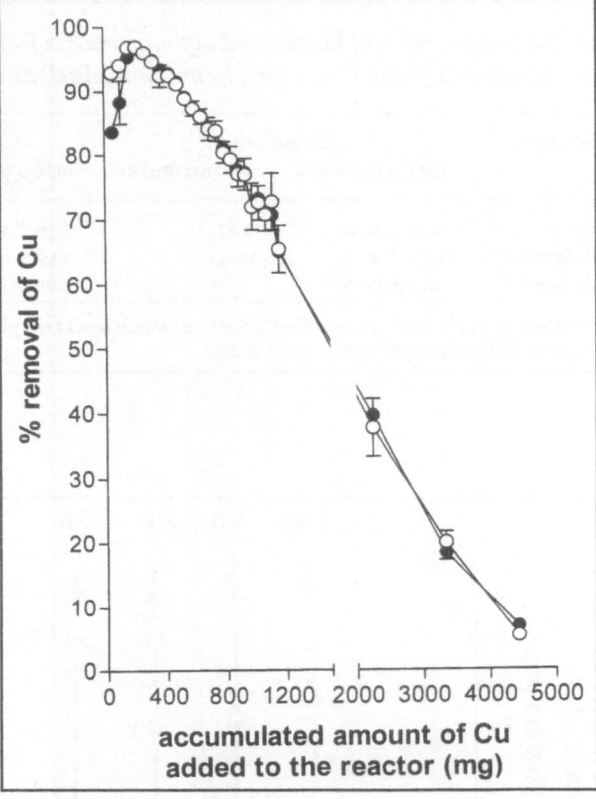

Fig. 2.8. Residual copper concentrations during repeated dosing of the reactor with 1.8 L copper solution (Cu concentration remained in solution at the end of 3 hour exposure of each dose was shown; ●: immobilized algal beads ○: alginate blank beads).

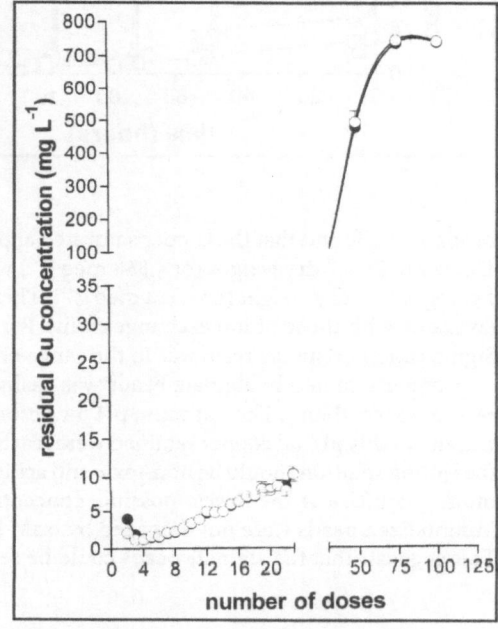

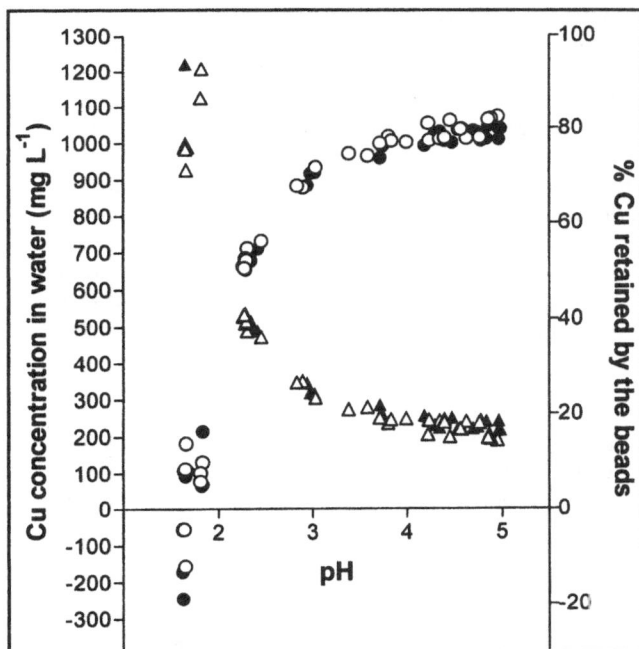

Fig. 2.9. Desorption of copper from copper-saturated immobilized beads at different pH (● indicate Cu concentrations eluted in acid solution; ▲ indicate % Cu retained in beads; ○, △ represent alginate blank beads; solid symbols represent algal beads).

as the copper saturated in the beads could be desorbed in a small volume of acid solution. The study also indicates that pH 3 would be the lower pH limit for the alginate beads to effectively adsorb copper from solution. The optimum pH for adsorption of metals by immobilizing *Saccharomyces cerevisiae* in polyacrylamide beads was also found to be 3 to 5.[50] pH has been considered as the most important single parameter influencing the biosorption rate. Crist et al[22,25] suggested that the zero-point charge or isoelectric point of algal cells would be around pH 3. Above this pH, algal cells would have a net negative charge favoring the binding of positively charged metal ions. The ability to desorb and recover bound metals from immobilized beads is of great economic significance. The concentrated metals could be recycled and reused in manufacturing processes and the regenerated beads could be reused, which would reduce operational costs of the algal treatment system. Such metal adsorption-desorption cycles also make the immobilized algal system mechanically suitable for batch and continuous packed-bed reactors. At least 8 successive copper adsorption and desorption cycles by immobilized yeast biomass has been reported.[50] Our recent studies also found that the packed-bed alginate immobilized algal beads could be reused for more than 10 successive cycles without lowering the treatment efficiency of Cu and Ni from electroplating effluent.[42]

Conclusions

The present study shows that both free and alginate-immobilized unicellular green alga, *Chlorella vulgaris,* were capable of removing copper from solution. The amount of copper removal was related to cell densities—more copper was removed when high cell densities were used. However, the increase in cell density led to a lower specific uptake of copper by algal cells, suggesting the amount of algal biomass employed for metal treatment is an important parameter and must be determined. In addition to cell densities, the copper removal was affected by pretreatment of algal cells, and dead cells, especially heat killed ones, adsorbed more copper than living cells. Free cells at a high cell density, 5×10^8 cells mL^{-1}, removed more copper and at a faster rate than the alginate- immobilized cells. Most of the adsorbed copper in alginate beads was retained in alginate matrix, with less than 15% adsorbed by cells, indicating alginate blank beads had a comparable copper removal efficiency as the algal beads. The study also demonstrates that the immobilized beads had a large capacity to retain copper but at pH lower than 2, adsorbed metals would be desorbed, suggesting the metal saturated beads could be regenerated and reused.

Acknowledgments

The authors would like to thank Dr. Arthur Lau, of HKUST, and Mr. Cheung Sung Fung, of the City University of Hong Kong, for their assistance. The project was financially supported by the Croucher Foundation and the University Research Grant Council of Hong Kong.

References

1. Corradi MG, Corbi G, Bassi M. Hexavalent chromium induces gametogenesis in the freshwater alga *Scenedesmus acutus*. Ecotoxicol Environ Safety 1995; 30: 106-110.
2. Volesky B. Advances in biosorption of metals: selection of biomass types. FEMS Microbiol Rev 1994; 14:291-302.
3. Asku Z, Sag Y, Kutsal T. The biosorption of copper (II) by *C. vulgaris* and *Z. ramigera*. Environ Technol 1992; 13: 579-586.
4. EPD (Environmental Protection Department). Environment Hong Kong 1990. Hong Kong:Hong Kong Government Printer, 1990.
5. Patterson JW. Waste Water Treatment. USA: Science Publishers Inc., 1997.
6. Aksu Z, Kutsal T. A comparative study for biosorption characteristics of heavy metal ions with *C. vulgaris*. Environ Technol 1990; 11:979-987.
7. Trevors JT, Stratton GW, Gadd GM. Cadmium transport, resistance and toxicity in bacteria, algae and fungi. Can J Microbiol 1986; 32:447-464.
8. Gadd GM. Microbial control of heavy metal pollution. In: Fry JC, Gadd GM, Jones CW et al., eds. Microbial Control of Pollution. Cambridge: Cambridge University Press, 1992:59-88.
9. Volesky B. Removal and recovery of heavy metals by biosorption. In: Volesky B ed. Biosorption of Heavy Metals. Boca Raton:CRC Press, 1990:7-43.
10. Lombrana JI, Varona F, Mijangos F. Study of nickel sorption onto biological sludges: Possibilities for heavy metal removal treatments. Wat Air Soil Pollut 1995; 82:645-658.
11. Chatterjee S, Asthana RK, Tripathi AK et al. Metal removal by selected sorbents. Process Biochem 1996; 31:457-462.
12. Simmons P, Tobin JM, Singleton I. Considerations on the use of commercially available yeast biomass for the treatment of metal-containing effluents. J Ind Microbiol 1995; 14:240-246.
13. Lustigman B, Lee LH, Khalil A. Effects of nickel and pH on the growth of *Chlorella vulgaris*. Bull Environ Contam Toxicol 1995; 55:73-80.

14. Greene B, McPherson R, Darnall D. Algal sorbents for selective metal ion recovery. In: Patterson JW, Passino R, ed. Metals Speciation, Separation and Recovery. Chicago: Lewis Publishers, Inc., 1987:315-318.

15. Aksu et al. 1990 Asku Z, Sag Y, Kutsal T. A comparative study of the adsorption of chromium (VI) ions to *C. vulgaris* and *Z. ramigera*. Environ Technol 1990; 11:33-40.

16. Wilkinson SC, Goulding KH, Robinson PK. Mercury removal by immobilized algae in batch culture systems. J Appl Phycol 1990; 2:223-230.

17. Wehrheim B, Wettern M. Comparative studies of the heavy metal uptake of whole cells and different types of cell walls from *Chlorella fusca*. Biotechnol Tech 1994; 8:221-226.

18. Blumreisinger M, Meindl D, Loos E. Cell wall composition of Chlorococcal algae. Phytochem 1983; 22:1603-1604.

19. Aksu Z, Kutsal T. A bioseparation process for removing lead (II) ions from wastewater by using *C. vulgaris*. J Chem Tech Biotechnol 1991; 52:109-118.

20. Darnall DW, Greene B, Henzl MT et al. Selective recovery of gold and other metal ions from an algal biomass. Environ Sci Technol 1986; 20:206-208.

21. Greene B, Hosea M, McPherson R et al. Interaction of gold (I) and gold (III) complexes with algal biomass. Environ Sci Technol 1986; 20:627-632.

22. Crist RH, Oberholser K, Schwartx D et al. Interaction of metals and protons with algae. Environ Sci Technol 1988: 22:755-760.

23. Xue HB, Strumm W, Sigg L. The binding of heavy metals to algal surfaces. Wat Res 1988; 22:917-926.

24. Wilde EW, Benemann JR. Bioremoval of heavy metals by the use of microalgae. Biotechnol Adv 1993; 11:781-812.

25. Crist RH, Martin JR, Carr D et al. Interaction of metals and protons with algae. 4. Ion exchange *vs* adsorption models and a reassessment of scatchard plots: ion exchange rates and equilibrium compared with calcium alginate. Environ Sci Technol 1994; 28:1859-1866.

26. Harris PO, Ramelow GJ. Binding of metal ions by particulate biomass derived from *Chlorella vulgaris* and *Scenedesmus quadricula*. Environ Sci Technol 1990; 24:220-228.

27. Ting YP, Lawson F, Prince IG. Uptake of cadmium and zinc by the alga *Chlorella vulgaris*: II Multi-ion situation. Biotechnol Bioengn 1991; 37:445-455.

28. Avery SV, Tobin JM. Mechanisms of Sr uptake by laboratory and brewing strains of *Saccharomyces cerevisiae*. Appl Environ Microbiol 1992; 58:3883-3889.

29. Urriutia Mera M, Kemper M, Doyle R etal. The membrane induced proton motive force influences the metal binding ability of *Bacillus subtilus* cell walls. Appl Environ Microbiol 1992; 58:3837-3844.

30. Brady JM, Tobin JM. Adsorption of metal ions by *Rhizopus arrhizus* biomass: characterization studies. Enzyme Microbiol Technol 1994; 16:671-675.

31. Robinson PK, Wilkinson SC. Removal of aqueous mercury and phosphate using gel entrapped *Chlorella* in packed bed reactors. Enzyme Microbiol Technol 1994; 16:802-807.

32. Tolley MR, Strachan LF, Macaskie LE. Lanthanum accumulation from acidic solutions using a *Citrobacter* sp. immobilized in a flow-through bioreactor. J Ind Microbiol 1995; 14:271-280.

33. Wilkinson SC, Goulding KH, Robinson PK. Mercury accumulation and volatilization in immobilized algal cell systems. Biotechnol Lett 1989; 11:961-964.

34. Garnham GW, Codd GA, Gadd GM. Accumulation of cobolt, zinc, and maganese by the estuarine green microalge *Chlorella salina* immobilized in alginate microbeads. Environ Sci Technol 1992; 26:1764-1770.

35. Sharma OP. Textbook of Algae. New Delhi: McGraw-Hill, 1986: 161pp.
36. Tam NFY, Wong YS. Wastewater nutrient removal by *Chlorella pyrenoidosa* and *Scenedesmus* sp. Environ Pollut 1989; 58:19-34.
37. Tam NFY, Wong YS. The comparison of growth and nutrient removal efficiency of *Chlorella pyrenoidosa* in settled and activated sewages. Environ Pollut 1990; 65:93-108.
38. Tam NFY, Lau PS, Wong YS. Wastewater inorganic N and P removal by immobilized *Chlorella vulgaris*. Wat Sci Technol 1994; 30:369-374.
39. Tam NFY, Wong YS. Feasibility of using Chlorella pyrenoidosa in the removal of inorganic nutrients in primary settled sewage. In: Phang SM, Lee YK, Borowitzka MA, Whitton BA, eds. Algal Biotechnology in Asia-Pacific Region. University of Malaya, Malaysia, 1994:291-299.
40. Lau PS, Tam NFY, Wong YS. Effect of algal density on nutrient removal from primary settled wastewater. Environ Pollut 1995; 89:59-66.
41. Lau PS, Tam NFY, Wong YS. Wastewater nutrients removal by *Chlorella vulgaris*: Optimization through acclimation. Environ Technol 1996; 17:183-189.
42. Lau PS, Tam NFY, Wong YS. An algal biosystem for industrial wastewater treatment. In: Proceedings of Asian Industrial Technology Congress Vol. 1. Hong Kong: The Operation and Organizing Committee, AITC, 1997:116-120.
43. Harrison PJ. Phytoplankton stains. In: Lobban CS, Chapman DJ, Kremer BP, eds. Experimental Phycology: A Laboratory Manual. Cambridge: Cambridge University Press, 1988:23-26.
44. Pirszel J, Oawlik B. Skowronski T. Cation-exchange capacity of algae and cyanobacteria: a parameter of their metal sorption abilities. J Ind Microbiol 1995; 14:319-322.
45. Sloof JE, Viragh A, Van der Veer B. Kinetics of cadmium uptake by green algae. Wat Air Soil Pollut 1995; 83:105-122.
46. Ting YP, Teo WK, Soh CY. Gold uptake by *Chlorella vulgaris*. J Appl Phycol 1995; 7:97-100.
47. Munda IM, Hudnik V. Trace metal content in some seaweeds from the Northern Adriatic. Bot Mar 1991; 34:241-249.
48. Malea P., Haritonidis S., Straits I. Bioaccumulation of metals by Rhodophyta species at Antikyra Gulf (Greece) near an aluminium factory. Bot Mar 1994; 37:505-513.
49. Sigma. Sigma Biochemicals, Organic Compounds for Research, and Diagnostic Reagents 1994. Sigma Chemical Company;1994.
50. Wilhelmi BS, Duncan JR. Reusability of immobilized *Saccaromyces cerevisiae* with successive copper adsorption - desorption cycles. Biotechnol Lett 1996; 18:531-536.

CHAPTER 3

Biosorption of Heavy Metals by Microalgae in Batch and Continuous Systems

Zümriye Aksu

Introduction

The removal of toxic or economically important heavy metal ions from wastewaters is of great importance from an environmental and industrial viewpoint. The biosorption of heavy metal ions by algae is a promising property with a potential for industrial use. The focus of this chapter is on investigating the biosorption of single- and multi-metal ions to free and immobilized microalgae in batch and continuous systems. There will be a discussion of heavy metal pollution, the usage and advantages of free and immobilized algal cells for heavy metal removal, biosorption mechanisms and developing isotherms in single- and multi-metal systems. Mathematical description of biosorption of a single metal ion to algae in a batch and in a continuous fixed bed column will also be presented and discussed.

Heavy Metal Pollution

Environmental pollution due to technological development is one of the most important problems of this century. Heavy metals such as copper, nickel, lead, chromium, cadmium, etc., in wastewaters are hazardous to the environment. Due to their toxicity, their effect on our ecosystem presents possible human health risk. The existence of these ions in water may cause toxic and harmful effects to living organisms in water and also to the consumers of it.[1-5]

All of the reported metals are widely used materials in daily life. Acid mine drainage, plating and brass plating industries involve undesirably high levels of Copper (II). In zinc and brass plating, in viscose rayon yarn and fiber production and in metal processing industries, zinc ion concentrations are higher than values allowed by water standards. Mining and metal processing industries, dye and textile industries, and petroleum refining cause iron(II) and iron(III) pollution. Lead(II) pollution results from dye and textile, and ceramic and glass industries,

Wastewater Treatment with Algae, edited by Yuk-Shan Wong and Nora F.Y. Tam.
© Springer - Verlag and Landes Bioscience 1998.

as well as, petroleum refining, storage battery manufacture and tetraethyl lead production. Besides these ions, chromium(VI), now known to be carcinogenic, is also a common pollutant in the environment. Mining and metallurgy of chromium, chromium coating and electroplating, and dye and leather industry wastes contain undesirable amounts of chromium(VI) ions. Due to the recently discovered extreme acute toxicity of cadmium, this metal has now joined lead(II) and mercury(II) as one of the "big three" heavy metals posing the greatest hazard to human health. Cadmium(II), well recognized for its negative effect on the environment where it accumulates throughout the food chain, is used in a wide variety of industrial processes, e.g., alloy preparation, metal plating and electronics. A great deal of these aforementioned metal ions are also frequently encountered together in industrial waste waters such as from plants producing machine parts and chemicals. As mandated by various water standards, heavy metal ion levels in waste waters must be controlled and reduced to desired values.[1-10]

The removal of toxic heavy-metal contaminants from industrial wastewaters is one of the most important environmental issues to be solved today. Waste water treatment, including heavy metal ions, generally depends on the capacity and the type of plant, the characteristics and flux of waste water and the treatment method and materials. Chemical precipitation, ion exchange, reverse osmosis and solvent extraction are the most commonly used procedures for removing metal ions from dilute aqueous streams. However, these high-technology processes have significant disadvantages, such as incomplete metal removal, expensive equipment and monitoring systems requirements, high reagent or energy requirements and/or generation of toxic sludge or other waste products that require disposal. Such processes also may be ineffective or extremely expensive when initial heavy metal concentrations are in the range of 10-100 mg/l. New technologies are required that can reduce heavy metal concentrations to environmentally acceptable levels at affordable costs.[1-5, 7,11-13]

The Use of Algal Cells for Heavy Metal Ions Removal

The search for new and innovative treatment technologies has focused attention on the effect of heavy metal toxicity on and uptake by microorganisms. In the area of waste water treatment, there appears to be a great potential in exploiting the intrinsic capability of microorganisms to clean up waters. Using microorganisms as biosorbents for heavy metals offers a potential alternative to existing methods for detoxification and recovery of toxic or valuable metals from industrial waste waters. Many yeasts, algae, bacteria and various aquatic flora are known to be capable of concentrating metal species from dilute aqueous solutions and accumulating them within the structure of the microorganism. Based upon this property, more economic, practical and efficient techniques are being developed for the treatment of industrial waste waters.[2,3,7-19]

Metal accumulative bioprocesses generally fall into one of two categories: biosorptive uptake by nonliving, non-growing biomass or biomass products and bioaccumulation by living cells. Bioaccumulative uptake forms the principle for waste detoxification processes using, for example, biological fluidized beds employing continually growing biofilm. The continually self-replenishing system can be left to run continuously for extended periods if the problem of metal toxicity to the growing cell is overcome by the use of metal-resistant organisms. This cannot be expressed simply in terms of metal accumulated per unit of biomass since both are continually increasing. Continued metal uptake is dependent upon biomass growth which may be sensitive to other co-pollutants; the system operates in a nondefined and empirical way which can be extremely difficult to model in mathematical, predictive terms. The living cells do, however, have the potential for mutant isolation or generic recombination to improve the metal-accumulative strain. The potential for metal recovery from intracellular sites is limited and these processes are generally restricted to treatment of low metal content wastes, with sacrificial biomass. Maintaining a viable organism during metal recovery can be very difficult.[2-5,9,10,15,19-22]

Dead (heat killed, dried, acid and/or otherwise chemically treated) cells have been shown to accumulate heavy metal ions to the same or greater extent than growing or resting cells. The main reason for this is the inhibition of microbial growth when the concentration of metal ion in industrial effluent is too high or when significant amounts of metal ions are adsorbed by the microorganisms. The use of dead microbial cells is more advantageous for water treatment in that dead organisms are not affected by toxic wastes. Nonliving biomass doesn't need to be provided with nutrients and can be used for many process cycles. Dead cells may be stored or used for extended periods at room temperature without putrefection occurring. This kind of adsorption is also termed as "biosorption." Biosorption is quite rapid, occurs to a high degree and is frequently selective. The mechanisms associated with metal biosorption by microorganisms are complex, dependent on the metal ion and the biological system and include extracellular and intracellular metal binding. The biosorption of heavy metal ions is also a very beneficial property of most algae and algae have been successfully used as a sorption agent for heavy metals.[2,4,11,15,17,19,20,23]

Biosorption of heavy metal ions by algae is better than precipitation in terms of ability to adjust to changes in pH and heavy metal concentrations, and better than ion exchange and reverse osmosis in terms of sensitivity to the presence of suspended solids, organics, and the presence of other heavy metals. Also, only ion exchange can compete with biosorption in terms of residual heavy metal concentrations. Overall, there are some potential advantages of biosorption processes compared with conventional heavy metal removal methods, including:

• Use of naturally abundant renewable biomaterials that can be cheaply produced;

• Ability to treat large volumes of wastewater due to rapid kinetics;

• High selectivity in terms of removal and recovery of specific heavy metals;

• Ability to handle multiple heavy metals and mixed waste;

• High affinity, reducing residual metals to below 1 ppb in many cases;

• Less need for additional expensive reagents which typically cause disposal and space problems;

• Operation over a wide range of physicochemical conditions including temperature, pH, and presence of other ions (including Calcium (II) and Magnesium (II));

• Relatively low capital investment and low operational costs;

• Greatly improved recovery of bound heavy metals from the biomass;

• Greatly reduced volume of hazardous waste produced.[2,19,20,24]

Use of Algal Cells for Multi-Metal Ion Removal

While much research has been carried out on the uptake of single species of metal ions, little attention seems to have been given to the study of multi-metal ion systems. Despite the fact that single toxic metallic species rarely exist in nature and wastewaters and the presence of a multiplicity of metals often gives rise to interactive effects, insufficient attention seems to have been paid to this problem. The examination on the effects of divalent cations in various combinations is more representative of the actual environmental problems faced by organisms than are single metal studies. This recognition results from the realization that environmental loadings of cations from anthropogenic sources rarely involve single cation contributions, and if they do, the introduced cation will interact with a host of chemicals native to the receiving system. Thus, organisms potentially impacted by these toxicants face a multiple rather than a single toxicant insult.

In general, a mixture of heavy metals can produce three possible types of behavior: synergism, antagonism and noninteraction. These terms are defined as follows:

• Synergism: The effect of the mixture is greater than that of each of the individual effects of the constituents in the mixture.

• Antagonism: The effect of the mixture is less than that of each of the individual effects of the constituents in the mixture.

• Noninteraction: The effect of the mixture is no more or no less than that of each of the individual effects of the constituents in the mixture.

Each of these responses has been shown by a number of microorganisms, including algae.[11,12,17,21,22,26,27]

Use of Immobilized Algae for Heavy Metal Biosorption

Much of the bioremoval literature deals with artificially immobilized biomass. Researchers have recognized that immobilizing nonliving or living biomass in a granular or polymeric matrix may improve biomass performance, biosorption capacity and facilitate separation of biomass from metal-bearing solution. Biosorptive metal accumulation by immobilized nonliving biomass have advantages such as:

• Growth-independent, nonliving biomass, not subject to toxicity limitations. No necessity for nutrients in feed solution, or problems of disposal of surplus nutrients or metabolic products;

• Immobilization processes are clear-cut; they allow higher biomass concentrations and column operations;

• Process not governed by physiological constraints;

• Choice of immobilization technique not governed by toxicity limitations or thermal inactivation;

• Very rapid and efficient metal uptake; the biomass behaves as an ion exchanger;

• Metals can be desorbed readily, and recovered. Immobilized systems may be well suited for non-destructive recovery, and after metal loading the metal may be concentrated in a small volume of solid material or desorbed into a small volume of eluant for recovery, disposal or containment;

• Unlike certain membrane systems, there may be only very limited clogging in continuous-flow systems involving immobilized biomass;

• The system can be mathematically defined.

Some of the disadvantages of biosorption by immobilized non-living biosorbent follow:

• Early saturation: when metal-interactive sites are occupied, metal desorption is necessary prior to further use. Irrespective of the metal value;
   • Adsorptive uptake is sensitive to pH, metal speciation;
   • There is no potential to biologically alter the metal valency state, e.g., to give less soluble forms;
   • There is no potential for degradation of organometallic species;
   • The potential for biological process improvement is limited since the cells are not metabolizing: production of adsorptive agent during pre-growth;
   • Diffusion limitations are created, which result in many of the surface sites on the biomass only slowly becoming available to the metal ions.

In the specific case of algae immobilization, several polymeric matrices have already been studied, sodium alginate being the one with greatest accumulated knowledge. However, a series of other supports such as glutaraldehyde, agar, cellulose-acetate, with lower cost, are being tried. These gels, along with polyacrylamide, ethyl acrylate-ethylene glycol dimethacrylate, polysulfone and silica gels can be highly efficient in small-scale systems where diffusion limitations are minimized. Such gels can be hardened, formed into beads and used in packed-bed, fluidized-bed and air-lift bioreactors that resemble the column processes used with ion exchange resins and activated carbon systems. The use of countercurrent processes can substantially reduce the amount of biomass required in the process. There are some problems with these immobilizing agents, however. For example, the calcium-alginate system has a disadvantage in that it is prone to damage by cation replacement or chelation—an important attribute when accumulating metal cations. Moreover, alginate systems are unstable at high pH. The immobilization of algae in a polyacrylamide matrix has also been used to demonstrate the potential for removal and selective recovery of metals from waste streams. Polyacrylamide immobilization has been shown to be superior to calcium-alginate, glutaraldehyde, agar or cellulose-acetate immobilization because this immobilization process only slightly decreases metal biosorption properties of the biomass. Polyacrylamide immobilization has an advantage in that it is not prone to damage by cation replacement or chelation, as calcium-alginate systems are an important attribute when accumulating metal cations. However, polyacrylamide is also far from ideal as an immobilizing medium for algal cells. Polyacrylamide gel, although providing a good model system, would be inappropriate for large-scale use due to the expense and toxicity of the gel precursors, and the lack of mechanical strength due to high water concentrations(80-95%). The other polymeric matrices have similar problems.[2,11,12,14,16,18,20,24,27-30,34]

Biosorption Mechanisms of Algae

Biosorption involves a combination of active and passive transport mechanisms starting with the diffusion of the metal ion to the surface of the microbial cell. Once the metal ion has diffused to the cell surface, it will bind to sites on the cell surface which exhibit some chemical affinity for the metal. This step contains a number of passive accumulation processes and may include adsorption, ion exchange, coordination, complexation, chelation and microprecipitation. Generally, such metal ion adsorption is fast, reversible, and not a limiting factor in bioremoval kinetics when dealing with dispersed cells. Biosorption is often followed by a slower metal binding process in which additional metal ion is bound, often irreversibly. This slow phase of metal uptake can be due to a number of mechanisms, including covalent bonding, surface precipitation, redox reactions, crystallization on the cell surface or, most often, diffusion into the cell interior and binding to proteins and other intracellular sites. Sometimes this slow, irreversible uptake requires metabolic energy (light, substrates), indicating an active transport system.

Microalgae can sequester heavy metal ions by the same adsorption and absorption mechanisms as other microbial biomass. The mechanism of binding metal ions by inactivated algal biomass may depend on the species and ionic charges of metal ion, the algal organism (Differences between algal species in the magnitude of change in metal ion binding capacity upon modification may be due to the wide variation in cell structure, depending upon the algal division, genera and species), and the chemical composition of the metal ion solution. It is important to understand the chemical nature of the metal binding process to use algal biomass effectively for water purification and metal reclamation. The first structure of an algal cell encountered by a heavy metal is the algal cell wall. The biosorption of metal ions by the algal biomass arises from the coordination of the ions to different functional groups on the algal cell. Algal cell walls contain polysaccharides as basic building blocks. The polysaccharides of the cell wall could provide binding amino, carboxylic, sulfhydryl, phosphate and thiol groups as well as the sulphate. The amino groups in the proteins on the cell wall and the nitrogen and oxygen of the peptide bond also could be available for bonding metallic ions. Such bond formation could be accompanied by displacement of protons dependent in part on the extent of protonation as determined by the pH. Metallic ions could also be electrostatically bonded to unprotonated carboxyl oxygen and sulphate. In addition to these functional binding groups, polysaccharides often have ion exchange properties.

Metal ion uptake by algae has been suggested as taking place in two stages. The first stage, known as physical adsorption or ion exchange at the cell surface, is a very rapid, reversible reaction, also called passive uptake and occurs a short time after the microorganism comes into contact with the metal. The subsequent stage, usually referred to as active uptake, is slower, is related to the metabolic activity and involves membrane transport of the metal ions into the cytoplasm of the cell.[2-6,8,11,13,14,22,29,32-36]

Developing Biosorption Isotherms for Single- and Multi-Metal Ion Systems

Equilibrium studies on biosorption provide information about the capacity of the adsorbent or the amount required to remove a unit mass of pollutant under system conditions. Typically, bioremoval studies described in the literature involve the use of 1 to 5 mg/ml dry weight of algal biomass mixed with solutions containing known concentrations of the heavy metal ion of interest. The mixture is shaken for a few minutes and then centrifuged or filtered, and the amount of metal left in solution (and/or present in the biomass) is determined. The most useful presentation of the data is in terms of the residual (equilibrium) metal concentration left in solution after binding (C_{eq}, typical units mg metal/liter) vs. the amount of metal bound to the biomass (q_{eq}, usually determined by difference, typical units mg metal/g of dry weight). The equilibrium biosorptive uptake capacity (q_{eq}) is a function of metal equilibrium(residual) concentration (C_{eq}). An adsorption isotherm is characterized by certain constants, the values of which express the surface properties and affinity of the adsorbent, and can also be used to compare biosorptive capacity of biomass for different metal ions. Such a plot (binding isotherm) is usually fitted to either the Langmuir or the Freundlich equation. Both the non-competetive Langmuir and Freundlich isotherm models have been shown to be suitable for describing the short-term and monocomponent adsorption of heavy metal ions by algal cells. The non-competitive Langmuir model, the most simple one used for adsorption phenomena of one component, has a theoretical basis which relies on a postulated chemical or physical interaction (or both) between solute and vacant sites on the adsorbent surface. The well known expression of the non-competitive Langmuir model is given by Eq. 1.

$$q_{eq} = \frac{Q^\circ b C_{eq}}{1 + b C_{eq}} \qquad 1)$$

where Q° is the maximum amount of metal ion bound at high C_{eq}, and b is a constant related to the affinity of the binding site. We can say the Q° value represents a practical limiting biosorption capacity when the surface is fully covered with metal ions and assists in the comparison of biosorption performance, particularly in cases where the sorbent biomass did not reach its full saturation in experiments. In an other way of speaking, Q° is the maximum value of q_{eq}. Q° and b can be determined from q_{eq}^{-1} and C_{eq}^{-1} plot. The intercept on such a plot is $1/Q^\circ$ and the slope is b/Q°. The model, which is based on assumptions of surface adsorption, showed a good fit where passive biosorption prevailed.

While the Freundlich model doesn't describe the saturation behavior of the sorbent, Q° is the monolayer saturation at equilibrium. The Freundlich equation is empirical, but more often than not fits the data better than the Langmuir equation, suggesting (as expected) that metal binding sites are not equivalent and/or independent. The non-competetive Freundlich equation is given below by Eq. 2.

$$q_{eq} = K_F \, C_{eq}^{1/n} \qquad 2)$$

where K_F and n are the non-competitive Freundlich constants characteristic of the system. K_F and n are indicators of adsorption capacity and adsorption intensity, respectively. Eq. 2. can be linearized in logarithmic form and Freundlich constants can be determined.

Both models were developed for a single-layer metal adsorption. However, the Freundlich model physically provides a more realistic description of metal adsorption by organic matter because it accounts for different binding sites. The Freundlich equation generally agrees well with the Langmuir equation at experimental data over moderate ranges of concentration. Both of them are characteristic of adsorption from aqueous solutions.[2-6,8,9,18,36-39,45]

One of the difficulties in describing the adsorption of metal ions from wastestreams is that wastewaters contain not one, but many metal ions. When several components are present, interference and competition phenomena for adsorption sites occur and lead to a more complex mathematical formulation of the equilibrium. Several isotherms have been proposed to describe equilibrium and competitive adsorption for such a system. These isotherms range from simple models related to the individual isotherm parameters only, to more complex models related to the individual isotherm parameters and to correction factors. One of them is a modified Langmuir model based on the same hypotheses as for the single-component Langmuir model and also assumes identical saturation capacities for all components. The extension of the basic Langmuir model to competitive adsorption of multicomponent mixtures is written as:

$$q_{eq_i} = \frac{Q_i^\circ b_i C_{eq_i}}{1 + \sum_{j=1}^{N} b_j C_{eq_j}} \qquad 3)$$

where the Q°_i and b_i, are derived from the corresponding individual Langmuir isotherm equations.

Sheindrof et al[42] derived a Freundlich-type, multicomponent isotherm model, where the coefficients relating to isotherms could be determined from monocomponent isotherm data except for the biosorption competition coefficients, which had to be determined experimentally. The Freundlich model restricted to multicomponent mixtures could be described as follows:

$$q_{eq_i} = K_{F_i} C_i \left[\sum_{j=1}^{N} a_{ij} C_j \right]^{n_i} \qquad 4)$$

where the K_{F_i} and n_i are derived from the corresponding individual Freundlich isotherm equations and a_{ij}'s are the competition coefficients for N metal ion species.[12,17,26,40-44]

Biosorption of Heavy Metals by Microalgae in Batch Systems

A great deal of the studies are performed in batch systems with single metal ion species. These processes are conceptually simple. A suitable microbial biomass is added to, cultivated in, or otherwise contacted with aqueous solution containing a metal ion. The contacting process is allowed to proceed for a sufficient time for the biomass to sequester the metal ions. Then the biomass is separated from the liquid phase and the metal-containing biomass is either regenerated (by eluting the metal as a concentrated solution) or disposed of in an environmentally acceptable manner.

Basic factors play an extremely important role in the mobility and bioremoval of metallic elements. These factors are the specific surface properties of the organism and the physicochemical influence of the environment. In practice, there are many variables and parameters to consider in the design and operation of such a process:

• Biomass selection and pretreatment (e.g., strain selection, cultivation methods, pretreatment(s), physiological state)

• Containment and contacting (e.g., immobilization scheme, pH, biomass concentration, contact time)

• Separation and recovery (e.g., separating biomass from wastewater after treatment, eluting bound metals from biomass);

• Disposal of spent biomass; and, of course;

• Economic considerations for all of the above steps and the overall process.

Although thousands of species of algae have been identified during the last 200 years, very few have been investigated to determine their absolute or relative abilities to sequester toxic heavy metal ions. In the few, limited studies where species have been compared, results often reveal major differences in metal binding efficacy between species, and even among strains of a single species, for any given metal and/or set of physicochemical conditions. It is also apparent that algae and other microorganisms are much more efficient at removing some metal ions than others. It is well established that several marine and fresh water algae are able to take up various heavy metals selectively from aqueous media and to accumulate these metals within their cells. The vast majority of studies involving metal binding by algae have been conducted using freshwater green algae, principally *Chlorella vulgaris, Chlamydomonas reinhardtii, Chlorella homosphaera, Chlorella pyrenoidosa, Chlorella fusca;* brown marine algae with high alginate content, *Sargassum natans, Focus vesiculosus, Ascophyllum nodosumand, Lamineria japonica* and nitrogen-fixing blue-green algae. These species are relatively easy to grow in culture and readily attainable from numerous culture collection. Also some of these algae are produced commercially in large quantities, primarily for use in the health food market. In addition to the lack of sufficient comparisons of marine and freshwater algae, there is a paucity of data in the literature on the metal uptake abilities of freshwater filamentous diatoms or sheath-forming filamentous blue-green algae, types that are likely to bind metals efficiently and also be amenable to liquid separation techniques.

Microalgae use light as an energy source, facilitating the maintenance of metabolism in the absence of organic carbon sources, and electron acceptors required by bacteria or fungi. Thus the use of metabolically active microalgal systems may be more readily achieved. Also, microalgae cultures can be cultivated in open ponds or in large-scale laboratory culture, providing a reliable and consistent supply of biomass for such studies and eventual scale-up work. The medium alga is grown in and the age or growth phase of the culture appear to also be significant factors influencing metal binding efficacy.

After algal biomass has been cultivated or otherwise obtained, various pretreatments including heat shock, salt shock and acid treatment of the biomass are possible before it is combined with the metal-containing solution in a bioremoval process. Differences in adsorption profiles have also been demonstrated by using various strands of algae and/or freeze drying or heat killing the algal biomass to reduce the matrix effects from living algae.

Temperature, adsorption pH, initial metal ion concentration, biomass concentration, and concentrations of other interfering ions are the environmental influences which are important in the biosorption of heavy metal ions. The binding of most metals to algae by biosorption is enchanced as temperature is increased. Perhaps the most important single parameter influencing the biosorption rate and capacity is pH. Crist et al[31] suggested that zero-point charge, or isoelectric point, would be found at pH 3. Above this, algal cells would have a net negative charge. The ionic state of such ligands as carboxyl, phosphate, imidazole and amino groups will be such as to promote reaction with metal ions. This would lead to electrostatic attractions between positively charged cations such as copper(II), lead(II), zinc(II), cadmium(II), nickel(II) and negatively charged binding sites, hence the rapid rise in binding efficiency between pH 5.0 and 5.5. As the pH is lowered, however, the overall surface charge on the algal cells will become positive and the interaction of metal ions such as chromium(VI), and selenium(VI) in anionic forms, with the algal cells will be primarily electrostatic in nature. Since some metals (e.g., mercury(II), silver(I)) exhibit significant binding over a broad pH change, there must be an additional type of binding that is not electrostatic. From the studies, it can be concluded that although some generalities occur in the literature, optimal pH for bioremoval is unpredictable and highly dependent on the type(s) of algae used and other conditions. In practical applications of biosorption the pH could be adjusted to be more favorable to the removal of the ions of interest. Biosorption can be accomplished with a high yield by increasing metal ion concentrations up to 200-250 mg/l, therefore algal biosorption can be used successfully with both low and high metal ion concentrations in wastewaters. Increasing the mass of the algae sample also seems to be the best method to increase the amount of the metals removed in solution. However, saturating a solution with alga can produce other problems concerning decanting and sample handling. Adjustment of the concentrations of other interfering ions is more problematic. In this regard, microalgae appear to offer a distinctive advantage over ion exchange processes, in that metal binding is not very sensitive to calcium(II) and magnesium(II) ions, avoiding the need for water softening prior to treatment. Although there has been considerable study of metal interactions in microalgae biosorption, few predictions are possible at present. Thus, studies relevant to bioremoval process development must be carried out, whenever possible, with actual or simulated waste solutions, something seldom reported in the literature.

Ideally, the bioremoval process will result in the recovery and reuse of the toxic metal(s) bound by the biomass. Bound metals can be removed from algal biomass with a wash or stripping step and then subsequently be rebound. This can often be achieved by eluting with a metal chelator, a low or high pH solution, or a salt solution, to reduce metal ion binding. In some cases, it is even possible to elute and recover several metals sequentially and selectively.[2-5,8,10,11,13,21,22,24,29,31-36]

Metal ion toxicity and biosorption of multi-metal ions on the algae are also affected by the kind of microorganism, the specific surface properties of the cells and the physicochemical parameters of solution such as pH, species of metals, metal combination and levels of metal ion concentrations. Even though optimum adsorption conditions of heavy metal ions to microorganisms are known, there is a need to understand how combinations of metal ions affect the ecological processes of the biomass and the ability of biomass to accumulate heavy metal ion combinations from the surrounding.

When the combined effects of the two metals together on the same alga is found to be antagonistic; such as the combined effect of copper(VI) and chromium(VI), and cadmium(II) and zinc(II) on *Chlorella vulgaris*, it is postulated that the effective binding sides available for single-metal uptake are reduced, depending on the equilibria between adsorption competition from all the cations. Different cations have different affinities to cell binding sites, hence single-metal uptake is inhibited to varying degrees. The order of inhibition from other cations is in the same series as for their electro-negativities. In contrast, copper(II) and nickel(II) uptake is found to be synergistic for the same alga, The most logical reason for this synergistic action is claimed to be screening or competition for the binding sites on the cellular surface, resulting in the metal ions mutually ameliorating their individual toxic effects.[7,25,26]

Mathematical Modeling of Batch Biosorption

Kinetics of Batch Biosorption

Several mathematical models are available to describe the unsteady-state biosorption of metal ions onto various micro algae in a batch mode. These models use equilibrium isotherms in addition to differential equations describing mass transfer and have ranged in complexity from single-component models to more complex models that include radial concentration profiles. The more complex models provide information on the concentration profile within an individual bead, but add an additional level of complexity.

A nonequilibrium lumped-interior model that uses an overall mass-transfer coefficient that accounts for both film and pore diffusion resistances is given here to describe a monocomponent biosorption in the batch system. Assuming it is limited by external surface, the rate of adsorption can be modelized according to the mass transfer equation:

$$-\frac{dC}{dt} = k_1 a (C - C_{eq}) \qquad 5)$$

where C is the residual metal ion concentration in the solution, a is the specific surface area of the alga and k_1 mass transfer coefficient. C_{eq}; equilibrium metal concentration left in solution after binding is described by the Langmuir or the Freundlich equilibrium equation given in Eqs. 1 and 2. These equations can be easily solved.[12,38,43,46]

Mathematical Description of Batch Biosorption as a Separation Process

The single metal ion biosorption in a batch reactor can be considered as a single-staged equilibrium operation and it depends on two basic constraints, that of equilibrium (given in Eqs. 1 and 2) and that of a mass balance. The mass balance for single metal ion is given by:

$$V_o C_o + X_o q_o = V_o C_{eq} + X_o q_{eq} \qquad 6)$$

$$V_o (C_o - C_{eq}) = X_o (q_{eq} - q_o) \qquad 7)$$

$$-\frac{V_o}{X_o}(C_{eq} - C_o) = (q_{eq} - q_o) \qquad 8)$$

where:

C_o: Initial (or feed) metal ion concentration
q_o: Amount of metal ion adsorbed per unit weight of algae before adding the metal-bearing solution
V_o: Volume of metal-bearing solution
X_o: Amount of sorbent (dried algae)

Eq. 8 belongs to a straight line and the line passes through (C_o, q_o) and (C_{eq}, q_{eq}) with ($-V_o/X_o$) slope as the operation line of this stage. The relation between q_{eq} and C_{eq} is given by the Freundlich or the Langmuir equation, so the single-staged batch operation can be shown in a figure on the same coordinates by drawing an operation line and equilibrium curve according to Eq. 8 and the Freundlich or the Langmuir equation (Eq. 1 or Eq. 2), respectively. V_o/X_o for desired purification or

C_{eq} and q_{eq} values at a given V_o/X_o can be determined from this figure for a given initial (or feed) metal ion concentration. If calculation is required, Equation 8 can be rearranged using the Freundlich equation as:

$$-\frac{V_o}{X_o}(C_{eq} - C_o) = (K_F C_{eq}^{1/n} - q_o) \qquad 9)$$

The amount of metal adsorbed per unit weight of algae at the beginning (q_o) is equal to 0.0, so Eq. 9 can be written as:

$$\frac{V_o}{X_o} = \frac{K_F C_{eq}^{1/n}}{C_o - C_{eq}} \qquad 10)$$

As K_F and n can be found from experimental data, Eq. 10 also gives V_o/X_o for desired purification or C_{eq} at a given V_o/X_o for a given initial (or feed) metal ion concentration.

The removal of a given amount of metal can be accomplished with greater economy of adsorbent if the solution is treated with separate small batches of adsorbent rather than in a single batch, with filtration between each stage. Biosorption of multi-metal ions on the algae in a batch system (or in a series of batch systems) can also be mathematically modelled with the provided effective multicomponent equilibrium constants in the same manner.[6,36-38,46]

Biosorption of Heavy Metals by Microalgae in Continuous Systems

A major consideration with any biosorption scheme is the separation of liquid and solids after batch or counter current contacting. Centrifugation of filtration, as routinely used in the laboratory, are not generally practical in industrial processes; thus, continuous systems such as continuous stirred tank reactors, fluidized bed, moving bed and packed bed columns must be used. For continuous operation, the most convenient configuration is that of a packed column, much like that used for ion exchange. Continuous packed bed biosorption has a number of advantages related to its process engineering. It is a simple, high yielded operation and relatively easily scaled up from a laboratory scale procedure, the stages in the separation protocol can be automated and high degrees of purification can often be achieved in a single step process.

Continuous immobilized or free algae packed into column system is simply a piece of pipe stood on its end and filled with biosorbent beads. Fluid containing the metal ion of interest flows into one end of the pipe, and out the other end. Initially, most of the metal ion is adsorbed, so that the solute concentration in the effluent is low. As biosorption continues, the effluent concentration rises, slowly at first, but then abruptly. When this abrupt rise or "breakthrough" occurs, the flow is stopped.

Much of the information needed to evaluate column performance is contained in plots of effluent concentration as a function of time or throughput volume. The general position of the breakthrough curve along the volume axis depends on the capacity of the column with respect to the feed concentration and flow rate. The maximum capacity of the column for a given feed concentration is equal to the area behind the breakthrough curve. The amount of solute that remains in the effluent is the area under this curve. The breakthrough curve would be a step function for favorable separations, i.e., there would be an instantaneous jump in the effluent concentration from zero to the feed concentration at the moment the column capacity was reached.

It appears from the results that metal uptake in batch systems is dependent on metal concentration. However, in the column experiments 99% removal of metal from solution is obtained before the breakthrough points are reached. Immobilized or free algae column systems behave according to theoretical equilibrium plate theory and are therefore likely to be capable of removing metal cations more completely from solution than would a free cell suspension, which would permit only a single equilibration. The limitation of the efficiency of bioaccumulation is therefore only determined by the physical size of the column. The theoretical plate lengths depend on such factors as the biomass type, the diffusion coefficient in the immobilized biomass, the metal ion species and concentration and the flow rate.[12,14,16,19,20,30,38,39,43,45,46]

Mathematical Description of Continuous Fixed Bed

In continuous fixed bed biosorption, the concentration in the fluid phase and the solid phase change with time as well as with the position in the bed. In order to obtain the scale-up parameters for the design of the processing equipment and to determine the optimum size and operating conditions, mathematical models describing the removal process must be developed. One of the mathematical models is a distributed parameter model taking into account liquid film and pore diffusion resistances to mass transfer which is more suitable for cases where the biosorbent particles are spherical, of constant density and size.

In the biosorption in a continuous fixed bed, the mass balance for the mobile phase is described by Eq. 11:

$$\varepsilon \frac{\partial C}{\partial t} = -v \frac{\partial C}{\partial z} + E \frac{\partial^2 C}{\partial z^2} - (1-\varepsilon) \frac{\partial q}{\partial t} \qquad \text{11)}$$

where:
C: Unadsorbed metal ion concentration in the fluid stream,
e: Void fraction in the bed,
v: Superficial velocity,
E: The axial dispersion coefficient,
t: Time,
z: The axial coordinate.

As in the earlier equation, the left-hand side represents accumulation in the liquid, now taken per differential volume in the bed. The first term on the right-hand side corresponds to the amount of solute flowing in minus that flowing out and the last term on the right-hand side gives the metal ion transferred from the liquid into the adsorbent. The second term on the right-hand side represents dispersion in the bed. Such dispersion leads to mixing of solute and solvent even in the absence of any adsorption and when the fluid phase is liquid and the flow is assumed plug flow, usually E=0.

The second key equation is a mass balance on the sorbed metal ion.

$$(1-\varepsilon)\left(\frac{\partial q}{\partial t}\right) = r \qquad \qquad 12)$$

where r is the adsorption rate per bed volume (has dimensions of mass adsorbed per volume per time). As before, the adsorption rate r depends on the mechanism responsible for adsorption. This mechanism may be controlled by mass transfer from the bulk solution to the surface of the adsorbent. Alternatively, it may be controlled by diffusion and reaction within the adsorbent particles. In either case, the major assumption is made that the adsorption rate is linear:

$$r = k_1 a (C - C_{eq}) \qquad \qquad 13)$$

Eqs. 12 and 13 also describe the monocomponent biosorption process in a batch biosorption unit as mentioned previously. The fourth and final key equation is an isotherm equation which gives the equilibrium between the biosorbent and solution concentrations, as reported previously. The solution of Eqs. 11-13 and the isotherm equation is nonlinear and coupled and so must be found numerically. The complete solutions are not discussed in detail here. For multicomponent adsorption with competition between species, the appropriate forms of Eqs. 11-13 with or without axial dispersion can be rewritten and solved numerically.[12,30,38,43,45,46]

Conclusions

One major conclusion is that biosorption by algae is a technically efficient and economically feasible technology for removing and recovering single- and multimetal ions from solutions in both batch and continuous systems. Selection of the best biosorption process will depend on the particular requirements, biomass, scale, processing methods, contacting environments, waste compositions, and other factors of a site-specific nature.

Microalgae biosorption technologies are still being developed and much more work is required. Some practical applications have been achieved, and the fundamentals look promising: microalgae have the potential to remove metal ions to very low concentrations and to accumulate large amounts of specific toxic elements. However, very little comparative or comparable information, especially economic analyses, are available. Economics will be very case-specific, and will depend on the properties of the biomass in both affinity and capacity for heavy metals of interest in the multi-ion environment of actual wastewaters. This makes the selection of strains suitable for a specific task of major importance. It is important to study the metal accumulating characteristics of many algal species to identify possible individual differences and exploit them.

From a commercial perspective, the objective is to develop a proprietary, low cost, biosorption process that can be easily marketed. Thus far, the approach has been to use commercially available microbial biomass and to develop patentable or proprietary methods, usually an immobilization process, for contacting of the aqueous solution with metabolically inert (dead) microbial biomass. Although higher adsorption rates and capacities for selective or simultaneous removal of metal ions can be carried out in a batch reactor by adjusting the pH of waste water and diluting waste water to the lower levels of metal ion concentrations, the use of either immobilized or non-immobilized algae in continuous columns for the removal of toxic or re-coverable metal ions from waste streams appears much more promising. Immobilizing algae in polymers may lead to a whole new class of biosorbents, which will function like more traditional ion exchange or chelating resins. Immobilization processes currently used are far from optimal, however. This is a general problem in many other fields of biotechnology. The major requirement is to develop immobilization systems which are more open to hydraulic flow yet still provide large contact areas, and which have good mechanical properties. The beads must be mechanically strong, chemically resistant and easy to regenerate and they must not swell and contract during use, as do many ion exchange resins. The production of the beads must be examined in hopes of increasing the yield in the desired size range, improving porosity, and increasing capacity. One suggestion made is that naturally immobilized algal systems could be considered. Algal colonies and flocs are also capable of achieving high cell densities while minimizing diffusion limitations in continuous systems.

In summary, application of biosorption by the algae in selective and simultaneous removal of metal ions from wastewaters offers great potential for large-scale exploitation.

References

1. Patterson JW. Waste Water Treatment. Science Publishers Inc. USA 1977.
2. Wilde EW, Benemann JR, Bioremoval of heavy metals by the use of microalgae. Biotech Adv 1993; 11:781-812.
3. Aksu Z, Kutsal T. A comparative study for biosorption characteristics of heavy metal ions with *C. vulgaris*. Environ Tech 1990; 11:979-987.
4. Aksu Z, Sag Y, Kutsal T. The biosorption of copper(II) by *C. vulgaris*. and *Z. ramigera*. Environ Tech 1992; 13:579-586.
5. Aksu Z, Sag Y, Kutsal T. A comparative study of the adsorption of chromium(VI) ions to *C. vulgaris* and *Z. ramigera*. Environ Tech 1990; 11:33-40.
6. Aksu Z, Kutsal T. A bioseparation process for removing lead(II) ions from wastewater by using *C. vulgaris*. J Chem Tech Biotech 1991; 52:109-118.
7. Panchanadikar VV, Das RP. Biosorption process for removing lead(II) ions from aqueous effluents using *Pseudomonas sp*. Intern J Environ Studies 1994; 46:243-250.
8. Holan ZR, Volesky, B Prasetyo I. Biosorption of cadmium by biomass of marine algae. Biotech Bioeng 1993; 41:819-825.
9. Volesky B, May H, Holan ZR. Cadmium biosorption by *Saccharomyces cerevisiae*. Biotech Bioeng 1993; 41:826-829.
10. Leborans GF Novillo A. Toxicity and bioaccumulation of cadmium in *Olisthodiscus luteus* (Raphidophyceae). Water Res 1996; 30(1):57-62.
11. Harris PO, Ramelow GJ. Binding of metal ions by particulate biomass derived from *Chlorella vulgaris* and *Scenedesmus quadricauda*. Environ Sci Tech 1990; 24:220-227.
12. Trujillo EM, Jeffers TH, Ferguson C, Stevenson HQ. Mathematically modelling the removal of heavy metals from a waste water using immobilized biomass. Environ Sci Tech 1991; 25:1559-1565.
13. Wehrheim B, Wettern M. Biosorption of cadmium, copper and lead by isolated mother cell walls and whole cells of *Chlorella fusca*. Appl Microbiol Biotech 1994; 41:725-728.
14. Brady D, Duncan JR. Bioaccumulation of metal cations by *Saccharomyces cerevisiae*. Appl Microbiol Biotech 1994; 41:149-154.
15. Brady D, Stoll A, Duncan JR. Biosorption of heavy metal cations by non-viable yeast biomass. Environ Tech 1994; 15:429-438.

16. Sag Y, Nourbaksh M, Aksu Z, Kutsal T. Comparison of Ca-alginate and immobilized *Z ramigera* as sorbents for copper(II) removal. Process Biochemistry 1995; 30(2):175-181.

17. Sag Y, Kutsal T. Fully competitive biosorption of chromium(VI) and iron(III) ions from binary metal mixtures by *R. arrhizus*. Use of the competitive Langmuir model. Process Biochemistry 1996; 31:573-585.

18. Mattuschka B, Straube G. Biosorption of metals by a waste biomass. J Chem Tech Biotech 1993; 58:57-63.

19. Spinti M, Zhuang H, Trujillo EM. Evaluation of immobilized biomass beads for removing heavy metals from wastewaters. Water Environ Res 1995; 67(6):943-952.

20. Macaskie LE. An immobilized cell bioprocess for the removal of heavy metals from aqueous flows. J Chem Tech Biotech 1990; 49:357-379.

21. Rachlin JW, Grosso A. The growth response of the green alga *Chlorella vulgaris* to combined divalent cation exposure. Arch Environ Contam Toxicol 1993; 24:16-20.

22. Ting YP, Lawson F, Prince IG. Uptake of cadmium and zinc by the alga *C. vulgaris*: Part 1: Individual ion species. Biotech Bioeng 1989; 34:990-999.

23. Simmons P, Singleton I. A method to increase silver biosorption by an industrial strain of *Saccharomyces cerevisiae*. Appl Microbiol Biotech 1996; 45:278-285.

24. Nestle NFEI, Kimmich R. NMR imaging of heavy metal absorption in alginate, immobilized cells, and kombu algal biosorbents. Biotech Bioeng 1996; 51:538-543.

25. Ting YP, Lawson F, Prince IG. Uptake of cadmium and zinc by the algae *C. vulgaris*: Part II: Multi-ion species. Biotech and Bioeng 1991a; 37:445-455.

26. Açikel Ü. Investigation of adsorption of heavy metal ion combinations in industrial wastewaters to *Chlorella vulgaris*, a green alga. Master Thesis. Hacettepe University, Ankara, Turkey 1996.

27. da Costa ACA, Leite SGF. Metals biosorption by sodium alginate immobilized *Chlorella homosphaera* cells. Biotech Lett 1992; 13(8):559-562.

28. Jang LK, Lopez SL, Eastman SL, Pryfogle P. Recovery of copper and cobalt by biopolymer gels. Biotech Bioeng 1991; 37:266-273.

29. Strong JRP, Madgwick JC, Ralph BJ. Metal binding polysaccharide from the alga, *Klebshormidium fluitans*. Biotech Lett 1991; 4:239-242.

30. Chen D, Lewandowsky Z, Roe F, Surapaneni P. Diffusivity of copper(II) in calcium alginate gel beads. Biotech Bioeng 1993; 41:755-760.

31. Crist HR, Oberholser K, Shank N et al. Nature of bonding between metallic ions and algal cell walls. Environ Science Tech 1981; 15:1212-1217.

32. Khummongkol D, Canterford GS, Fryer C. Accumulation of heavy metals in unicellular algae. Biotech Bioeng 1982; 24:2643-2660.

33. Gardea-Torresdey JL, Becker-Hapak MK, Hosea JM, Darnall DW. Effect of chemical modification of algal carboxyl groups on metal ion binding. Environ Sci Tech 1990; 24:1372-1378.

34. Darnall DW, Greene B, Henzl MT et al. Selective recovery of gold and other metal ions from an algal biomass. Envir Sci Tech 1986; 20(2):206-208.

35. Becker EW. Limitations of heavy metal removal from wastewater by means of algae. Water Res 1983; 17(4):459-466.

36. Aksu Z, Özer D, Ekiz H, Kutsal T, Çağlar A. Investigation of biosorption of chromium(VI) on *Cladophora crispata* in two staged batch reactor. Environ Tech 1996; 17:215-220.

37. Belter PA, Cussler EL, Hu WS. Bioseparations-Downstream Processing for Biotechnology. Wiley-Interscience, New York, 1988.

38. Humphrey AE Millis NF. Biochemical Engineering. 2nd Ed. Academic Press 1973.

39. Ruthven DM. Adsorption Kinetics-Adsorption: Science and Technology. In: AE Rodrigues et al, eds. NATO ASI Series, Series E: Applied Sciences, Kluwer Academic Publishers 1989; 158:87-114.
40. Bellot JC, Condoret JS. Modelling of liquid chromatography equilibria. Process Biochemistry 1993; 28:365-376.
41. Khan AR, Al-Wheab IR, Al-Haddad A. A generalized equation for adsorption isotherms for multi-component organic pollutants in dilute aqueous solution. Environ Tech 1996; 17:13-23.
42. Sheindrof C, Rebhun M, Sheintuch M. A Freundlich type multicomponent isotherm. J Colloid Interface Sci 1981; 79:136-141.
43. Tan HK, Spinner IH. Multicomponent ion exchange column dynamics. The Can J Chem Eng 1994; 72:330-341.
44. Aksu Z, Asikel U, Kutsal T. Application of multi-comparent adsorption isotherms to simultaneous biosorption of iron(III) and chromium(VI) on *C. vulgaris*. J Chem Tech Biotech, in press 1997.
45. Tsezos M, Deutschmann AA. The use of a mathematical model for the study of the important parameters in immobilized biomass biosorption. J Chem Tech Biotech 1992; 53:1-12.
46. Guibal E, Lorenzelli R, Vincent T, Cloirec P. Application of silica gel to metal ion sorption: static and dynamic removal of uranyl ions. Environ Tech 1995; 16:101-114.

CHAPTER 4

Microalgal Removal of Organic and Inorganic Metal Species from Aqueous Solution

Simon V. Avery, Geoffrey A. Codd and Geoffrey M. Gadd

Introduction

Microalgae are important primary producers, ubiquitous in fresh, brackish and marine waters and in terrestrial ecosystems. Microalgal population densities vary; under appropriate conditions (e.g., abundance of Fe- and P-containing nutrients) microalgal blooms may form.[1] Dense microalgal populations may be detrimental to water quality, although they may also be exploited for human benefit. For example, in shallow oxidation ponds or waste stabilization ponds, oxygen production from microalgal photosynthesis may support bacterial oxidation of raw primary sewage or secondary sewage products.[2] The specific exploitation of microalgae in biotechnology is particularly attractive as these organisms have relatively simple nutritional requirements (e.g., solar energy and carbon dioxide) and their cultivation does not necessitate addition of large amounts of expensive substrates. Thus, microalgae have been used as "biocatalysts" for the commercial production of vitamins, carotenoids, pigments, polysaccharides and single-cell protein.[3] More recent applications have also included decontamination of wastewaters, e.g., those containing phosphorus-, sulfur- and nitrogen-rich pollutants.[3] The potential role and/or use of microalgae in toxic metal removal is another area that is currently generating considerable interest.

Pollution resulting from the continuing accidental and controlled release of toxic metals into the environment has led to concern over the interactions of metals with biological systems in recent years.[4] One of the main problems associated with toxic metal pollution is that unlike organic chemicals, which can be broken down to innocuous products such as carbon dioxide and water, metals persist in the environment. Certain metals can be converted to organic forms, which in some cases involves biologically-mediated processes.[5] However, many organometallic compounds, e.g., organotins, can be more toxic to biota than their inorganic forms.[6] Aquatic ecosystems are often the ultimate destination for heavy metal pollutants

Wastewater Treatment with Algae, edited by Yuk-Shan Wong and Nora F.Y. Tam.
© Springer - Verlag and Landes Bioscience 1998.

and both marine and freshwater microorganisms may become exposed to elevated localized metal concentrations.[7] This situation has been exacerbated by the application of metal-containing formulations to control algal and cyanobacterial growth.[8] Microalgae are commonly found in metal-contaminated freshwater habitats.[9] Differential selection of metal-tolerant species during heavy metal pollution of water bodies can lead to an altered structure of indigenous algal communities.[10] Microalgae represent the lowest trophic level in many aquatic food chains. Thus, the transfer of accumulated metals through food webs from algae to higher organisms, which has been clearly demonstrated,[11] is of considerable concern. However, as a result of their high capacity for metal accumulation,[12-14] microalgae may also help reduce the problems associated with metal release into the environment.

Accumulation of heavy metals by algae generally comprises two phases. A rapid initial biosorptive phase is complete within 5-10 min, is reversible and metabolism-independent (being unaffected by metabolic inhibitors). A slower second phase of accumulation is often irreversible and metabolism-dependent.[10,15] The relative importance of these two processes is dependent on both the metal and the algal species. Biosorption has been noted as the major component of uptake for several metals in *Chlorella* spp.,[12,16] including cadmium, whereas the reverse was true for cadmium uptake in *Eremosphaera viridis*[17] and caesium uptake in *Chlorella emersonii*.[18] In other cases metal uptake via the two mechanisms may be of a similar order.[15,19] Metal biosorption by algae can generally be represented by Freundlich and Langmuir isotherms, indicating a linear relationship between the metal concentration in solution and that bound to the cell.[20,21] Algal cell walls provide many potential metal-binding sites associated with both polysaccharides (e.g., cellulose) and proteins.[19] Both ionic and covalent bonding are involved in metal biosorption to carboxyl, amine and sulfydryl groups.[13,22-25] Nakajima et al[12] reported that the binding capacity for a range of metals was similar in live and dead *Chlorella vulgaris*. However, in certain cases additional binding sites may become available in dead cell-preparations.[26] Whole microalgal cells are known to accumulate more metal than their isolated cell walls.[21] In addition to intracellular accumulation as soluble ions, internal metal sequestration sites may include metal-binding polypeptides, such as phytochelatins.[27,28]

Environmental parameters strongly affect the speciation of dissolved metals and hence their availability to aquatic microorganisms.[29] Precipitation of metals at high pH can lead to reduced microalgal metal accumulation. Conversely, alterations in cytoplasmic-membrane potential and decreased competition from H^+ in alkaline medium may enhance metal uptake in many cases.[30,31] High concentrations of other monovalent and divalent cations (including other heavy metals) can also affect metal uptake by microalgae.[19,32]

Investigations into the potential application of microalgae for decontamination of metal-laden solutions have arisen as a consequence of the high metal-uptake capacities and generally cheap availability of these organisms.[13,21,33-35] Their ease of growth and independence from organic nutrition offer some advantages over several other microbial biosorption systems.[10] Microalgae also display considerably higher metal biosorption capacities than macroalgae, which can be partly attributed to greater surface area to volume ratios.[36] Furthermore, metal adsorption on microalgal surfaces is higher than that on abiotic organic and inorganic adsorbing materials (e.g., activated carbon, silica gel),[37] including clay particles.[14] Microalgal metal removal efficiencies are highly variable.[38,39] This variation is primarily related to differences in the chemical properties of studied metals, although biosorption capacities of different microalgae for single metals can also vary considerably.[39] Clearly, any variation in experimental conditions may also potentially affect metal uptake and a range of manipulations may be adopted to maximize metal removal efficiency. Immobilization in alginate (for improving the physical characteristics of the biomass)[3] generally has little effect on metal biosorption by dead cells (although the alginate matrix may add to total metal-binding), but can result in stimulated active metal uptake by live microalgae.[40] Thus, by packing algal biomass into a permeable polymeric matrix, Barkley[41] achieved very high (~99%) extraction of mercury from contaminated groundwater. Biosorption processes commonly employ batch or continuous-column configurations.[13] Alternative systems for metal removal from polluted waters may use algae contained in lagoons or meander systems. However, the effectiveness of the latter systems hinges on algal activity and may be susceptible to metal toxicity effects.[13] Microalgae are also amenable to metal desorption for biomass recycling and/or for recovery of valuable metals.[42] Recovery of metals bound to microalgal cell surfaces can be achieved by washing with acidic solutions[38,40] or with solutions rich in metal-complexing ligands.[13,25] In addition, metal displacement by supplementation with other cationic species can be effective.[40] Greene et al[43] found that column-packed *Chlorella pyrenoidosa* could be subjected to 50 cycles of gold binding and stripping (using acidic thiourea solutions) with no loss in column efficiency. Similar results were obtained for removal and concentration of Cu(II), a common contaminant of electroplating wastes and ground waters.[13] It is envisioned that microalgal biosorption may act in conjunction with, or possibly replace, existing technologies for metal removal. The latter currently rely on metal precipitation and/or sorption from the medium using exchange resins, which can be expensive.[3,34]

Thus, some information on the removal of polyvalent inorganic metal ions by microalgae is now available. However, information relating to other groups of metals, which may not fall within the traditional "heavy metal" group defined arbitrarily by Duxbury[44] but which nevertheless pose serious environmental problems, is scarce. Caesium is a monovalent radionuclide that has been the focus of particular environmental concern since the Chernobyl disaster in 1986, when large amounts of ^{137}Cs were released into the atmosphere. Other sources of environmental Cs have included nuclear weapons testing and accidental and controlled discharge from nuclear installations.[45] Current technologies for Cs removal from industrial waste effluents (e.g., zeolite ion-exchange) are expensive and relatively non-specific.[33] ^{137}Cs has a relatively long half-life (30.2 years) and is particularly mobile in biological systems because of its high water solubility and similarity to the biologically-essential cation K^+.[46] As a consequence of this latter property, Cs^+ is readily accumulated by microalgae and other microorganisms via active (metabolism-dependent) K^+ transport systems.[46,47] At high external Cs^+ concentrations the displacement of cellular K^+ by Cs^+, but inability of Cs^+ to functionally substitute for K^+, results in Cs^+ toxicity.[18,48] Microalgal monovalent cation transport systems generally exhibit a greater affinity for K^+ than Cs^+. Thus, Cs^+ bioconcentration factors are commonly low. However, the relative affinities of monovalent cation transporter(s) for Cs^+ and K^+ do display some variability and in certain cases Cs^+ uptake may occur preferentially.[46,49] For further enhancing Cs^+ removal, a means of stimulating active Cs^+ (K^+) influx would be desirable. Interestingly, among the mechanisms employed by halotolerant microorganisms to enable rapid adaptation to increases in external salinity is enhanced K^+ accumulation.[50] Thus, the single K^+ (and Cs^+) transport system known to exist in the halotolerant microalga *Chlorella salina*[47] made this organism the ideal candidate for our investigation into NaCl-stimulated microalgal Cs^+ removal, described here.

In common with caesium, very little information is currently available on organometal removal by microalgae. The organotins are one group of highly toxic organic metalloid species that have been widely used in industry, principally for their biocidal properties.[51] The toxicity of organotins depends partly on the number of organic groups attached to Sn and is positively correlated with both their molecular total surface area and their lipid solubility.[6,51] The widespread use of tributyltin (Bu_3Sn) compounds in antifouling paints (e.g., for ship hulls) has led to particular concern over their fate in aquatic ecosystems.[51,52] The toxicity of organotins towards microalgae has been characterized.[53-55] However, very few studies have investigated organotin uptake in the algae, despite accumulation being a prerequisite for toxicity and despite the likely role of algae in enhancing environmental organotin degradation.[56-58] For other organisms, it has been suggested that the organic moieties of organotins become associated with the surfaces of biological membranes rather than penetrating them.[6,51] Thus, over 50% of total Bu_3SnCl biosorption by the fungus *Aureobasidium pullulans* was observed to occur almost instantaneously when the cells and biocide were mixed.[59] Furthermore, accumulation of Bu_3SnCl by a *Pseudomonas* sp. isolate was not influenced by the metabolic activity of the cells and was attributed to adsorption at the cell surface.[60] Previous evidence has indicated that microalgae may outcompete bacteria for Bu_3SnCl biosorption under certain circumstances, which may be attributable to strong adsorption interactions with the algal cell surface.[61] In this chapter, in addition to studies on Cs^+, we characterize organotin (particularly Bu_3SnCl) biosorption by the freshwater microalga *Chlorella emersonii*. The potential application of microalgae in the removal of these model inorganic and organic metal species from aqueous solution is discussed in light of the results presented.

Methods

Axenic cultures of *Chlorella emersonii* 211 8b and *Chlorella salina* Kufferath CCAP 211/25 were grown at 22°C in 100 ml BG-11 freshwater medium[49] or 100 ml MN seawater medium,[47] respectively. Cultures were incubated in 250-ml Erlenmeyer flasks with rotary aeration at 150 cycles min^{-1} and a light intensity of 120 µE $m^{-2} s^{-1}$.

For metal uptake experiments, cells from the late exponential growth phase were collected by centrifugation (1,200 g, 10 min), then washed and resuspended in 10 mM TAPS buffer, pH 8.0 (*C. salina*) or MES, pH 5.5 (*C. emersonii*). Where necessary, NaCl was included in the buffer at the desired concentration. Cells were suspended to a density of approximately $10^6 ml^{-1}$ [or approximately 0.5 mg (dry weight) ml^{-1}]. Cell numbers were determined using a modified Fuchs-Rosenthal hemocytometer. Dry weights were determined after drying to a constant weight at 105°C in tared foil cups.

Cs^+ uptake was initiated by the addition of CsCl, with ^{137}Cs (Amersham) added as a tracer to a final activity of 0.93 - 7.40 kBq ml^{-1}. At intervals, 200 µl samples were removed and harvested by centrifugation (8,000 g, 30 s) through a layer comprising 40% (v/v) Dow-Corning 550 silicone oil and 60% (v/v) bis-3,3,5-trimethylhexylphthalate (Fluka) in 500 µl Beckman PRO22 plastic tubes, using an

Eppendorf 5412 microcentrifuge. The bottom of each tube was cut off and placed in 5 ml Ecoscint A scintillation fluid (National Diagnostics) for 24 h before measuring radioactivity using a Packard Minaxi tri-carb 4000 scintillation counter. To determine radioactivity in the external medium, 200 µl samples of supernatant were taken following centrifugation of cells (1,200 g, 10 min) and analyzed in the same way.

For organotin uptake, 10-30 µl of an appropriate organotin stock solution was added to 1 ml of *C. emersonii* suspension to give the desired organotin concentration. To assess the effect of pH on biosorption of Bu_3SnCl, cells were treated as above but were washed and finally suspended in 10 mM MES (pH 4.5, 5.5 and 6.5) or 10 mM TAPS (pH 7.5 and 8.5) buffers containing 1% (v/v) ethanol. At time intervals, cells were harvested by centrifugation (8,000 g, 30 s) and the supernatant analyzed for residual organotin concentration. Organotin concentrations were measured using a Metrohm (CH-9101 Herisau, Switzerland) polarograph (663 VA stand and 626 polarorecord) as described previously.[62]

Production of *C. salina*-containing calcium-alginate microbeads (approximate diameter, 0.5 mm) for Cs^+ uptake experiments was performed using a microbead maker (obtained from Dr. Gudmund Skjak-Braek, Institute of Biotechnology, University of Trondheim, Norway) as described previously.[53] 7.5 ml packed beads were washed and resuspended in 20 ml 10 mM TAPS + 0.5 M NaCl, pH 8, in 100 ml conical flasks. Cs^+ uptake was initiated and analyzed (using supernatants) as described above.

Results and Discussion

Cs^+ *Accumulation by* Chlorella salina

The effect of NaCl on Cs^+ accumulation by the halotolerant microalga *C. salina* was examined. After 2 h incubation in the presence of 0.5 M NaCl, Cs^+ accumulation was approximately 28-fold higher than that of cells incubated in the absence of NaCl (Fig. 4.1). LiCl had one-half of the stimulatory effect of NaCl on Cs^+ accumulation. Mannitol, at 0.05 or 1.0 M, had negligible stimulatory effect on Cs^+ accumulation (Fig. 4.1). Thus, the effect was specific to NaCl and LiCl, and not attributable to a general osmotic response. The initial rate of active Cs^+ uptake (20-120 min, after cell-wall binding) by *C. salina* was found to increase with NaCl concentration (results not shown). Cs^+ influx (at 50 µM CsCl) was lowest at 50 µM NaCl [approximately 0.3 nmol Cs^+ h^{-1} $(10^6$ cells$)^{-1}$], the lowest concentration examined, and comparable to that of the freshwater species *C. emersonii* in the absence of NaCl; NaCl inhibited Cs^+ uptake by *C. emersonii*.[18] Cs^+ uptake by *C. salina* increased to a maximum level of approximately 4 nmol Cs^+ h^{-1} $(10^6$ cells$)^{-1}$ between 25 and 500 mM NaCl, but was reduced at 1 M NaCl. The results indicate that in contrast to *E. coli*,[64] the K^+ transport system that mediates Cs^+ uptake by *C. salina* is under NaCl-control. In certain diatoms the operation of K^+/Na^+ symports has been described,[65] although any reciprocal stimulation of Na^+ uptake by Cs^+ in *C. salina* (to support a role of Na^+ symport) was not discernible in this study due to a very rapid accumulation of $^{22}Na^+$ (results not shown). It should be stressed that other mechanisms besides a symport may also explain the current observations. Ehrenfeld and Cousin[66] have postulated that K^+/Na^+ exchange in *Dunaliella tertiolecta* counteracts the large passive influx of Na^+ that follows hypertonic shock. However, Pick et

Eppendorf 5412 microcentrifuge. The bottom of each tube was cut off and placed in 5 ml Ecoscint A scintillation fluid (National Diagnostics) for 24 h before measuring radioactivity using a Packard Minaxi tri-carb 4000 scintillation counter. To determine radioactivity in the external medium, 200 µl samples of supernatant were taken following centrifugation of cells (1,200 g, 10 min) and analyzed in the same way.

For organotin uptake, 10-30 µl of an appropriate organotin stock solution was added to 1 ml of *C. emersonii* suspension to give the desired organotin concentration. To assess the effect of pH on biosorption of Bu_3SnCl, cells were treated as above but were washed and finally suspended in 10 mM MES (pH 4.5, 5.5 and 6.5) or 10 mM TAPS (pH 7.5 and 8.5) buffers containing 1% (v/v) ethanol. At time intervals, cells were harvested by centrifugation (8,000 g, 30 s) and the supernatant analyzed for residual organotin concentration. Organotin concentrations were measured using a Metrohm (CH-9101 Herisau, Switzerland) polarograph (663 VA stand and 626 polarorecord) as described previously.[62]

Production of *C. salina*-containing calcium-alginate microbeads (approximate diameter, 0.5 mm) for Cs^+ uptake experiments was performed using a microbead maker (obtained from Dr. Gudmund Skjak-Braek, Institute of Biotechnology, University of Trondheim, Norway) as described previously.[63] 7.5 ml packed beads were washed and resuspended in 20 ml 10 mM TAPS + 0.5 M NaCl, pH 8, in 100 ml conical flasks. Cs^+ uptake was initiated and analyzed (using supernatants) as described above.

Results and Discussion

Cs^+ Accumulation by Chlorella salina

The effect of NaCl on Cs^+ accumulation by the halotolerant microalga *C. salina* was examined. After 2 h incubation in the presence of 0.5 M NaCl, Cs^+ accumulation was approximately 28-fold higher than that of cells incubated in the absence of NaCl (Fig. 4.1). LiCl had one-half of the stimulatory effect of NaCl on Cs^+ accumulation. Mannitol, at 0.05 or 1.0 M, had negligible stimulatory effect on Cs^+ accumulation (Fig. 4.1). Thus, the effect was specific to NaCl and LiCl, and not attributable to a general osmotic response. The initial rate of active Cs^+ uptake (20-120 min, after cell-wall binding) by *C. salina* was found to increase with NaCl concentration (results not shown). Cs^+ influx (at 50 µM CsCl) was lowest at 50 µM NaCl [approximately 0.3 nmol Cs^+ h^{-1} (10^6 cells)$^{-1}$], the lowest concentration examined, and comparable to that of the freshwater species *C. emersonii* in the absence of NaCl; NaCl inhibited Cs^+ uptake by *C. emersonii*.[18] Cs^+ uptake by *C. salina* increased to a maximum level of approximately 4 nmol Cs^+ h^{-1} (10^6 cells)$^{-1}$ between 25 and 500 mM NaCl, but was reduced at 1 M NaCl. The results indicate that in contrast to *E. coli*,[64] the K^+ transport system that mediates Cs^+ uptake by *C. salina* is under NaCl-control. In certain diatoms the operation of K^+/Na^+ symports has been described,[65] although any reciprocal stimulation of Na^+ uptake by Cs^+ in *C. salina* (to support a role of Na^+ symport) was not discernible in this study due to a very rapid accumulation of $^{22}Na^+$ (results not shown). It should be stressed that other mechanisms besides a symport may also explain the current observations. Ehrenfeld and Cousin[66] have postulated that K^+/Na^+ exchange in *Dunaliella tertiolecta* counteracts the large passive influx of Na^+ that follows hypertonic shock. However, Pick et

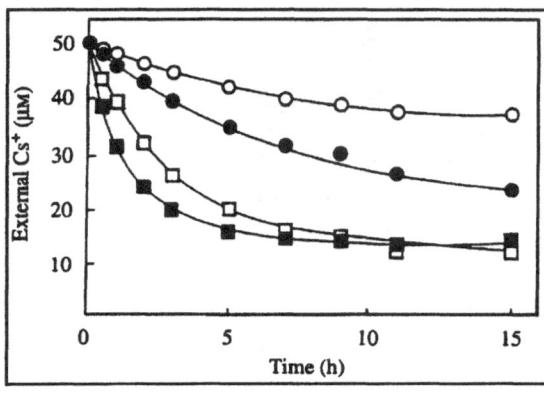

Fig. 4.2. Influence of cell density on Cs⁺ accumulation by *C. salina*. Cells were suspended in 10 mM TAPS buffer + 0.5 M NaCl, pH 8, supplemented with 50 μM CsCl. The graph shows Cs⁺ remaining in solution at cell densities of 4 x 10⁵ (○), 1 x 10⁶ (●), 4 x 10⁶ (□) and 1 x 10⁷ (■).ml⁻¹. Mean values from three replicate determinations are shown. SEM values are smaller than the dimensions of the symbols. Figure adapted from Avery et al. J Gen Microbiol 1993; 139:2239-2244.

highest cell density examined (1×10^7 cells ml⁻¹), accumulation of Cs⁺ was complete after approximately 7 h at which stage approximately 28% of Cs⁺ remained in solution.

In order to assess the effect of alterations in biomass physical properties on Cs⁺ removal-efficiency, Cs⁺ uptake was compared in freely suspended cells and in cells immobilized in calcium-alginate. In both cases, approximately 70% of total Cs⁺ was removed from solution after 15 h incubation in the presence of 50 μM CsCl (Fig. 4.3). However, approximately one-half of that was attributable to binding of Cs⁺ to the calcium-alginate matrix (determined by incubation of cell-free calcium-alginate beads alone). The results indicated that *C. salina* accumulated approximately 46% less Cs⁺ when cells were present in an immobilized state than when in free suspension. This contrasts with the enhanced active uptake of other metals (Co^{2+}, Zn^{2+} and Mn^{2+}) in immobilized *C. salina* described by Garnham et al.[40] However, a reduced relative importance of metabolism-dependent Cs⁺ uptake, compared to cell-surface Cs⁺ adsorption, has been reported in a range of immobilized fungi.[46,71] Immobilization is known to have negligible influence on the photosynthesis of *C. salina*, but results in a 30% decrease in the organism's rate of respiration.[40] Rb⁺ and Cs⁺ influx, in *Chlorella pyrenoidosa* and *C. emersonii*, respectively, are independently supported by cyclic photophosphorylation and oxidative phosphorylation.[18,72] Thus, it is likely that a reduction in respiration would also result in a decline in Cs⁺ uptake in immobilized *C. salina*.

The efficacy of repeated incubations of *C. salina* at high NaCl for removal of Cs⁺ from solution was investigated (Table 4.1). Approximately 72% Cs⁺ was removed after 15 h incubation of *C. salina* in buffer in the presence of 50 μM CsCl. Approximately 95% of accumulated Cs⁺ was released when cells were subsequently washed in buffers of lower osmotic strength, containing 0-50 mM NaCl. A similar effect of hypo-osmotic shock on cellular K⁺ has been reported for another halotolerant microalga, *Chlorococcum submarinum*.[73] Such treatment is known to have very little effect on the growth or structural integrity of *C. salina*.[69] At higher NaCl concentrations, a lower proportion of cellular Cs⁺ was released during washing and negligible Cs⁺ loss resulted at 500 mM NaCl. Re-incubation of the cells for a second 15 h period, in the original Cs⁺-containing buffer, resulted in little or no further Cs⁺ accumulation. In all of the above cases, further accumulation of Cs⁺ did not occur during a third incubation period (Table 4.1). The inclusion of

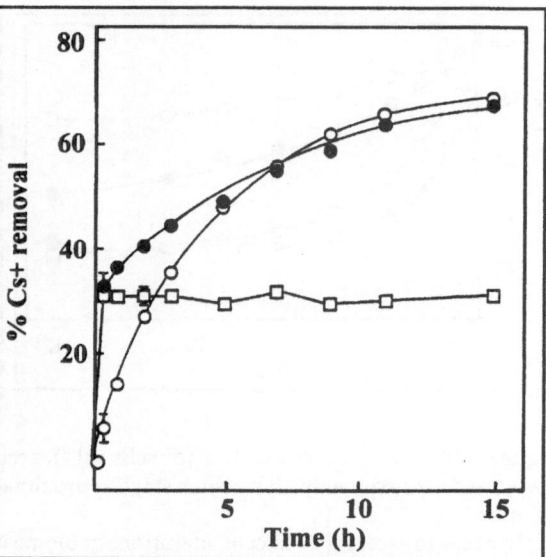

Fig. 4.3. Effect of immobilization on Cs$^+$ accumulation by *C. salina*. Cells (free and immobilized) were suspended to 1 x 10^6 ml^{-1} in 10 M TAPS buffer + 0.5 M NaCl, pH 8, supplemented with 50 μM CsCl. The graph shows % Cs$^+$ removal in the presence of free cells (O), cells immobilized in beads (●) and cell-free beads (□). Mean values from three replicate determinations are shown ±SEM where these values exceed the dimensions of the symbols. Figure adapted from Avery et al. J Gen Microbiol 1993; 139:2239-2244.

increasing amounts of KCl in washing solutions containing 500 mM NaCl resulted in decreased levels of Cs$^+$ release. However, the greatest percentage removal of Cs$^+$, approximately 85.1% after three successive 15 h incubations, occurred when cells were washed in buffer containing 500 mM NaCl and 200 mM KCl. Washing cells in other combinations of NaCl, KCl and mannitol resulted in lower final levels of Cs$^+$ removal. The percentage removal of Cs$^+$ was increased when cells were incubated in the presence of 1 and 10 μM CsCl (Table 4.1). After incubation periods totalling 45 h, >87% of the initial Cs$^+$ was accumulated at these lower CsCl concentrations. Further enhancement of Cs$^+$ removal was achieved by using either fresh cells or fresh buffer for the second incubation period (results not shown). The latter results suggested that the reduced uptake evident during the second incubation period may have been partly due to increased competition from K$^+$ released by cells in exchange for Cs$^+$. In addition, as some cellular Cs$^+$ was retained during washing, subsequent accumulation of Cs$^+$ to a final level equal to that obtained prior to washing only required reduced uptake. The maximal level of Cs$^+$ uptake reported here [85.1% removal from 50 μM Cs$^+$, ~42.6 nmol (10^6 cells)$^{-1}$] corresponds to approximately 66 mg Cs$^+$ (g dry wt)$^{-1}$. This is lower than the value of 150 mg metal (g dry wt)$^{-1}$ suggested by Brierley et al[74] as being a threshold below which biological metal-removal becomes non-viable, although it must be stated that environmental considerations did not take a priority in this assessment. Although no improvement in Cs$^+$ uptake efficiency resulted from cell immobilization, the system did display a number of advantages, including non-destructive treatment for complete recovery of accumulated Cs$^+$, over many other living-cell systems. Furthermore, despite the probable influence of external K$^+$ on Cs$^+$ accumulation following successive incubation periods, competition by K$^+$ is low in *C. salina*[47] compared to certain other microorganisms[46] and the influence of naturally abundant ions like Ca^{2+} and Mg^{2+} was negligible.

Table 4.1. Uptake-efflux cycles for removal/recovery of Cs+ by C. salina

CsCl conc. (μM)	Addition to washing solution*	Percentage removal of Cs+ after: 1 cycle†	2 cycles	3 cycles	% recovery of accumulated Cs+ at 15 h
50	0 - 50 mM NaCl	71.5 ± 1.2	69.7 ± 0.7	69.5 ± 0.6	94.8 ± 1.7
50	200 mM NaCl	71.5 ± 1.2	76.0 ± 0.0	76.0 ± 0.2	67.9 ± 0.2
50	500 mM NaCl	71.5 ± 1.2	75.6 ± 0.0	76.3 ± 0.1	4.0 ± 0.0
50	500 mM NaCl + 50 mM KCl	71.5 ± 1.2	76.9 ± 0.4	77.2 ± 0.8	82.4 ± 0.0
50	500 mM NaCl + 200 mM KCl	71.5 ± 1.2	81.8 ± 0.2	85.1 ± 0.1	63.9 ± 0.2
50	500 mM NaCl + 350 mM KCl	71.5 ± 1.2	63.6 ± 0.6	61.4 ± 0.5	29.3 ± 0.3
50	500 mM NaCl + 500 mM KCl	71.5 ± 1.2	58.6 ± 0.9	52.3 ± 0.3	12.4 ± 0.1
50	250 mM NaCl + 250 mM KCl	71.5 ± 1.2	54.8 ± 1.2	51.9 ± 0.0	64.5 ± 0.7
50	200 mM KCl	71.5 ± 1.2	45.5 ± 0.2	32.3 ± 4.9	11.3 ± 0.0
50	1 M mannitol	71.5 ± 1.2	52.7 ± 0.8	48.2 ± 0.1	3.4 ± 0.0
1	200 mM NaCl	79.3 ± 0.0	85.1 ± 0.2	87.8 ± 0.0	-
10	200 mM NaCl	79.4 ± 0.1	84.9 ± 7.4	87.6 ± 0.0	-

*Cells were washed at the end of each cycle (using 10 mM TAPS, pH 8, with the specified supplements) and then returned to the original Cs+ containing buffer for the next cycle.
†Each cycle represents 15 h incubation in Cs+-containing buffer, followed by washing.

Organotin Biosorption by Chlorella emersonii

Organotin compounds were selected as model organometal species for microalgal biosorption studies. Levels of biosorption differed for the various organotins examined but generally increased with the molecular mass of the triorganotins, being maximal for triphenyltin chloride (Ph_3SnCl) (Fig. 4.4). In contrast, during studies with the microalga *Scenedesmus obliquus*, Bu_3SnCl uptake was found to be greater than Ph_3SnCl uptake.[75] A similar trend to that observed in the present study has already been established for triorganotin toxicity.[51,53,54] This correlation between levels of biosorption/toxicity and the total surface area and lipid solubility of trisubstituted tins supports the theory that triorganotin compounds exert their toxic effects primarily through interactions with membrane lipids.[6,51,55,76] Conversely, although biosorption of monobutyltin chloride ($BuSnCl_3$) was lower than for the other butyltins, Bu_3SnCl uptake was 40% lower than that observed for dibutyltin chloride (Bu_2SnCl_2) in *C. emersonii*. Furthermore, biosorption of tripropyltin chloride (Pr_3SnCl) was lower than $BuSnCl_3$ biosorption (Fig. 4.4). Thus, the lower toxicity of mono- and di-substituted tins is not a result of reduced uptake. It is likely that mono- and di-organotins possess a genuine lesser ability to inhibit cellular processes. The lowest levels of uptake were evident for the low molecular mass trisubstituted organotins, trimethyltin chloride (Me_3SnCl) and triethyltin chloride (Et_3SnCl).

In view of the high levels of uptake and well-known environmental toxicology of Bu_3SnCl,[51] subsequent experiments in the present study focused on this triorganotin. Unlike in cyanobacteria, where Bu_3SnCl uptake can be largely accounted for by rapid adsorption to the cell surface,[62] but consistent with

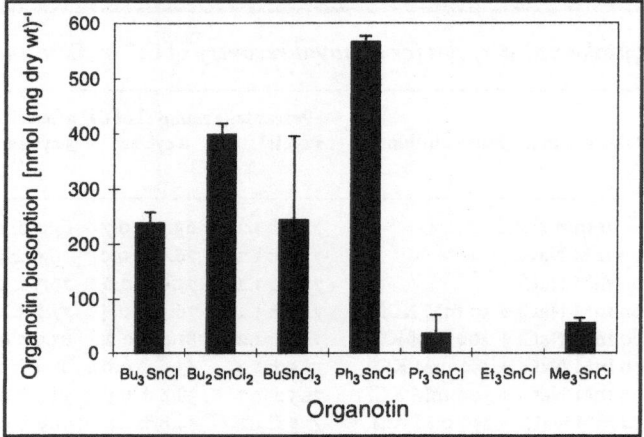

Fig. 4.4. Biosorption of differing organotin compounds by *C. emersonii*. Cells were suspended in 10 mM MES buffer, pH 5.5, supplemented with 0.5 mM of the appropriate organotin chloride. Cells were harvested after 30 min incubation. Mean values from three replicate determinations are shown ±SEM: Bu, butyl; Ph, phenyl; Pr, propyl; Et, ethyl; Me, methyl. Figure adapted from Avery et al. Appl Microbiol Biotechnol 1993; 39:812-817.

observations in higher algae,[58,77] biosorption of Bu_3SnCl by *C. emersonii* was found to be prolonged (results not shown). A rapid initial phase of uptake, to approximately 145 nmol Bu_3SnCl (mg dry wt)$^{-1}$, occurred within 5 min and was followed by a slower, apparently linear, uptake. The level of Bu_3SnCl biosorbed by *C. emersonii* after 2 h was approximately 350 nmol (mg dry wt)$^{-1}$. It is unlikely that the latter phase represented metabolism-dependent intracellular accumulation as the high concentration used (0.5 mM) was far greater than that necessary to exert toxicity.[53,54] Gadd et al[59] ascribed a similar disappearance of Bu_3SnCl from the incubation medium, in the period following cell surface adsorption in *A. pullulans*, partly to abiotic volatilization of the organotin from the medium as well as to further permeation of the cells by the organotin with increased membrane disruption and cell death. Increases in uptake following biomass killing have also been reported for inorganic heavy metals, and were attributed to an increased availability of metal-binding sites.[26,78] The additional membrane-bound organelles and other intracellular components, that would be exposed following partial cell lysis of *C. emersonii* but that are not found in prokaryotic cyanobacteria, may account for some of the observed differences in accumulation of Bu_3SnCl in these organisms.[62]

    pH and salt concentration are two highly variable physicochemical parameters of aqueous *C. emersonii* habitats; their effects on Bu_3SnCl biosorption were examined here. Bu_3SnCl uptake by *C. emersonii* varied with pH, although a clear trend was difficult to discern (Fig. 4.5a). Generally, biosorption appeared to increase with pH. Uptake levels were approximately 33% lower at pH 4.5 and 6.5 than at pH 5.5, 7.5 and 8.5. At pH 8.5, Bu_3SnCl uptake levels were greater for *C. emersonii* than those reported for two cyanobacterial species, which showed a marked decline in Bu_3SnCl uptake at pH > 6.5.[62] In aqueous solution of low pH, Bu_3SnCl exists as the

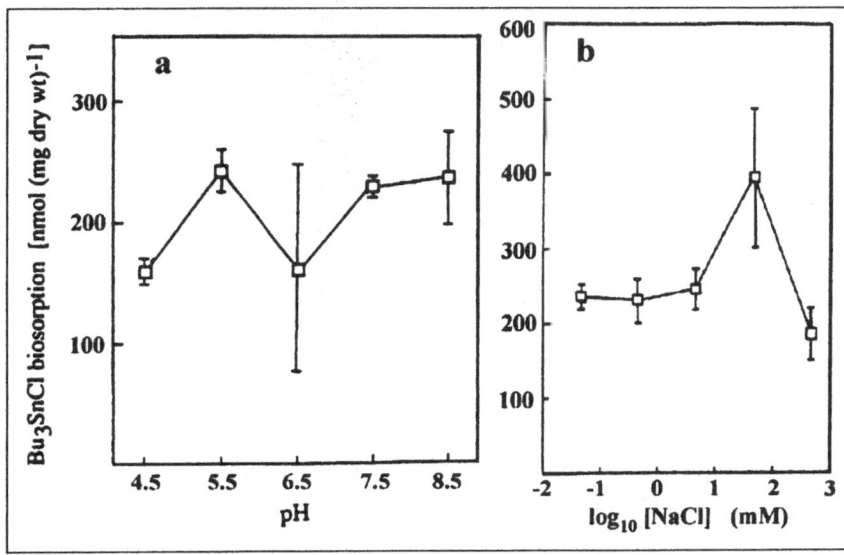

Fig. 4.5. Influence of pH and NaCl on Bu₃SnCl biosorption by *C. emersonii*. Cells were suspended in 10 mM of the appropriate buffer (see text), adjusted to (a) the specified pH or (b) the specified NaCl concentration, pH 5.5. Cells were harvested after 30 min incubation. Mean values from three replicate determinations are shown ± SEM where these exceed the dimensions of the symbols. Figure adapted from Avery et al. Appl Microbiol Biotechnol 1993; 39:812-817.

hydrated butyltin cation $[Bu_3Sn(H_2O)_2]^+$ whereas at pH > 6.5 simple neutral hydroxides of organotins predominate.[79,80] It is probable that reduced biosorption of Bu_3SnCl at pH values lower than pH 5.5 is due to competition from H^+. Any effect of H^+ at higher pH values would probably not be relevant in view of the non-cationic nature of Bu_3SnCl under these conditions. The results presented here and elsewhere[62] suggest that cyanobacterial Bu_3SnCl uptake is more dependent on the organotin possessing positive charge than microalgal Bu_3SnCl uptake. This difference may be a consequence of the lesser contribution of cell surface adsorption to total Bu_3SnCl uptake by *C. emersonii*. Indeed, hydrophobic mechanisms (e.g., membrane permeation in *C. emersonii*) have been shown to be more important in total organotin uptake when the organotin exists as a neutral species.[51] It should also be noted that differences in the composition of cell walls (which are based on mucopolymers in cyanobacteria and on cellulose in microalgae) can affect biosorptive processes.[5]

NaCl also influenced Bu_3SnCl biosorption by *C. emersonii*. Whereas little difference in uptake was evident between 0.05 and 5.0 mM NaCl, a further increase in NaCl concentration to 50 mM was concomitant with an increase in Bu_3SnCl biosorption from 245 to 395 nmol (mg dry wt)$^{-1}$ (Fig. 4.5b). However, Bu_3SnCl uptake was considerably lower [at approximately 180 nmol Bu_3SnCl (mg dry wt)$^{-1}$] at 0.5 M NaCl, a concentration approximately matching that of seawater. These observations may partly explain why Wuertz et al[81] detected tributyltin-resistant bacteria in tributyltin-polluted freshwater sites but not in similar estuarine sites, where Bu_3SnCl-microbe interactions may be reduced. Our results were also in agreement with the decreased toxicity of Bu_3SnCl towards yeast at high salinities reported previously.[82,83] Reduced Bu_3SnCl uptake/toxicity at high NaCl may result from interaction of Na^+ or Cl^- with cell binding sites or with Bu_3SnCl, e.g., by competition for uptake by Na^+, by reduced solubility of the organotin due to formation of covalent tributyltin chlorides, or because of alterations in the membrane lipid composition in response to salt stress.[51,83] It is not clear why increased Bu_3SnCl uptake occurred between 5 and 50 mM NaCl. It is possible that, as for neutral Bu_3SnOH species (at pH >6.5), the formation of neutral Bu_3SnCl at intermediate NaCl concentrations may facilitate uptake by hydrophobic mechanisms.[51]

Biosorption of Bu_3SnCl at varying Bu_3SnCl concentrations is represented as Langmuir adsorption isotherms (Fig. 4.6). The amount of Bu_3SnCl biosorbed per unit weight of biomass (q_e) at fixed temperature is expressed as a function of the concentration of Bu_3SnCl remaining in solution at equilibrium (C). Bu_3SnCl biosorption increased with external Bu_3SnCl concentration (Fig. 4.6a). The relationship was curvilinear, indicating that some saturation of cellular binding sites occurred at high Bu_3SnCl concentrations. Interestingly, despite Bu_3SnCl uptake being lower in *C. emersonii* than in cyanobacteria at low Bu_3SnCl concentrations, suggesting greater uptake-affinity in the latter,[62] cyanobacterial cells were more readily saturated with Bu_3SnCl. Thus, at higher concentrations, Bu_3SnCl uptake was greater in the present study with *C. emersonii*. The q_e value for *C. emersonii* after 15 min incubation in the presence of 3 mM Bu_3SnCl was approximately 680 nmol Bu_3SnCl (mg dry wt)$^{-1}$ (Fig. 4.6a). When the data for q_e and C were transformed to reciprocal values, a curved relationship was evident. This suggested a more complex Bu_3SnCl bonding pattern to *C. emersonii* than to cyanobacteria[62] and to the fungus *Aureobasidium pullulans*,[59] where reciprocal Langmuir plots were linear. The curved plot obtained here was probably a consequence of the biphasic nature of Bu_3SnCl uptake by *C. emersonii*, although it should be noted that the involvement of multiple cell-surface binding mechanisms has been suggested for the uptake of other (inorganic) metal species by *C. salina*.[15] Extrapolation of the plot in Figure 4.6b to the y axis gave a theoretical maximal value for q_e (that would occur on complete saturation of the cell surface) of approximately 1050 nmol Bu_3SnCl (mg dry wt)$^{-1}$. This value was approximately 2-fold higher than those obtained for cyanobacteria under the same conditions.[62] Moreover, the value was more than 2-fold higher [at 342 mg Bu_3SnCl (g dry wt)$^{-1}$] than the threshold proposed by Brierley et al[74] to be required for the economic viability of a biosorption process. Thus, *C. emersonii* appears to have considerable potential for application in organotin removal.

Concluding Remarks

The present investigation demonstrates microalgal removal of inorganic monovalent and organic metal species, and thus builds on previous reports of polyvalent cationic metal removal by microalgae. The novel method for markedly enhancing the Cs^+ removal capabilities of *Chlorella salina* circumvented problems associated with poor Cs^+ bioconcentration encountered in previous studies.[33] Levels of cellular Cs^+ accumulation here were higher than any described previously over such a short time-scale.[33] Moreover, that the system depended on Na^+ was particularly advantageous, as competition by Na^+ (and K^+) is the major drawback

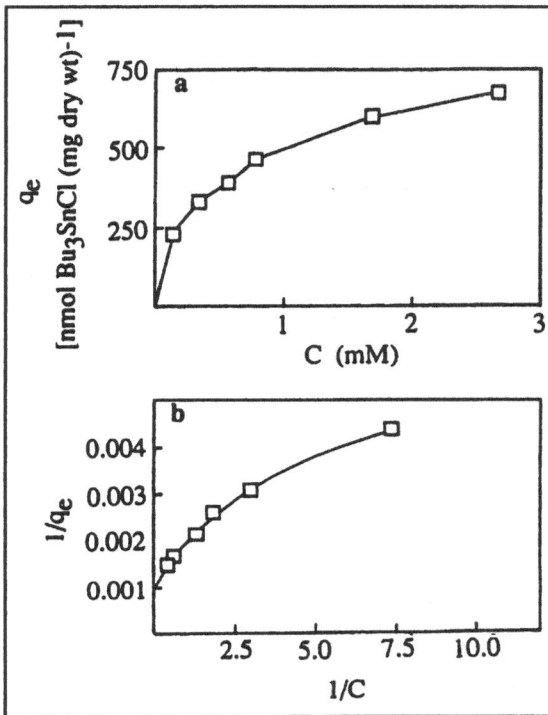

Fig. 4.6. Langmuir adsorption isotherms of Bu₃SnCl biosorption by *C. emersonii*. Cells were suspended in 10 mM MES buffer, pH 5.5, with the appropriate concentration of Bu₃SnCl. (a) The amount of Bu₃SnCl taken up per unit weight of biomass (q_e) is expressed as a function of the Bu₃SnCl concentration remaining in solution (C) after 15 min equilibration with the biomass. (b) Reciprocal plot of the data presented in (a). Mean values from three replicate determinations are shown ± SEM where these exceed the dimensions of the symbols. Figure adapted from Avery et al. Appl Microbiol Biotechnol 1993; 39:812-817.

of conventional Cs⁺-removal strategies using zeolites;[84] it is noteworthy that K⁺ is a weak competitor of Cs⁺ uptake in *C. salina*.[47] The many other advantages of the present system (discussed above) suggest genuine potential for microalgal Cs⁺ removal. In view of the wide reported variation in microbial Cs⁺ accumulation capacities[33] and ease of screening by autoradiography,[85] as well as potential further uptake-enhancement by physiological[18] and molecular[86,87] manipulations, future development of the present system may lead to further improved removal-efficiency.

The influence of metal speciation on biosorption processes was exemplified here for Bu₃SnCl uptake at varying pH and NaCl concentration; generally, effects of such parameters are related to either the organism's physiology (e.g., Na⁺ on Cs⁺ uptake by *C. salina*) or, as appeared to be the case for Bu₃SnCl, the metal's environmental chemistry. Effects may also depend on whether uptake is principally via active or passive mechanisms. Our results underscore the need to consider external physicochemical parameters in any proposed organotin remediation strategy involving microalgae. The present investigation presents a preliminary characterization of microalgal organotin uptake, thus complementing previous studies on toxicity and biodegradation[53,56] for which uptake is clearly a prerequisite.

In addition to high environmental organotin concentrations arising through leaching from antifouling paints on ship hulls, an increasing number of reports have revealed elevated organotin concentrations in sewage sludges and other sinks for industrial discharges in recent years;[51] thus, conventional treatments do not effectively remove organotins from certain effluents. The high levels of Bu_3SnCl uptake by *C. emersonii* reported here suggest that microalgal treatment may represent one alternative strategy for organotin removal. The metabolism-independence of microalgal Bu_3SnCl biosorption should be advantageous for any application in Bu_3SnCl removal. The dependence of uptake on membrane-lipid composition (i.e., hydrophobicity) is one feature that could be readily manipulated[88] and exploited to maximize Bu_3SnCl uptake efficiencies in the future.

Acknowledgments

SVA gratefully acknowledges receipt of a NERC research studentship for his work conducted while at the University of Dundee. GMG and GAC gratefully acknowledge financial support from the Natural Environment Research Council (NERC GR3/7290); GMG also gratefully acknowledges receipt of a NATO Linkage Grant (ENVIR.LG.950387).

References

1. Fitzgerald GP. The biotic relationships within water blooms. In: Rosowski JR, Parker BC, eds. Selected Papers in Phycology. London: Academic Press, 1971: 26-32.
2. Oswald WJ. Microalgae and waste-water treatment. In: Borowitzka MA, Borowitzka LJ, eds. Micro-algal Biotechnology. Cambridge: Cambridge University Press, 1988:305-328.
3. Vílchez C, Garbayo I, Lobato MV et al. Microalgae-mediated chemicals production and waste removal. J Chem Technol Biotechnol 1997; in press.
4. Gadd GM. Interactions of fungi with toxic metals. New Phytol 1993; 124:25-60.
5. Gadd GM. Microbial control of heavy metal pollution. In: Fry JC, Gadd GM, Herbert RA et al, eds. Microbial Control of Pollution. Cambridge: Cambridge University Press, 1992:59-88.
6. Cooney JJ, Wuertz S. Toxic effects of tin compounds on microorganisms. J Indust Microbiol 1989; 4:375-402.
7. Clark RB. Marine Pollution. Oxford: Oxford Science Publications, 1989.
8. Evans LV. Marine algae and fouling: a review with particular reference to ship-fouling. Botanica Marina 1981; 24:167-182.
9. Whitton BA. Zinc and plants in rivers and streams. In: Nriagu JO, ed. Zinc in the Environment. Part II. New York: John Wiley & Sons, 1980:364-400.
10. Reed RH, Gadd GM. Metal tolerance in eukaryotic and prokaryotic algae. In: Shaw AJ, ed. Heavy Metal Tolerance in Plants: Evolutionary Aspects. Boca Raton: CRC Press, 1990:105-118.
11. King SF. Uptake and transfer of cesium-137 by *Chlamydomonas*, *Daphnia* and bluegill fingerlings. Ecology 1964; 45:852-859.
12. Nakajima A, Horikoshi T, Sakaguchi T. Studies on the accumulation of heavy metal elements in biological systems. XVII. Selective accumulation of heavy metal ions by *Chlorella regularis*. Eur J Appl Microbiol Biotechnol 1981; 12:76-83.
13. Greene B, Darnall DW. Microbial oxygenic photoautotrophs (cyanobacteria and algae) for metal-ion binding. In: Ehrlich HL, Brierley CL, Brierley JA, eds. Microbial Mineral Recovery. New York: McGraw-Hill, 1990:277-302.
14. Garnham GW, Codd GA, Gadd GM. Uptake of cobalt and cesium by microalgal-and cyanobacterial-clay mixtures. Microb Ecol 1993; 25:71-82.
15. Garnham GW, Codd GA, Gadd GM. Kinetics of uptake and intracellular location of cobalt, manganese and zinc in the estuarine green alga *Chlorella salina*. Appl Microbiol Biotechnol 1992; 37:270-276.

16. Garnham GW, Codd GA, Gadd GM. Accumulation of zirconium by microalgae and cyanobacteria. Appl Microbiol Biotechnol 1993; 39:666-672.

17. Geisweid HJ, Urbach W. Sorption of cadmium by the green microalgae *Chlorella vulgaris, Ankistrodesmus braunii* and *Eremosphaera viridis*. Zeit Pflanzenphysiol 1983; 109:127-141.

18. Avery SV, Codd GA, Gadd GM. Replacement of cellular potassium by caesium in *Chlorella emersonii*: differential sensitivity of photoautotrophic and chemo-heterotrophic growth. J Gen Microbiol 1992; 138:69-76.

19. Trevors JT, Stratton GW, Gadd GM. Cadmium transport, resistance and toxicity in bacteria, algae, and fungi. Can J Microbiol 1986; 32:447-464.

20. Khummongkol D, Canterford GS, Fryer C. Accumulation of heavy metals in unicellular algae. Biotechnol Bioeng 1982; 24:2643-2660.

21. Wehrheim B, Wettern M. Biosorption of cadmium, copper and lead by isolated mother cell-walls and whole cells of *Chlorella fusca*. Appl Microbiol Biotechnol 1994; 41:725-728.

22. Crist RH, Oberholser K, Shank N et al. Nature of bonding between metallic ions and algal cell walls. Environ Sci Technol 1981; 15:1212-1217.

23. Cho DY, Lee ST, Park SW et al. Studies of the biosorption of heavy-metals onto *Chlorella-vulgaris*. J Environ Sci Health Part A – Environ Sci Engin 1994; 29:389-409.

24. Crist RH, Martin JR, Carr D et al. Interaction of metal and protons with algae. 4. Ion exchange vs adsorption models and a reassessment of Scatchard plots; ion-exchange rates and equilibria compared with calcium alginate. Environ Sci Technol 1994; 28:1859-1866.

25. Crist DR, Crist RH, Martin JR et al. Ion exchange systems in proton-metal interactions with algal cell walls. FEMS Microbiol Rev 1994; 14:309-314.

26. Mang S, Tromballa HW. Uptake of cadmium by *Chlorella fusca*. Zeit Pflanzenphysiol 1978; 90:293-302.

27. Gekeler W, Grill E, Winnacker E-L et al. Algae sequester heavy metals via synthesis of phytochelatin complexes. Arch Microbiol 1988; 150:197-202.

28. Wikfors GH, Neeman A, Jackson PJ. Cadmium-binding polypeptides in microalgal strains with laboratory-induced cadmium tolerance. Mar Ecol Prog Ser 1991; 79:163-170.

29. Hughes MN, Poole RK. Metal speciation and microbial growth–the hard (and soft) facts. J Gen Microbiol 1991; 137:725-734.

30. Peterson HG, Healey FP, Wagemann R. Metal toxicity to algae: a highly pH-dependent phenomenon. J Can Sci Halicut Aquat 1984; 41:974-979.

31. Skowronski T. Influence of some physicochemical factors on cadmium uptake by the green alga *Stichococcus bacillaris*. Appl Microbiol Biotechnol 1986; 24: 423-425.

32. Singh SP, Yadava V. Cadmium uptake in *Anacystis nidulans*: Effect of modifying factors. J Gen Appl Microbiol 1985; 31:39-48.

33. Avery SV. Microbial interactions with caesium–implications for biotechnology. J Chem Technol Biotechnol 1995; 62:3-16.

34. Cai XH, Traina ST, Logan TJ et al. Applications of eukaryotic algae for the removal of heavy-metals from water. Molec Mar Biol Biotechnol 1995; 4:338-344.

35. Khoshmanesh A, Lawson F, Prince IG. Cadmium uptake by unicellular green microalgae. Chem Engin J Biochem Engin J 1996; 62:81-88.

36. Pirszel J, Pawlik B, Skowronski T. Cation-exchange capacity of algae and cyanobacteria: a parameter of their metal sorption abilities. J Indust Microbiol 1995; 14:319-322.

37. Fehrmann C, Pohl P. Cadmium adsorption by the non-living biomass of micro-algae grown in axenic mass culture. J Appl Phycol 1993; 5:555-562.
38. Roy D, Greenlaw PN, Shane BS. Adsorption of heavy-metals by green-algae and ground rice hulls. J Environ Sci Health Part A-Environ Sci Engin 1993; A28: 37-50.
39. Brady D, Letebele B, Duncan JR et al. Bioaccumulation of metals by *Scenedesmus, Selenastrum* and *Chlorella* algae. Water SA 1994; 20:213-218.
40. Garnham GW, Codd GA, Gadd GM. Accumulation of cobalt, zinc and manganese by the estuarine green microalga *Chlorella salina* immobilized in alginate microbeads. Environ Sci Technol 1992; 26:1764-1770.
41. Barkley NP. Extraction of mercury from groundwater using immobilized algae. Air Waste Mgmt Assoc 1991; 41: 1387-1393.
42. Cordery J, Wills AJ, Atkinson K et al. Extraction and recovery of silver from low-grade liquors using microalgae. Miner Engin 1994; 7:1003-1015.
43. Greene B, Hosea M, McPherson R et al. Interaction of gold(I) and gold(III) com-plexes with algal biomass. Environ Sci Technol 1986; 20:627-632.
44. Duxbury T. Ecological aspects of heavy metal responses in microorganisms. In: Marshall KC, ed. Advances in Microbial Ecology. Vol. 8. New York: Plenum Press, 1985:185-235.
45. Avery SV. Fate of radiocaesium in the environment: distribution between the abiotic and biotic components of aquatic and terrestrial ecosystems. J Environ Radioact 1996; 30:139-171.
46. Avery SV. Caesium accumulation by microorganisms: Uptake mechanisms, cat-ion competition, compartmentalization and toxicity. J Indust Microbiol 1995; 14:76-84.
47. Avery SV, Codd GA, Gadd GM. Transport kinetics, cation inhibition and intrac-ellular location of accumulated caesium in the green microalga *Chlorella salina*. J Gen Microbiol 1993; 139:827-834.
48. Perkins J, Gadd GM. Caesium toxicity, accumulation and intracellular localiza-tion in yeasts. Mycol Res 1993; 97: 717-724.
49. Avery SV, Codd GA, Gadd GM. Caesium accumulation and interactions with other monovalent cations in the cyanobacterium *Synechocystis* PCC 6803. J Gen Microbiol 1991; 137:405-413.
50. Gilmour D. Halotolerant and halophilic microorganisms. In: Edwards C, ed. Mi-crobiology of Extreme Environments. Milton Keynes: Open University Press, 1990:147-177.
51. Fent K. Ecotoxicology of organotins. Crit Rev Toxicol 1996; 26:1-117.
52. Huggett RJ, Unger MA, Seligman PF et al. The marine biocide tributyltin. En-viron Sci Technol 1992; 26:232-237.
53. Wong PTS, Chau YK, Kramar O, Bengert GA. Structure-toxicity relationship of tin compounds. Can J Fish Aquat Sci 1982; 39:483-488.
54. Avery SV, Miller ME, Gadd GM et al. Toxicity of organotins towards cyanobacterial photosynthesis and nitrogen-fixation. FEMS Microbiol Lett 1991; 84:205-210.
55. Huang GL, Dai SG, Sun HW. Toxic effects of organotin species on algae. Appl Organomet Chem 1996; 10:377-387.
56. Maguire RJ, Wong PTS, Rhamey JS. Accumulation and metabolism of tri-*n*-butyltin cation by a green alga, *Ankistrodesmus falcatus*. Can J Fish Aquat Sci 1984; 41:537-540.
57. Reader S, Pelletier E. Biosorption and degradation of butyltin compounds by the marine diatom *Skeletonema costatum* and the associated bacterial community at low temperature. Bull Environ Contam Toxicol 1992; 48:599-607.

58. Stlouis R, Pelletier E, Marsot P et al. Distribution and effects of tributyltin chloride and its degradation products on the growth of the marine alga *Pavlova lutheri* in continuous-culture. Water Res 1994; 28:2533-2544.

59. Gadd GM, Gray DJ, Newby PJ. Role of melanin in fungal biosorption of tributyltin chloride. Appl Microbiol Biotechnol 1990; 34:116-121.

60. Blair WR, Olson GJ, Brinckman FE et al. Accumulation and fate of tri-*n*-butyltin cation in estuarine bacteria. Microb Ecol 1982; 8:241-251.

61. Molander S, Dahl B, Blanck H et al. Combined effects of tri-*n*-butyltin (TBT) and diuron on marine periphyton communities detected as pollution-induced community tolerance. Arch Environ Contam Toxicol 1992; 22:419-427.

62. Avery SV, Codd GA, Gadd GM. Biosorption of tributyltin and other organotin compounds by cyanobacteria and microalgae. Appl Microbiol Biotechnol 1993; 39:812-817.

63. Avery SV, Codd GA, Gadd GM. Salt-stimulation of caesium accumulation in the euryhaline green microalga *Chlorella salina*: potential relevance to the development of a biological Cs-removal process. J Gen Microbiol 1993; 139:2239-2244.

64. Macaskie LE. The application of biotechnology to the treatment of wastes produced from the nuclear fuel cycle: Biodegradation and bioaccumulation as a means of treating radionuclide-containing streams. Crit Rev Biotechnol 1994; 11:41-112.

65. Brown LM, Hellebust JA. Ionic dependence of deplasmolysis in the euryhaline diatom *Cyclotella cryptica*. Can J Microbiol 1978; 56:408-412.

66. Ehrenfeld J, Cousin J-L. Ionic regulation of the unicellular green alga *Dunaliella tertiolecta*: response to hypertonic shock. J Membr Biol 1984; 77:45-55.

67. Pick U, Ben-Amotz A, Karni L et al. Partial characterization of K⁺ and Ca²⁺ uptake systems in the halotolerant alga *Dunaliella salina*. Plant Physiol 1986; 81:875-881.

68. Gadd GM. Fungi and yeasts for metal accumulation. In: Ehrlich HL, Brierley CL, Brierley JA, eds. Microbial Mineral Recovery. New York: McGraw-Hill, 1990: 249-275.

69. Garnham GW, Codd GA, Gadd GM. Effect of salinity and pH on cobalt biosorption by the estuarine microalga *Chlorella salina*. Biol Metals 1991; 4:151-157.

70. Meikle AJ, Gadd GM, Reed RH. Manipulation of yeast for transport studies: critical assessment of cultural and experimental procedures. Enz Microb Technol 1990; 12:865-872.

71. De Rome L, Gadd GM. Use of pelleted and immobilized yeast and fungal biomass for heavy metal and radionuclide recovery. J Indust Microbiol 1991; 7: 97-104.

72. Springer-Lederer H, Rosenfeld DL. Energy sources for the absorption of rubidium by *Chlorella*. Physiol Plantarum 1968; 21:435-444.

73. Blackwell JR, Gilmour DJ. Physiological response of the unicellular green alga *Chlorococcum submarinum* to rapid changes in salinity. Arch Microbiol 1991; 157:86-91.

74. Brierley JA, Goyak GM, Brierley CL. Considerations for commercial use of natural products for metal recovery. In: Eccles H, Hunt S, eds. Immobilization of Ions by Biosorption. Chichester: Ellis Horwood, 1986:105-117.

75. Huang GL, Bai ZP, Dai SG et al. Accumulation and toxic effect of organometallic compounds on algae. Appl Organomet Chem 1993; 7:373-380.

76. Eng G, Tierney EJ, Bellama JM et al. Correlation of molecular total surface area with organotin toxicity for biological and physicochemical applications. Appl Organomet Chem 1988; 2:171-175.

77. Mouhri K, Marsot P, Pelletier E et al. Effects of tributyltin chloride on the growth and metabolism of the marine diatom *Phaeodactylum-tricornutum* (Bohlin). Oceanolog Acta 1995; 18:363-370.

78. Avery SV, Tobin JM. Mechanisms of strontium uptake by laboratory and brewing strains of *Saccharomyces cerevisiae*. Appl Environ Microbiol 1992; 58:3883-3889.
79. Blunden SJ, Chapman A. Organotin compounds in the environment. In: Craig PJ, ed. Organometallic Compounds in the Environment. Harlow, U.K.: Longman, 1986:111-159.
80. Shoukry MM. Equilibrium study of tributyltin(IV) complexes with amino acids and related compounds. Bull Soc Chim Fr 1993; 130:117-120.
81. Wuertz S, Miller CE, Pfister RM et al. Tributyltin-resistant bacteria from estuarine and freshwater sediments. Appl Environ Microbiol 1991; 57:2783-2789.
82. Cooney JJ, De Rome L, Laurence OS et al. Effects of organotin and organolead compounds on microorganisms. J Indust Microbiol 1989; 4:279-288.
83. Laurence OS, Cooney JJ, Gadd GM. Toxicity of organotins towards the marine yeast *Debaryomyces hansenii*. Microb Ecol 1989; 17:275-285.
84. Kesraoui-Ouki S, Cheeseman CR, Perry R. Natural zeolite utilisation in pollution control: a review of applications to metal effluents. J Chem Tech Biotechnol 1994; 59:121-126.
85. Tomioka N, Uchiyama H, Yagi O. Isolation and characterization of cesium-accumulating bacteria. Appl Environ Microbiol 1992; 58:1019-1023.
86. Bossemeyer D, Schlösser A, Bakker EP. Specific cesium transport via the *Escherichia coli* Kup (TrkD) K^+ uptake system. J Bacteriol 1989; 171:2219-2221.
87. Sentenac H, Bonneaud N, Minet M et al. Cloning and expression in yeast of a plant potassium ion transport system. Science 1992; 256:663-665.
88. Cossins AR, ed. Temperature Adaptation of Biological Membranes. London: Portland Press, 1994.

CHAPTER 5

Bioaccumulation and Biotransformation of Arsenic, Antimony, and Bismuth Compounds by Freshwater Algae

Shigeru Maeda and Akira Ohki

Introduction

Arsenic, which is ubiquitous in the Earth's crust, ranks 20th among the elements in abundance. Arsenic is widely but sparsely distributed in nature. Mostly the element is associated with igneous and sedimentary rocks, particularly with sulfidic ores. Arsenic enters aquatic environments indirectly from industrial and other air emissions, such as smelting operations and fossil-fuel combustion, and enters directly from localized effluent discharges.[1]

The chemical form of arsenic in seawater is mainly inorganic. However, arsenic may be methylated by microorganisms to form volatile methylated arsines and nonvolatile organoarsenic compounds, followed by the release of arsenic compounds into water. Marine organisms can accumulate arsenic directly from water and through the food chain.[2]

Antimony occurs in nature mainly as Sb_2S_3 (stibnite, antimonite) and Sb_2O_3 (valentinite) which is a decomposition product of stibnite. These kinds of antimony are commonly found in ores of copper, silver, and lead. It should be noted for antimony that Sb_2O_3 has been used for flameproofing (flame retardants) of textiles, paper, and plastics. Thus, when wastes which contain such flame retardants are incinerated, the antimony will be emitted into the atmosphere.[2,3]

Most bismuth occurs as bismuth sulfides which are associated with lead, copper, and silver. Bismuth is used for low-melting metals, pharmaceutical compounds, and glass and ceramic products.[2]

Freshwater algae which can bioaccumulate the metals are needed for wastewater treatment of arsenic, antimony, and bismuth. In this chapter, we describe the bioaccumulation and biotransformation of arsenic, antimony, and bismuth compounds by freshwater algae.

Wastewater Treatment with Algae, edited by Yuk-Shan Wong and Nora F.Y. Tam.
© Springer - Verlag and Landes Bioscience 1998.

Background Literature

A number of books and review articles have dealt with arsenic species and their transformation in the environment. Particular mention should be made of one book, "Arsenic" from the Committee on Medical and Biological Effects of Environmental Pollutants, the National Research Council, Washington, DC.[1] The third Spurenelement-Symposium resulted in a useful volume, "Arsen," edited by Anke et al,[4] a symposium sponsored by the Chemical Manufacturers Association and the National Bureau of Standards gave "Arsenic: Industrial, Biomedical, Environmental Perspective" edited by Lederer and Fensterheim,[5] and the first arsenic symposium sponsored by the Japanese Arsenic Scientists Society (JASS) provided "Arsenic: Chemistry, Metabolism and Toxicity" edited by Ishinishi et al.[6] Subsequent JASS symposia resulted in successive volumes, as special issues of Applied Organometallic Chemistry, of "Natural and Industrial Arsenic" edited by Irgolic et al[6,7] and by Maeda and Craig[8,9] and of "Environmental and Industrial Arsenic" edited by Maeda and Craig.[10,11] Fowler[12] and Nriagu[13,14] have edited books devoted to the biological and environmental effects of arsenic. Other books contain chapters of interest by Brinckman and Bellama[15] and some useful reviews have appeared by Phillips and Depledge,[16] Cullen and Reimer,[17] Phillips,[18] Maeda and Sakaguchi[19] and Maeda.[2]

Arsenic compounds in marine environments have been described in many of those books and reviews; whereas a few have mentioned those in freshwater environments.

Generally, the arsenic content in freshwater algae is lower than that in marine algae.[18] However, some freshwater algae accumulate arsenic to a large degree. Lunde cultivated three freshwater algae (green alga: *Chlorella pyrenoidosa*; blue green alga: *Oscillatoria rubescens*; diatom: *Phaeodactylum tricornutum*) and three marine algae (green alga: *Chlorella ovalis*; diatoms: *Phaeodactylum* sp. and *Skeletonema costatum*) in a freshwater and a seawater media, respectively, which contained radioactive arsenic ions.[20] Enrichment of 240 to 2800 times and that of 710 to 2900 times in the arsenic concentration were observed in the freshwater algae and the marine algae, respectively.

So far a few studies which relate to biotransformation of arsenic by freshwater algae have been performed. Lunde measured arsenic content in the lipid phase extracted from freshwater algae (*Chlorella ovalis, Phaeodactylum tricornutum*, and *Oscillatoria rubescens*) using neutron activation technique.[21] Bioaccumulation of arsenic from enriched cultures containing 1 to 3 µg/L of arsenic into the lipid phase took place at concentrations of 0.5 to 5 µg/g-dry cell. These results suggest that the arsenic in lipids extracted from algae is chemically bound as can be seen in case of marine organisms.

Baker et al isolated four freshwater algae (*Ankistrodesmus* sp., *Scenedesmus* sp., *Chlorella* sp., and *Selenastrum* sp.) which biomethylated inorganic arsenic into methylarsonic acid, dimethylarsinic acid, and trimethylarsine oxide in lake water and Bold's basal medium.[22] A blue-green alga, *Rizoclonium* sp., was exposed to [74]As-arsenate for one week and then extracted with hot ethanol.[23] The principal arsenic compounds appeared to be present as lipid- and water-soluble "lipid-related" compounds.

The transformation of antimony and bismuth in the environment has also been reported by Maeda.[2] However, thus far few studies have been carried out on the bioaccumulation of antimony and bismuth compounds by freshwater algae.

The authors have studied the bioaccumulation and biotransformation of arsenic, antimony, and bismuth compounds by freshwater algae, such as *Chlorella*, for the last 15 years. We describe the results of those studies.

Bioaccumulation of Arsenic by Freshwater Algae

The authors have collected soil and water samples containing microalgae from arsenic-polluted places, such as old mines and near geothermal electric power plants. Those algae were inoculated in a modified Detmer medium and cultured, and then repeatedly inoculated in modified Detmer media containing arsenic for which the concentration was increased stepwise (1, 10, 50, 100, 500, 1000, and 2000 mg/L as elemental arsenic with appropriate concentration of Na_2HAsO_4; abbreviated as As(V)). In this manner, some algae which showed arsenic-tolerance were screened.[24]

For such arsenic-tolerant algae, a green alga, *Chlorella vulgaris*, and blue-green algae, *Nostoc* sp., *Phormidium* sp., *Hydrocoleum* sp., and *Microchaete* sp. were isolated.[24,25] The algae were placed in a modified Detmer medium (300 mL) containing various concentrations of As(V) in an Erlenmeyer flask (500 mL). The culture system was aerated by germ-free moisture-saturated air and illuminated for 12 h under fluorescent light (2000-5000 lx) and in the dark for 12 h daily at room temperature (20-30°C). Figure 5.1 shows the influence of As(V) concentration in the medium upon the cell growth of *C. vulgaris*, *Nostoc* sp., and *Phormidium* sp.[25-27] For *C. vulgaris* and *Phormidium* sp., as the As(V) concentration was raised, the cell growth was enhanced up to 5×10^2 to 5×10^3 mg/L of As(V), and still multiplied even under 1×10^4 mg/L of As(V).

The determination of arsenic concentration in the algal cell was carried out. A constant volume of culture suspension was harvested and centrifuged, and the cell was rinsed twice with pure water and dried at 105°C for 2 h. The dry cell (10-20 mg) was placed in a 100 mL porcelain crucible, 2 mL of 50% $Mg(NO_3)_2$ was added, and the mixture was heated at 60°C for 12 h and then at 550°C for 6 h to be mineralized. After cooling, the mineralized sample was dissolved in a 5 mL of 10 N HCl solution, 40% KI solution (1 mL) was added, the solution was extracted twice with chloroform (5 mL), and the chloroform phase was then back-extracted with water (2 mL). The arsenic concentration was determined by graphite-furnace (flameless) atomic absorption spectrophotometry (Japan Jarrel Ash AA-890 with FLA-1000).

Figure 5.2 shows the growth curve of *C. vulgaris* and the arsenic concentration in the cell when the alga is cultured in a modified Detmer medium in the presence of 100 mg/L As(V). The most marked bioaccumulation of arsenic (7.2 mg/g-dry cell) was observed in the middle of log phase when the cell multiplication was the most remarkable. However, the bioaccumulation of arsenic was considerably retarded since the rate of arsenic excretion excelled that of the accumulation in the stationary phase.

Figure 5.3 presents the growth curve of *C. vulgaris* and the bioaccumulation of arsenic when the alga is exposed to 1.0 mg/L As(III).[28] The bioaccumulation of arsenic provided the maximum value at the beginning of the log phase and then rapidly decreased. It is proposed that As(III) is more toxic than As(V), so that excretion of the former takes place more rapidly than the latter.

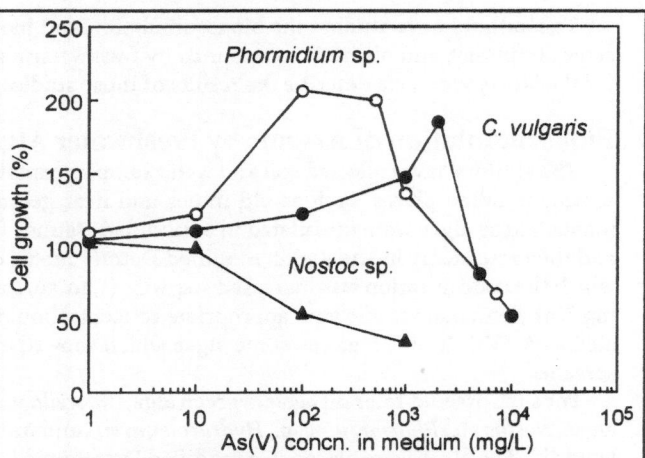

Fig. 5.1. Influence of As(V) concentration in the medium upon the cell growth of algae. Cell growth (%) reveals the ratio of the cell growth when the medium contains the concentration of As(V) to that in the absence of As(V).

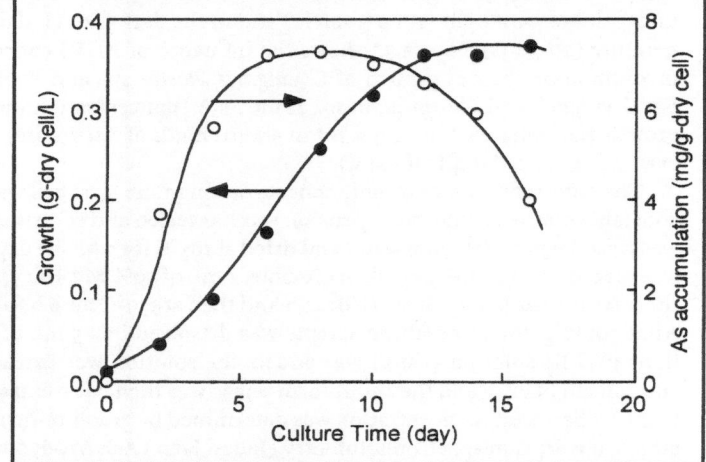

Fig. 5.2. Growth curve of *C. vulgaris* and the arsenic concentration in the cell when the alga is cultured in the presence of 100 mg/L of As(V).[25,29]

Bioaccumulation of Arsenic by
C. Vulgaris Using Raceway-Type Open-Culture Tank

To attain the practical use of arsenic bioaccumulation by algae, a raceway-type open-culture tank (50 L; Fig. 5.4) was prepared.[29] In this tank, air bubbles are provided by a compressor; this results in the circulation of a medium solution and thus, air supply and stirring for the medium are effectively attained. Figure 5.5 shows the growth curve of *C. vulgaris* and the bioaccumulation of arsenic when the alga is cultured in the presence of 10 mg/L of As(V) by use of the open-culture system.

Similar to the closed flask-culture system, which is germ-free, mentioned above, the bioaccumulation of arsenic exhibited the maximum value in the middle of log phase, which was followed by a decrease in accumulation. Compared to the closed system, the decrease in the stationary phase is remarkable, which suggests that the

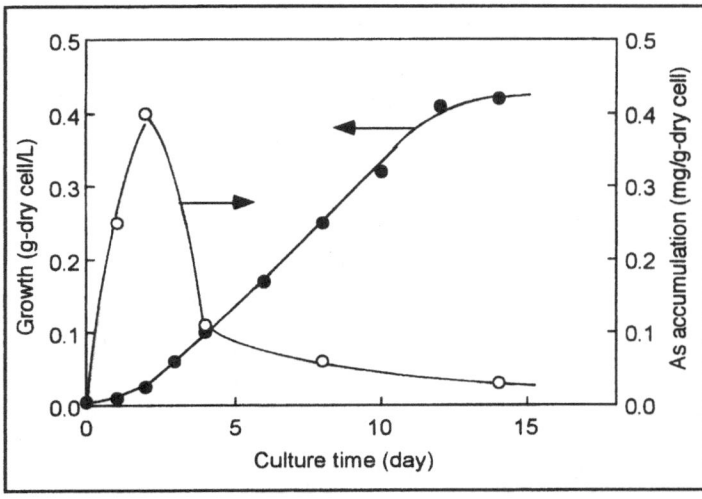

Fig. 5.3. Growth curve of *C. vulgaris* and the arsenic concentration in the cell when the alga is cultured in the presence of 1.0 mg/L of As(III).[28]

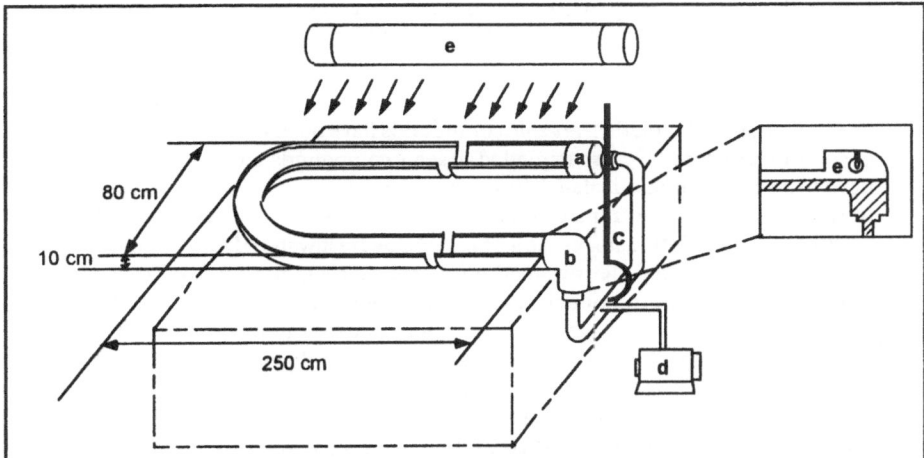

Fig. 5.4. A raceway-type open-culture tank (50 L).[29] a: Inlet; b: Outlet; c: Water level indicator; d: Compressor; e: Fluorescent light.

excretion of arsenic takes place more rapidly. Figure 5.5 also presents the arsenic concentration in the medium. Since it is proposed that the excretion of arsenic starts in late log phase, the minimum value of the arsenic concentration appears just behind the point where bioaccumulation shows the maximum value. By use of the open-culture system, the arsenic concentration in the aqueous phase can be decreased from 9.0 mg/L to 7.5 mg/L.

Biotransformation of Arsenic by Freshwater Algae: Exposure to Inorganic Arsenic

It is well known that some marine organisms highly accumulate arsenic from seawater.[2,6,17,19] Some portion of the accumulated arsenic, ranging from a few percent to about 100%, is biotransformed into organoarsenic compounds.

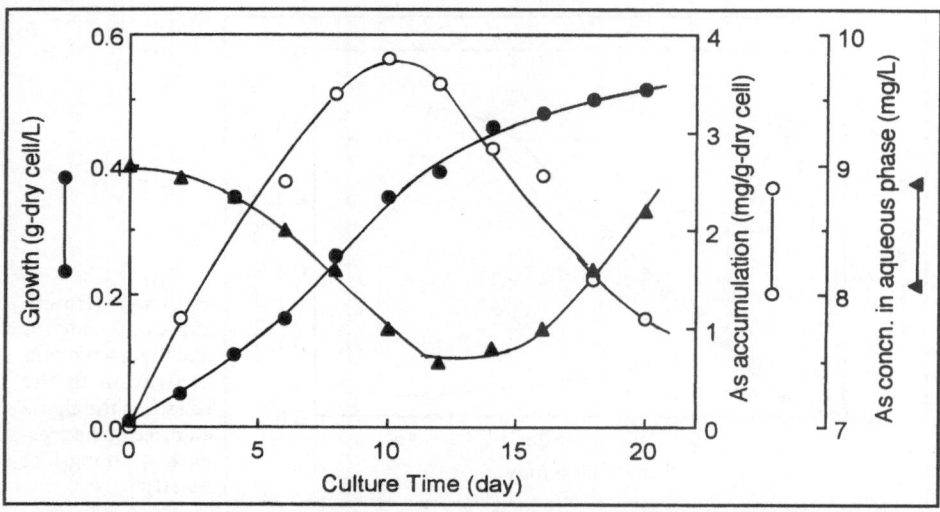

Fig. 5.5. Growth curve of *C. vulgaris* and the arsenic concentration in the cell when the alga is cultured in the presence of 9.0 mg/L of As(V) by use of the open-culture tank.[29]

Monomethyl- and dimethyl-arsenic compounds have been found in water extracts of some macroalgae.[30-32] Also trimethyl-arsenic compounds, such as arsenobetaine, have been detected and isolated from some marine organisms.[33-40] However, for freshwater algae the biotransformation of arsenic has scarcely been studied.

To probe the biotransformation of arsenic by freshwater algae, *C. vulgaris* was cultured in a modified Detmer medium containing 217 mg/L of As(V) until the stationary phase. The fractional determination of nonmethylated and methylated arsenic compounds in the whole cell was performed as follows.[41] The dried cells (10 mg) prepared in a similar manner as above were digested with 5 mL of 2 M NaOH solution at 90-95°C for 3 h. By this treatment, bonding of arsenic with intracellular compounds or groups except for methyl group are cleaved. Nonmethyl-(inorganic), monomethyl, dimethyl- and trimethyl-arsenic compounds (abbreviated as IA, MMA, DMA, and TMA, respectively) which exist in the cell are converted to arsenate, methanearsonate, dimethylarsinate, and trimethylarsine oxide, respectively. These four compounds were reduced with $NaBH_4$ to arsine, methylarsine, dimethylarsine, and trimethylarsine, respectively. The arsine gases generated were conducted with helium carrier gas into three drying tubes and then frozen out in a liquid-nitrogen-cooled U-trap. The U-trap was warmed to room temperature, resulting in the successive volatilization of each arsine gas in the order of its boiling point. The arsine gases were passed through a quartz tube atomizer and determined by an atomic absorption spectrometer (Japan Jarrel Ash AA-890).

The arsenic concentration in each biochemical component of algal cells was also carried out by the fractionation of the cell.[41] After reaching the stationary phase, the wet cells were homogenized with chloroform/methanol (2:1) using a homogenizer, the slurry was filtered under reduced pressure through a filter paper, and the residue was washed several times with the mixed solvent. The residue contained proteins, insoluble polysaccharides, and nucleic acids. The filtrate was combined with the wash and shaken with one-quarter of their total volume of water, the mixture was allowed to stand at room temperature overnight, and the top phase (water-soluble) and the bottom phase (lipid-soluble) were separated and evaporated to dryness. The whole cell, the residue (proteins and polysaccharides), and the two fractions were analyzed for total and methylated arsenic compounds.

The yield of the extraction and the concentration of total and methylated arsenic compounds in the whole cell and the biochemical components are summarized in Table 5.1.[41] Since the concentration of IA cannot be precisely obtained by the fractional determination mentioned above, the IA concentration presented in Table 5.1 is the difference between the total concentration and the sum of MMA, DMA, and TMA concentrations. The arsenic species detected in the whole cell and in the protein and polysaccharide fraction were almost IA; whereas considerable amount of DMA was detected in the lipid-soluble and water-soluble fractions. These results suggest that some part of arsenic accumulated is biomethylated in the interior of the cell into DMA.

Table 5.1. Extraction yields and arsenic concentrations in biochemical components of various algae[a,b]

Component	Yield (%)	As concentration (µg/g-dry wt)				
		Total As	IA	MMA	DMA	TMA
Chlorella vulgaris						
Whole cell	-	310	310	tr[d]	tr	tr
PP[c]	79	330	330	tr	nd[e]	nd
Lipid soluble	17	60	10	nd	50	nd
Water soluble	4	210	140	10	60	tr
Nostoc sp.						
Whole cell	-	860	860	nd	tr	nd
PP[b]	90	640	640	nd	nd	nd
Lipid soluble	8	60	nd	nd	60	nd
Water soluble	2	230	140	nd	90	nd
Phormidium sp.						
Whole cell	-	1950	1950	nd	nd	nd
PP[b]	94	3000	3000	nd	nd	nd
Lipid soluble	5	110	100	nd	10	nd
Water soluble	1	nd	nd	nd	nd	nd

a Modified from ref. 41
b Experimental conditions: see text
c Proteins and polysaccharides
d Trace
e Not determined

Biotransfermation of Arsenic by Freshwater Algae: Exposure to Organic Arsenic Compounds

To probe the biotransformation of organic arsenic compounds by freshwater algae, *C. vulgaris* was cultured in a modified Detmer medium containing methylated arsenic compounds (10 mg/L), methylarsonic acid, dimethylarsinic acid, and arsenobetaine.[42] The fractional determination of arsenic compounds in the whole cell was performed (Table 5.2). When the alga was exposed to an MMA compound (methylarsonic acid), DMA (24 µg/g-dry cell) was detected as well as MMA (27 µg/g-dry cell); whereas IA was not observed and a trace amount of TMA was present. When the alga was exposed to a DMA compound (dimethylarsinic acid) and a TMA compound (arsenobetaine), almost only DMA and TMA, respectively, were detected. These results suggest that *C. vulgaris* biomethylates MMA into DMA although further methylation from DMA to TMA scarcely takes place. Also, demethylation of methylated arsenic compounds, MMA, DMA and TMA, into less methylated compounds was hardly observed.

Table 5.3 summarizes the concentration of arsenic compounds in cell when a bacterium, *Klebsiella oxytoca*, which exhibits a high arsenic-tolerance, is exposed to four arsenic compounds.[43] Contrary to algae, such as *C. vulgaris*, *Nostoc* sp. and *Phormidium* sp., the bacterium scarcely biomethylates IA and MMA into further methylated species. When the bacterium was exposed to DMA and TMA, demethylation for one methyl group occurred somewhat. Similar results were obtained for other arsenic-tolerant bacteria, *Xanthomonas* sp. and *Arthrobacter* sp.[43]

Association Mode of Bioaccumulated Inorganic Arsenic

An arsenic-tolerant alga *C. vulgaris*, which had been cultured in a modified Detmer medium containing 1000 mg/L of As(V), accumulated arsenic at a level of 8.7 mg/g-dry cell.[44] The alga was fractionated with chloroform/methanol (2:1) in the same manner described previously. An extract residue (protein and polysaccharide fraction) containing 7.4 mg/g-dry wt of arsenic was obtained.

Table 5.2. Biotransformation of organic arsenic compounds by C. vulgaris[a,b]

Arsenic compounds in medium (10 mg/L)	Arsenic species accumulated in whole cell (µg/g-dry cell)				
	Total	IA	MMA	DMA	TMA
Methylarsonic acid	51	0	24	27	tr[c]
Dimethylarsinic acid	60	0	0	60	0
Arsenobetaine	58	0	0	tr	58

a Modified from ref. 42
b Experimental conditions: see text
c Trace

Table 5.3. Biotransformation of various arsenic compounds by K. oxytoca[a]

Arsenic compounds in medium (10 mg/L)	Arsenic species accumulated in whole cell (μg/g-dry cell)				
	Total	IA	MMA	DMA	TMA
Arsenic acid	33	33	tr[b]	tr	tr
Methylarsonic acid	51	nd[c]	51	tr	tr
Dimethylarsinic acid	3.1	nd	0.3	2.6	tr
Arsenobetaine	149	nd	nd	13	136

a Modified from ref. 43
b Trace
cNot determined

In order to investigate the association mode of inorganic arsenic with proteins in the residue, the barely soluble proteins in the residue were solubilized.[44] The residue (ca. 100 mg) was pulverized and mixed with 1% sodium dodecyl sulfate (SDS) solution (15 mL, pH 8.6) and allowed to stand at 40°C for 24 h. The suspension was centrifuged, the supernatant was concentrated by a rotary evaporator at a reduced pressure and a condensed protein solution was obtained. Solubilized arsenic-bound proteins were fractionated to their molecular weights by a gel-filtration chromatography by use of Sephadex G-75 (Pharmacia LKB Biotechnology; 40-120 μm diameter; fractional molecular weight ranging from 3000 to 80,000). The Sephadex G-75 column (2.0 cm i.d., 80 cm long) was preconditioned with an eluent solution containing 0.1% SDS and 10 mM Bicine buffer (pH 8.6). The clear aqueous protein solution was placed on the column and eluted with the eluent at a flow rate of 1 mL/min. The eluates were collected by a fraction collector (200 drops, ca 4.5 mL each) and the fraction was analyzed for arsenic and protein. The arsenic was determined by flameless atomic absorption spectrometry; while the protein was measured by spectrophotometry at 254 nm. The molecular weight was calibrated with standard poly(styrenesulfonate) samples (MW 6500, 16000, 31000). The plots of the retention volumes against the molecular weights of the standard samples provided a good linearity. In the same manner, arsenic-free proteins were separated from the extract residue of arsenic-free *C. vulgaris* cells.

The chromatogram (Fig. 5.6a) has two peaks (I and II) for protein and one peak for arsenic, and the chromatogram (Fig. 5.6b) has two peaks for protein only.[44] The fraction numbers of the two protein in the chromatogram (a) coincide with those in the chromatogram (b), respectively. The height of peak II in the former was larger than that in the latter. It was found that from the calibration curve that the solubilized protein had two types of proteins with molecular weights around 2 x 10^4 and 3 x 10^3, with the latter of which the accumulated arsenic was associated. Additionally, it appears that when arsenic is accumulated in the cell, the smaller protein which can bind arsenic increases relatively.

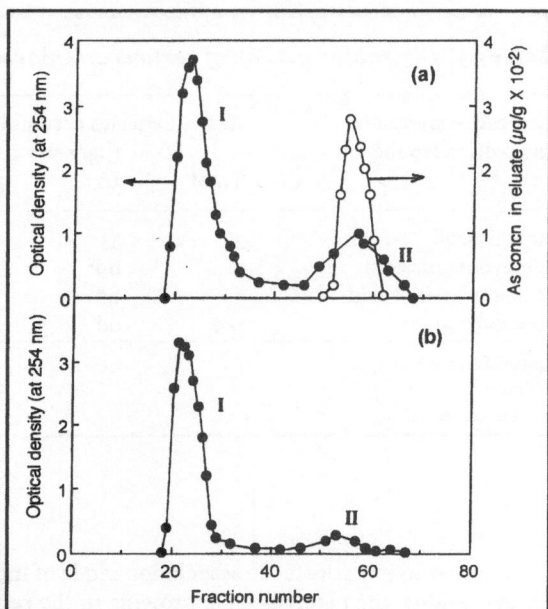

Fig. 5.6. Gel-filtration chromatograms for proteins from arsenic-holding (a) and arsenic-free (b) *C. vulgaris* cells.[44]

The eluates of peak II in Figure 5.6a and b were collected, dried, and analyzed for amino acids.[44] The dry powdered protein was mixed with 10 mL chloroform, the mixture was filtered on a 4.5 μm membrane filter and the filtrate was concentrated in a vacuum to give a white powder. The powder was dissolved with 6 M HCl and the protein solution was hydrolyzed by heating at 110 °C in a sealed tube for 22 h. The hydrolyzed amino acid solution was heated to dryness, the dry powder was dissolved with 0.5 mL of a citric acid buffer (pH 2.2), and the insoluble matter was removed by filtration on a 0.5 μm membrane filter. The filtrate was analyzed for amino acids using an automatic amino acid analyzer (JASCO 801-SC; detector, Hitachi 650-10S).

Table 5.4 presents the results of amino acid analysis for the protein obtained from the arsenic-exposed *C. vulgaris* cells and the arsenic-free cells.[44] Table 5.4 also shows the data obtained from a cadmium-bound protein fractionated in the same manner from *C. vulgaris* which had been cultured in a medium containing cadmium.[45] An interesting subject is whether a metallothionein-like protein is inductively biosynthesized in the *C. vulgaris* cell or not when the alga is exposed to arsenic. When the alga is exposed to cadmium, a cysteine-rich protein is synthesized, as shown in Table 5.4. However, the cysteine content (1.02 mol%) in the arsenic-bound protein is very close to that in the arsenic-free protein (0.76 mol%). It is proposed that metallothionein-like protein is not synthesized by *C. vulgaris* when the alga is exposed to arsenic. The alga may have another detoxifying process for arsenic, such as the methylation of arsenic.

Table 5.4. Amino acid analysis for the proteins obtained from the arsenic-exposed, cadmium-exposed, and control C. vulgaris *cells[a]*

Amino acid	Amino acid composition (mol %)		
	As-bound protein	Cd-bound protein	Metal-free protein
Glycine	22.1	0.84	11.1
Alanine	14.1	5.42	19.8
Valine	6.64	2.40	7.12
Leucine	1.53	4.88	2.97
Isoleucine	1.36	4.06	2.00
Serine	11.4	6.79	5.04
Threonine	3.92	2.74	6.01
Cysteine	1.02	8.31	0.76
Methionine	0	0.43	2.70
Aspartic acid	0	0.03	6.08
Glutamic acid	14.1	7.34	8.78
Arginine	3.41	5.71	5.74
Lysine	3.07	1.34	3.59
Histidine	2.90	1.13	1.52
Phenylalanine	1.87	2.93	5.60
Tyrosine	0.51	3.37	2.35
Tryptophane	0	1.54	0
Proline	12.1	42.1	8.71

a Modified from ref. 44

Biotransformation of Arsenic in Freshwater Food Chain

Many papers on biotransformation of arsenic through marine food chains have been published. For example, three trophic levels of marine organisms (phytoplankton: *Dunaliella marina*; zooplankton: *Artemia salina*; and shrimp: *Lysmata seticaudata*) were tested for their arsenic metabolism.[46] The experimental results led to a conclusion that organic forms of arsenic in marine food webs were derived from in vivo synthesis by primary producers and were efficiently transferred along a marine food chain. The shrimp, the highest trophic level in this food chain, could not form organic arsenic by itself. In this case, arsenate taken up from water was converted largely to arsenite. Similar conclusions were obtained from the experimental results on a phytoplankton-mussel (*Mytilus galloprovincialis*)-crab (*Carcinus marnas*) system,[47] a phytoplankton (*Fucus spiralis*)-grazer snail (*Littorina littoralis*)-carnivore snail (*Nucella laillus*) system,[48,49] and a phytoplankton (*Dunaliella tertiolecta*)-lobster (*Homarus americanus* juveniles) system.[50] More information about biotransformation of arsenic in the marine ecosystem is available in a review.[51]

Very few papers on biotransformation of arsenic in the freshwater ecosystem are available. Isensee et al examined the distribution of ^{14}C-labeled dimethylarsinic acid and dimethylarsine among freshwater organisms in a model ecosystem.[52] Fish, daphnia (*Daphnia magna*), snails and algae were exposed to dimethylarsinic acid and dimethylarsine for 3, 29, 32 and 32 days, respectively. The freshwater organisms represented parts of two food chains: water → algae → snails and water → diatoms, protozoa, and rotifers → daphnia → fish. Lower food chain organisms (algae and daphnia) bioaccumulated more dimethylarsinic acid and dimethylarsine, and the amounts accumulated indicate that dimethylarsinic acid and dimethylarsine do not show a high potential for biomagnifying in the environment.

First the authors examined the bioaccumulation and biotransformation of arsenic in a two-step freshwater food chain consisting of an autotroph (*C. vulgaris*) and a grazer (*Moina* sp.).[53] Arsenic-tolerant algae *C. vulgaris* and *Phormidium* sp. was cultured in a modified Detmer medium containing As(V). 250 *Moina macrocopa* (1.25 mg in dry mass) in 1 L aerated diluted modified Detmer medium (100 mL medium, 900 mL distilled water) were fed for seven days with the *C. vulgaris* cell or *Phormidium* sp. cell (about 18 mg dry mass per day: total 126 mg) which had bioaccumulated arsenic. The control group received arsenic-free bread yeast (Super camellia, dry yeast, Nissin Seifun Co., Japan). *Moina* sp., which had multiplied about 10-fold during seven days, was collected with a plankton net and rinsed with distilled water. A part of *Moina* sp. fed for seven days was analyzed for arsenic.

The bioaccumulation and biotransformation of arsenic when *Moina* sp. takes arsenic from water and through the two-step food chain are recorded in Table 5.5.[53,54] Table 5.5 also includes the data for arsenic-holding algae with which *Moina* sp. is fed. When *Moina* sp. was fed in water (a diluted modified Detmer medium) containing 1.0 mg/L of As(V), 10.3 µg/g-dry cell of arsenic which included 0.7 and 2.7 µg/g-dry cell of MMA and DMA, respectively, was bioaccumulated. When *Moina* sp. was fed in the food-chain in which *C. vulgaris* containing 2850 µg/g-dry cell of arsenic was used, 225 µg/g-dry cell of arsenic which included 21.5 and 15.6 µg/g-dry cell of MMA and DMA, respectively, was bioaccumulated. A similar result was obtained when *Moina* was fed with arsenic-holding *Phormidium* sp.

Table 5.5. Arsenic accumulation and metabolism in a two-step food chain[a]

Organisms	Accumulation route (As concn., mg/L)	Arsenic in organisms (µg/g-dry cell)				
		Total	IA	MMA	DMA	TMA
C. vulgaris	Water (100)	640	605	tr[b]	35.0	tr
Phormidium sp.	Water (100)	2900	2890	tr	5.2	tr
Moina sp.	Water (1.0)	10.3	6.9	0.7	2.7	tr
	Water (2.0)	17.9	13.6	1.1	3.3	tr
	Food (*C. vulgaris*)	225	188	21.5	15.6	tr
	Food (*Phormidium* sp.)	111	83.3	9.3	18.4	tr

a Modified from refs. 53 and 54
b Trace

The authors also examined the bioaccumulation and biotransformation of arsenic in a three-step freshwater food chain consisting of an autotroph (*C. vulgaris*), a grazer (*Moina* sp.), and a carnivore (a guppy, *Poecilia* sp.).[53] Four *Poecilia reticulata* (1.5 cm long and 10 mg dry mass) in the aerated diluted Detmer medium was fed for seven days with the arsenic-holding *Moina* sp. (about 0.125 mg dry mass per *Poecilia* sp. per day). The control group received "Tetrafin", a basic diet for goldfish (manufactured in Germany). The four *Poecilia* sp. were collected with a plankton net, rinsed with distilled water and analyzed for arsenic.

Table 5.6 is a record of the bioaccumulation and biotransformation of arsenic when *Poecilia* sp. takes arsenic directly from water and through the three-step food chain.[53] Total arsenic concentration in the cells when *Poecilia* sp. took arsenic directly from water was similar to that when *Poecilia* sp. was fed with the arsenic-holding *Moina* sp. in the three-step food chain. However, for TMA 4.6 µg/g-dry cell was detected in the *Poecilia* cell when the food chain was done, although the TMA concentration in the cell was quite small (0.8-1.1 µg/g-dry cell) for the direct uptake from water. When a goldfish, *Carassius* sp. was used instead of *Poecilia* sp., a similar result was obtained (Table 5.6).[54] It appears that as the trophic level is elevated in a food chain, the total arsenic accumulated in the cell decreases; while the relative proportion of methylated species increases.

Bioaccumulation of Antimony by Freshwater Algae

It seems reasonable to suppose that the arsenic-tolerant algae, such as *C. vulgaris*, also can bioaccumulate antimony, which is chemically similar to arsenic and is in the same group of the periodic table. To probe the bioaccumulation and biotransformation of antimony by *C. vulgaris*, the alga was cultured in a modified Detmer medium containing 10, 100, 4000 and 5000 mg/L of antimony (as elemental antimony with appropriate concentration of antimony potassium tartrate; abbreviated as Sb(III)).[55] Figure 5.7 shows the growth curve of *C. vulgaris* when the alga is exposed to various concentrations of Sb(III). The addition of 10 mg/L Sb(III)

Table 5.6. Arsenic accumulation and metabolism in a three-step food chain[a]

Organisms (Carnivore)	Accumulation route (As concn., mg/L)	Total	IA	MMA	DMA	TMA
			\multicolumn{4}{c}{Arsenic in organisms (µg/g-dry cell)}			
Poecilia sp.	Water (0.5)	6.8	5.0	0.6	0.1	1.1
	Water (1.0)	6.9	5.8	0.1	0.2	0.8
	Food[b]	5.6	0.9	tr[c]	0.1	4.6
Carassius sp.	Water (0.5)	33	32	tr	0.2	0.9
	Food[b]	37	24	6.9	1.3	5.3

a Modified from refs. 53 and 54
b The carnivore was fed with arsenic-holding *Moina* sp
c Trace

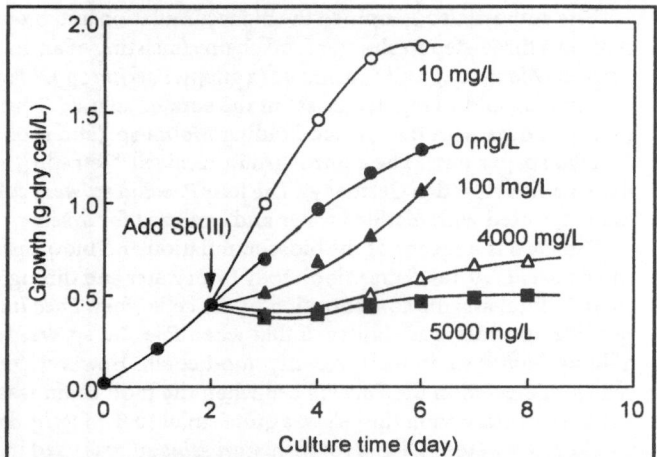

Fig. 5.7. Growth curve of *C. vulgaris* when a certain amount of Sb(III) is added after two days culture.[55]

enhanced the growth; while the growth was retarded at the Sb(III) level of 100 mg/L. Although the cells did not survive at 5000 mg/L of Sb(III), *C. vulgaris* could grow even at 4000 mg/L.

The total amount of antimony in the cell was determined. The algal cells were collected by centrifuging, washed twice or more with distilled water, and dried. The dried cells were dissolved in a concentrated nitric acid and heated at 80°C to homogeneous clear solution. The resulting pale-yellow transparent solution was mixed with 0.1 M tartaric acid, diluted to an appropriate volume, and subjected to flameless atomic absorption spectrometry.

Figure 5.8 shows the growth curve of *C. vulgaris* and the antimony concentration in the cell when the alga is cultured in a modified Detmer medium containing 50 mg/L Sb(III). The most marked bioaccumulation of antimony (12 mg/g-dry cell), which is somewhat higher than that for the As(V) exposure, appeared in the middle of log phase (six days culture). After seven days culture, the amount of antimony accumulated in the cell suddenly decreased although the cell growth increased. This result suggests that the antimony accumulated in the cell is rapidly excreted.

The valence of inorganic antimony excreted into the aqueous phase was determined by the cupferon-extraction method.[56] When *C. vulgaris*, which had bioaccumulated 372 μg/g-dry cell of antimony, was transferred into an antimony-free aqueous phase, the alga excreted a considerable amount of antimony (262 μg/g-dry cell), which consisted of Sb(III) (44%) and Sb(V) (56%). This result suggests that some part of Sb(III) accumulated is oxidized into Sb(V), which is less toxic than Sb(III), in the *C. vulgaris* cell.

The association mode of antimony with proteins was examined in a similar manner described above for arsenic. Figure 5.9 indicates a gel-filtration chromatogram of the solubilized antimony-bound proteins. Similarly to the chromatogram for arsenic (Fig. 5.6a), antimony was detected in the lower molecular-weight protein (MW = 3.0×10^4), which was analyzed for amino acids. As shown in Table 5.7, the arsenic-bound protein had an extraordinary high cysteine content, compared with the antimony-free protein. It is proposed that metallothionein-like protein is synthesized by *C. vulgaris* when the alga is exposed to antimony, which is quite different from the alga exposed to arsenic described before.

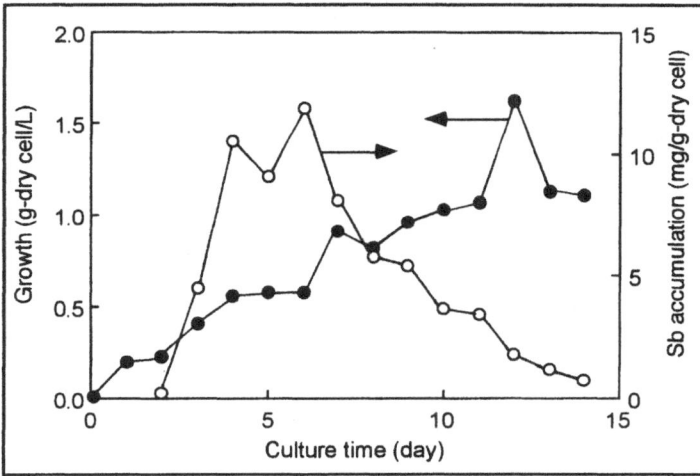

Fig. 5.8. Growth curve of *C. vulgaris* and the antimony concentration in the cell when the alga is cultured in the presence of 50 mg/L of Sb(III).[55]

Table 5.7. Amino acid analysis for the proteins obtained from the antimony-exposed C. vulgaris. *cells*

Amino acid	Amino acid composition (mol %)	
	Sb-bound protein	*Chlorella* in literature[a]
Glycine	18.0	10.6
Alanine	7.89	8.76
Valine	3.85	6.42
Leucine	4.12	8.67
Isoleucine	2.33	5.61
Serine	4.47	5.13
Threonine	18.3	5.22
Cysteine	6.36	0.96
Methionine	1.05	1.30
Aspartic acid	5.26	9.14
Glutamic acid	18.6	9.36
Arginine	2.59	5.82
Lysine	2.30	6.38
Histidine	0	1.54
Phenylalanine	1.89	4.42
Tyrosine	0.59	2.50
Tryptophane	0	1.08
Proline	2.47	7.05

a Literature data: average values for various *Chlorella* species not exposed to heavy metals (ref. 45)

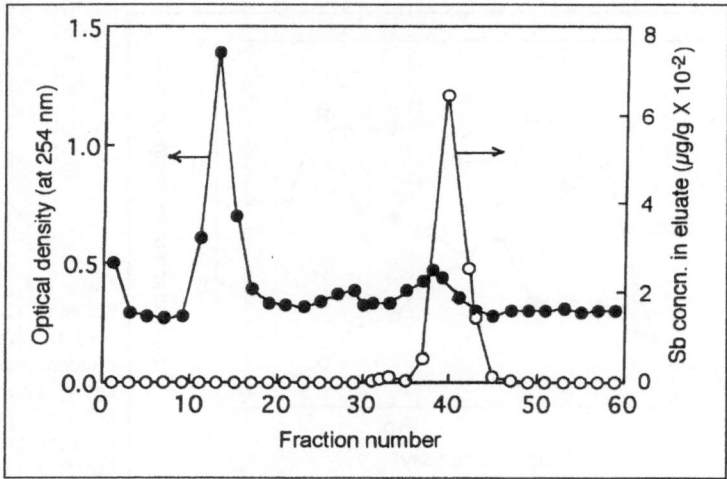

Fig. 5.9. Gel-filtration chromatograms for proteins from antimony-holding *C. vulgaris* cells.[55]

Bioaccumulation of Bismuth by Freshwater Algae

We also examined the bioaccumulation of bismuth by freshwater algae. Figure 5.10 indicates the growth curve of *C. vulgaris* and the bismuth concentration in the cell when the alga is cultured in a modified Detmer medium in the presence of 50 mg/L of bismuth (as elemental bismuth with appropriate concentration of bismuth potassium tartrate).

The total amount of bismuth in the cell was determined. The algal cells were collected by centrifuging, washed twice or more with distilled water, and dried. The dried cells were dissolved in a concentrated nitric acid and heated at 80 °C to homogeneous clear solution. The resulting solution was diluted to an appropriate volume, and subjected to flameless atomic absorption spectrometry.

As shown in Figure 5.10, the bioaccumulation of bismuth provided the maximum value at the beginning of log phase and then rapidly decreased, which is quite similar to the bioaccumulation of As(III) and Sb(III) (Figs. 5.3 and 5.8, respectively). It appears that *C. vulgaris* rapidly excrete these toxic compounds. The maximum value of bioaccumulation is 260 mg/g-dry cell of bismuth, which is much greater than those for As(V) and Sb(III).[57]

Conclusion

For the treatment of harmful heavy metals in wastewater by algae, freshwater algae which can bioaccumulate the metals are needed. From this point of view, we have collected freshwater algae from arsenic-polluted places, and the bioaccumulation and biotransformation of arsenic, antimony, and bismuth compounds by the algae have been studied.

Several algae, such as *C. vulgaris*, *Nostoc* sp., and *Phormidium* sp., which had been isolated from arsenic-polluted soil and water samples, exhibited a high tolerance against As(V). When *C. vulgaris* was cultured in a medium containing As(V), the bioaccumulation of arsenic occurred and then the excretion gradually took

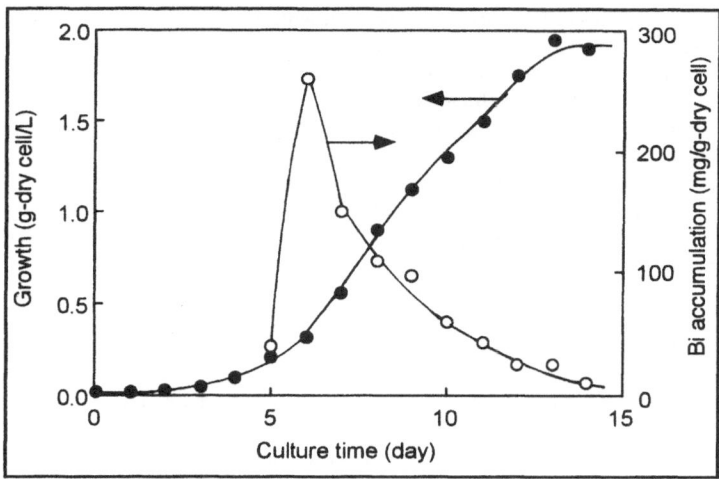

Fig. 5.10. Growth curve of *C. vulgaris* and the bismuth concentration in the cell when the alga is cultured in the presence of 50 mg/L of bismuth.[57]

place. Some portion of the arsenic accumulated in the cell was biotransformed into DMA, which considerably localized in the lipid and water soluble fractions. When *C. vulgaris* was exposed to As(III), Sb(III), and bismuth, the excretion occurred more rapidly compared with As(V). The association mode of inorganic arsenic with proteins was examined and it was found that metallothionein-like protein was not synthesized in the cell. On the contrary, when the alga was exposed to Sb(III) and Cd(II), metallothionein-like proteins were detected in the cell. In freshwater food chains, as the trophic level was elevated, the total arsenic accumulated in the cell decreased; while the relative proportion of methylated species increased.

By use of a raceway-type open-culture tank, the possibility of the practical use of arsenic bioaccumulation by freshwater algae has been suggested. To improve the efficiency of the metal recovery system using algae, it is important to control the excretion of metals which are once bioaccumulated. When we try to recover As(III), Sb(III) and bismuth by algae, this problem will be more essential because these metals are more rapidly excreted from algae than As(V), although the bioaccumulation of those metals is quite high. For As(III), this problem will be cancelled by means of the oxidation into As(V) as long as an effective bio- or chemical oxidation system, which can be coupled with the bio-recovery system, is developed. Obviously, more extended studies are needed to attain practical use of the bio-recovery system by freshwater algae.

References

1. National Research Council ed. Arsenic: Medical and Biologic Effects of Environmental Pollutants. Washington, DC: National Academy of Sciences, 1977.
2. Maeda S. Safety and environmental effects. In: Patai S ed. The Chemistry of Organic Arsenic, Antimony, and Bismuth Compounds. New York: John Wiley & Sons, 1994: 725-759.

3. Clarke LB, Sloss LL. Sources of trace elements in the atmosphere. IEA Coal Research 1992; 49:15-20.

4. Anke M, Schneider HJ, Brückner C eds. Arsen: 3rd Spurenelement-Symposium. Jena: Friedrich-Schiller-University, 1980.

5. Lederer WH, Fensterheim RJ eds. Arsenic: Industrial, Biomedical, Environmental Perspectives. New York: Van Nostrand Reinhold, 1983.

6. Ishinishi N, Okabe S, Kikuchi T eds. Hiso: Kagaku, Taisha, Dokusei (Arsenic: Chemistry, Metabolism, and Toxicity). Tokyo: Koseisha-Koseikaku, 1985.

7. Irgolic KJ, Kikuchi T, Maeda S, Craig PJ, eds. Natural and Industrial Arsenic. Appl Organomet Chem 1988; 2:283-404.

8. Irgolic KJ, Kikuchi T, Maeda S, Craig PJ, eds. Natural and Industrial Arsenic. Appl Organomet Chem 1990; 4:181-295.

9. Maeda S, Craig PJ, eds. Natural and Industrial Arsenic. Appl Organomet Chem 1992; 6: 307-420.

10. Maeda S, Craig PJ, eds. Environmental and Industrial Arsenic. Appl Organomet Chem 1994; 8: 165-283.

11. Maeda S, Craig PJ, eds. Environmental and Industrial Arsenic. Appl Organomet Chem 1996; 10:667-760.

12. Fowler BA, ed. Biological and Environmental Effects of Arsenic: Topics in Environmental Health. Amsterdam: Elsevier, 1983.

13. Nriagu JO, ed. Arsenic in the Environment. Part I: Cycling and Characterization. New York: John Wiley & Sons, Inc. 1994.

14. Nriagu JO ed. Arsenic in the Environment. Part II: Human Health and Ecosystem Effects, New York: John Wiley & Sons, Inc. 1994.

15. Brinckman FE, Bellama JM eds. Organometals and Organometalloids. Occurrence and Fate in the Environment. ACS Sym Ser Vol. 82. Washington DC; American Chemical Society, 1978.

16. Phillips, DJH, Depledge MH. Metabolic pathways involving arsenic in marine organisms: A unifying hypothesis. Mar Environ Res 1985; 17:1-12.

17. Cullen WR, Reimer KJ. Arsenic speciation in the environment. Chem Rev 1989; 89:713-764.

18. Phillips DJH. Arsenic in aquatic organisms: A review, emphasizing chemical speciation. Aqua Toxicol 1990; 16:151-186.

19. Maeda S, Sakaguchi T. Accumulation and detoxification of toxic metal elements by algae. In: Akatsuka I ed. Introduction to Applied Psychology. The Hague: SPB Academic Publishing bv, 1990:109-136.

20. Lunde G. The synthesis of fat and water soluble arseno organic compounds in marine and limnetic algae. Acta Chem Scand 1973; 27:1586-1594.

21. Lunde G. The analysis of arsenic in the lipid phase from marine and limnetic algae. Acta Chem Scand 1972; 26:2642-2644.

22. Baker MD, Wong PTS, Chau YK et al. Methylation of arsenic by freshwater green algae. Can J Fish Aquat Sci 1983; 40:1254-1257.

23. Nissen P, Benson AA. Arsenic metabolism in freshwater and terrestrial plants. Physiol Plant 1982; 54:446-450.

24. Maeda S, Kumamoto, T, Yonemoto M et al. Bioaccumulation of arsenic by freshwater algae and the application to the removal of inorganic arsenic from an aqueous phase. Part I. Screening of freshwater algae having high resistance to inorganic arsenic. Sep Sci Tech 1983; 18:375-385.

25. Maeda S, Nakashima S, Takeshita T et al. Bioaccumulation of arsenic by freshwater algae and the application to the removal of inorganic arsenic from an aqueous phase. Part II. By *Chlorella vulgaris* isolated from arsenic-polluted environment. Sep Sci Tech 1985; 20:153-161.

26. Maeda S, Kumeda K, Maeda M et al. Bioaccumulation of arsenic by freshwater algae (*Nostoc* sp.) and the application to the removal of inorganic arsenic from an aqueous phase. Appl Organomet Chem 1987; 1:363-370.

27. Maeda S, Fujita S, Ohki A et al. Takeshita T. Arsenic accumulation by arsenic-tolerant freshwater blue-green alga (*Phormidium* sp.). Appl Organomet Chem 1988; 2:353-357.

28. Maeda S, Kusadome K, Arima H et al. Uptake and excretion of total inorganic arsenic by the freshwater alga *Chlorella vulgaris*. Appl Organomet Chem 1992; 6:399-405.

29. Maeda S, Ohki A, Naka K. Approach to commercial-scale removal of arsenic from environmental waters by freshwater algae accumulating and transforming arsenic. Rep Asahi Glass Found 1991; 58:61-68.

30. Edmonds JS, Francesconi KA. Arseno-sugars from brown kelp (*Ecklonia radiata*) as intermediates in cycling of arsenic in a marine ecosystem. Nature (London) 1981; 289:602-604.

31. Morita M, Shibata Y. Chemical form of arsenic in marine macroalgae. Appl Organomet Chem 1990; 4:181-190.

32. Shibata Y, Jin K, Morita M. Arsenic compounds in the edible red alga, *Porphyra tenera*, and in *nori* and *yakinori*, food items produced from red algae. Appl Organomet Chem 1990; 4:255-260.

33. Edmonds JS, Francesconi KA. Methylated arsenic from marine fauna. Nature (London) 1977; 265:436.

34. Edmonds JS, Francesconi KA. The origin and chemical form of arsenic in the school whiting. Mar Pollut Bull 1981; 12:92-96.

35. Edmonds JS, Francesconi KA. Isolation and identification of arsenobetaine from the American lobster *Homarus americanus*. Chemosphere 1981; 10:1041-1044.

36. Edmonds JS, Francesconi KA. Trimethylarsine oxide in estuary catfish (*Cnidoglanis macrocephalus*) and school whiting (*Sillago bassenis*) after oral administration of sodium arsenate and as a natural component of estuary catfish. Sci Total Environ 1987; 64:317-323.

37. Edmonds JS, Francesconi KA. The origin of arsenobetaine in marine animals. Appl Organomet Chem 1988; 2: 297-302.

38. Norin H, Christakopoulas A. Evidence for the presence of arsenobetaine and another organoarsenical in shrimp. Chemosphere 1982; 11:287-298.

39. Norin H, Ryhage R, Christakopoulas A et al. New evidence for the presence of arsenocholine in shrimps (*Pandalus borealis*) by use of pyrolysis gas chromatography-atomic adsorption spectrometry/mass spectrometry. Chemosphere 1983; 12:299-315.

40. Kaise T, Watanabe S, Ito K et al. The study of organoarsenic compounds in fish and alga by extract mass measurement using fast atom bombardment mass spectrometry. Chemosphere 1987; 16:91-97.

41. Maeda S, Wada H, Kumeda K et al. Methylation of inorganic arsenic by arsenic-tolerant freshwater algae. Appl Organomet Chem 1987; 1:465-472.

42. Maeda S, Kusadome K, Arima H et al. Biomethylation of arsenic and its excretion by the alga *Chlorella vulgaris*. Appl Organomet Chem 1992; 6:407-413.

43. Maeda S, Ohki A, Miyahara K, et al. Metabolism of methylated arsenic compounds by arsenic-resistant bacteria (*Klebsiella oxytoca* and *Xanthomonas* sp.). Appl Organomet Chem 1992; 6:415-420.

44. Maeda S, Arima H, Ohki A et al. The association mode of arsenic accumulated in the freshwater alga *Chlorella vulgaris*. Appl Organomet Chem 1992; 6:393-397.

45. Maeda S, Mizoguchi M, Ohki A et al. Bioaccumulation of zinc and cadmium in freshwater alga, *Chlorella vulgaris*. Part II. Association mode of the metal and cell tissue. Chemosphere 1990; 21:965-973.

46. Wrench J, Fowler SW, Ünlü MY. Arsenic metabolism in a marine food chain. Mar Pollut Bull 1979; 10:18-20.

47. Ünlü MY. Chemical transformation and flux of different forms of arsenic in the crab *Carcinus maenas*. Chemosphere 1979; 5:269-275.

48. Klumpp DW. Accumulation of arsenic from water and food by *Littorina littoralis* and *Nucella lapillus*. Mar Biol 1980; 58:265-274.

49. Klumpp DW, Peterson PJ. Chemical characteristics of arsenic in a marine food chain. Mar Biol 1981; 62:297-305.

50. Cooney RV, Benson AA. Arsenic metabolism in *Homarus americanus*. Chemosphere 1980; 9:335-341.

51. Andreae MO. Biotransformation of arsenic in the marine environment. In: Lederer WH, Fensterheim RJ, eds. Arsenic: Industrial, Biochemical, Environmental Perspectives. New York: Van Nostrand Reinhold, 1983: 378-392.

52. Isensee AR, Kearney PC, Woolson EA et al. Distribution of alkyl arsenicals in model ecosystem. Envir Sci Technol 1973; 7:841-845.

53. Maeda S, Ohki A, Tokuda T et al. Transformation of arsenic compounds in a freshwater food chain. Appl Organomet Chem 1990; 4:251-254.

54. Maeda S, Inoue R, Kozono T et al. Arsenic metabolism in a freshwater food chain. Chemosphere 1990; 20:101-108.

55. Maeda S, Fukuyama H, Yokoyama E et al. Bioaccumulation of antimony by *Chlorella vulgaris* and the association mode of antimony in the cell. Appl Organomet Chem 1997; 11:393-396.

56. Luke CL, Campbell ME. Determination of impurities in germanium and silicon. Anal Chem 1953; 25:1588-1594.

57. Maeda S, Ohki A. unpublished data.

Metal Ion Binding by Biomass Derived From Nonliving Algae, Lichens, Water Hyacinth Root and *Sphagnum* Moss

Gerald J. Ramelow, Hua Yao and Wei Zhuang

Introduction

Many types of microorganisms are known to strongly bind metal ions under certain conditions.[1-8] Higher plants have also been studied.[9] Although several potential binding groups are thought to exist on the complex cell wall structures of proteins and polysaccharides, binding is thought to involve primarily COOH groups at pH values less than 2,[10] while diamine groups may be involved at pHs greater than 6.[11] Thus, binding of most metal ions is very pH dependent.[12] Generally, optimum binding is observed at a pH of around 5. Little binding is seen below pH values of 2 for most metal ions, but gold (in the form of the $AuCl_4^-$ ion) usually binds most strongly at a pH of 2. Darnall and coworkers found that certain algae were able to bind many metals, and especially precious metals such as gold, to a high degree.[13,14] Some metals such as silver and mercury seem to be less affected by pH than other metals.

The metal binding properties of algae and other microorgansims have been exploited by immobilizing biomass in inert polymeric materials such as silica gel,[13] styrene divinylbenzene[15] and polysulfone.[16] Thus, it might be possible to prepare adsorbent materials by immobilizing biomass in inert supports and use the sorbents thus prepared to either concentrate metal from aqueous streams or remove potentially harmful metal ions from waste streams.

In order to better utilize the metal-binding properties of microoganisms for practical purposes, it is important to: 1) survey the metal-binding properties of a wide variety of organisms to assess both the degree of binding of each metal by each organism with the aim of selecting eventually the organisms with the maximum ability to bind certain metals; 2) observe differences in metal-binding behavior

Wastewater Treatment with Algae, edited by Yuk-Shan Wong and Nora F.Y. Tam.
© Springer - Verlag and Landes Bioscience 1998.

between organisms with the ultimate aim of completely understanding the nature of the metal binding process; and 3) enhance the natural metal binding of biomass types by subjecting them to chemical and physical pretreatments.

Marine and freshwater algae are among the organisms that have been widely studied for metal-binding properties. Kuyucak and Volesky made a survey of several marine algae for ability to bind cobalt[17] and gold.[18] They found marked differences between algal species in their metal binding powers.

Living algal specimens have been used as monitors of marine pollution by metals.[19,20] However, the nature of the metal-binding process is not well understood. An understanding of metal binding by marine micro and macro algae is important first of all to elucidate the chemistry of dissolved metals and relationships between dissolved metals in the sea and marine algae. It has been found that nonliving biomass can also bind metals very efficiently. In fact, nonliving biomass is often better than living biomass in this regard.[21]

Not only is the understanding of the nature of metal binding by algae of importance to marine chemistry, but there are some very important potentially useful applications in other fields. If biomass can bind aqueous metal ions, then it has great potential for removing hazardous metals from waste streams and for reclaiming valuable metals. Of all biomass types, seaweed is extremely abundant in the world's seas and much of it ends up on beaches where it is considered a nuisance. It has been found that *Sargassum* sp. is an extremely good binder of metals.[22,23] Recent work has shown that algal biomass can be incorporated into polymeric materials for use as a metal sorbent.[1,13,15,16,23] One of the most attractive aspects of such sorbents is their low cost compared to currently available chelating resins.

In our laboratory the metal-ion binding properties of several marine algal species have been studied, including *Eisenia bicyclis, Sargassum fluitans, Ulva lactuca, Gracilaria conferta, Cladophora prolitera* and *Padina pavonica*.[22,24,25] Differences in metal preferences between the species have been observed. The effect of salts on metal sorption has also been studied. The aim has been to increase the percentage of metal adsorbed by first treating the dried algal biomass with chemicals such as 1 M HCl, 1 M HNO_3, 0.1 M NaOH and 1 M NaOH, acetone, detergent, 0.1% EDTA and warm (60°) water. Each of the treatments has shown some beneficial effect for some metals. No one treatment seems to be universally positive in increasing metal uptake capacity.

Lichens are another class of organisms that possess strong metal-ion binding properties. Lichens consist of an alga and a fungus growing in a symbiotic relationship. There are thousands of varieties of lichens found throughout the world. Lichens obtain their nourishment from rainwater and entrapment of atmospheric particles. They have been widely used as monitors of air pollution, not only because their presence or absence is a general air quality indicator, but also, their ability to metabolize metal-ions make them attractive for monitoring trace metal pollutants in the atmosphere.[26] The metal-ion binding properties of lichens have been studied and it has been found that nonliving lichen biomass is able to bind metal-ions to a greater degree than living lichens.[20]

In addition to algae and lichens, two other strong metal-ion sorbers are peat moss which is partially decomposed *Sphagnum* moss and water hyacinth (*Eichhornia crassipes*) root material. Both mosses[27] and water hyacinth plants[28] derive their nutrients directly from water, somewhat like lichens, so the metal-ion binding properties of these materials is not surprising. Water hyacinths have been used as water pollution monitors.[29,30]

Materials and Methods

Collection and Preparation of Biosamples

Marine samples tested consisted of six strains of seaweed algae and a sea plant. These included: a) *E. bicyclis* (Arame seaweed) purchased from Great Eastern Sun (Enka, NC); b) *S. fluitans* collected along the Louisiana Gulf of Mexico coast; c) *U. lactuca* supplied by Dr. Amir Neori of the National Center for Mariculture, Elat, Israel; d) *G. conferta* supplied by Dr. Michael Friedlander of the Israel Ocenaographic and Limnological Research, Ltd., Haifa, Israel; e) *C. prolitera* and *P. pavonica* supplied by Prof. Mustafa Unsal of the Institute of Marine Sciences, Erdemli, Turkey; and f) *Z. marina* collected along the Aegean coastline of Turkey. Dirt and other debris were removed from the specimens by rinsing with tap water and doubly-deionized water, then allowed to air dry for at least 48 hours.

Samples of two types of lichen common to southwest Louisiana, *Parmotrema praesorediosum* (Nyl.) Hale and *Ramalina stenospora* Mull. Arg., were collected from oak trees in Lake Charles, Louisiana. Additional lichens, tentatively identified as *Alectoria sarmentosa* (Ach.) Ach, *Letharia vulpina* (L.) Hue, and *Platismatia glauca* (L.) Culb. and Culb. were collected from pine trees in eastern Washington, northern Idaho and western Montana. Two others, *Cladina evansii* (Abb.) Hale and Culb., and *Cladonia abbreviatula* Merr., were collected on the ground in Destin, Florida. After identification and classification,[31] all lichen material was thoroughly washed in the laboratory with tap water and doubly-deionized water to remove dust and debris.

Sphagnum moss was purchased from a local garden store as partially decomposed peat moss. It originated from a Canadian peat bog. It was ground and sieved in the same manner as the lichen materials. Water hyacinth plants (*E. crassipes*) were collected from a waterway near Highway 27 between Lake Charles and Cameron, Louisiana. The roots were carefully removed from the plants, washed in tap water to remove dirt and debris, then dried in an oven for 48 hours at 80°C.

All biomaterials were oven dried at 80°C for 48 hours, and then ground to a fine powder in a laboratory grinder and sieved through numbers 35 (32 mesh), 60 (60 mesh), 120 (115 mesh), and 230 (250 mesh) sieves. The powder obtained from each sieve was collected and stored separately. Generally, 60-115 mesh-sized powder was used for the metal-ion uptake experiments and the 250 mesh and smaller fraction was used for immobilization in polymers.

Pretreatment of Biomass

Several of nine different treatments (0.1 M HNO_3, 1 M HNO_3, 1 M HCl, 0.1 M NaOH, 1 M NaOH, 0.1% EDTA, acetone, detergent and 60°C water) were used on each biomass powder to evaluate the effect on the Pb(II), Cu(II), Zn(II), Cd(II), Cr(III), Mn(II), Ni(II), Co(II), Ag(I) and Au(III)-binding capacity at pH values between 2 and 6. Five-gram amounts of biomaterial were mixed with 100 mL of each treating solution for 15 minutes. Three treatments (1 M NaOH, 1 M HNO_3 and 1 M HCl) were carried out on three marine algae at 15, 30 and 60 minute durations and at 25°C and 60°C to evaluate the effects of treatment time and temperature.

Metal uptake after each treatment was compared with biomass washed only in cold (25°C) water. Most treatments were carried out by stirring the biomass samples with treating agents at a room temperature of 25°C. The biomass was then filtered through paper, washed with doubly-deionized water and 0.05 M ammonium acetate solution (pH 6.5), and rinsed with doubly-deionized water. Finally, the biomass was dried in an oven at 60°C for 48 hours prior to metal uptake experiments.

Effect of pH on Metal Ion Uptake by Non-Immobilized Biomass

The uptake of metal ions by nonliving biomass derived from algae, lichens, peat moss and water hyacinth root as a function of pH was studied by a batch technique. One-hundred-milligram amounts of dried and treated biomass were mixed with 20 mL of five different 5 ppm mixed-metal standards each prepared in 0.05 M ammonium acetate whose pH had been adjusted to known values in the range of 2 to 6 with nitric acid and ammonium hydroxide, to give a lichen concentration of 5 mg/mL. The solutions were shaken for 1 hour (Heto Rotamix RS shaker), after which time the solutions were filtered through paper and the filtrates analyzed for residual metal ion content.

All metal analyses were carried out by atomic absorption spectrometry using air-acetylene flame atomization (Perkin-Elmer 372 or 5000 atomic absorption spectrometer). Commercial metal standards (1000 mg/L) were used to prepare all standard solutions for uptake experiments and to calibrate the spectrophotometer. An Orion Model 301 pH meter equipped with an Orion combination pH electrode was used to measure the pH of all solutions. Buffers of pH 4 and 7 (Sargent-Welch) were used to calibrate the pH meter.

Immobilization of Biomass in Polymers

The strong pH effect for many metal ions makes it possible to use biomass derived from algae and other biomaterials as ion exchange agents in a manner similar to chelating resins. In order to do so the biomass must be immobilized in some type of inert supporting material such as a polymer. Several types of polymeric materials were previously evaluated for immobilizing biomass, including styrene/divinylbenzene, ethyl acrylate/ethylene glycol dimethacrylate, ethyl acrylate/methylmethacrylate/ethylene glycol dimethacrylate, polysulfone and silica gel.[22] It was found that styrene/divinylbenzene and polysulfone were the best polymers for incorporating biomass, and that they compared favorably with a commercially available chelating resin. A biomass weight percentage of 25 was found to have greater sorption capacity than either 10 or 50%.

In the present study polysulfone was selected as the polymer for immobilizing biomass of peat moss because polysulfone is very easy to work with. Biomass was incorporated in polysulfone by first dissolving polysulfone in warm (60°C) chloroform. Then an amount of biomass calculated to give a weight percentage of 25 was added to the solution and stirred until a homogeneous suspension was obtained. This suspension was then added slowly to methanol to precipitate the polysulfone/biomass polymer. The polymer was filtered and air dried to give a light, porous polymer.

Laboratory Studies on Metal-Ion Binding by Biomass Immobilized in Polysulfone

Peat moss was immobilized in polysulfone as described previously to give a biomass weight percentage of 25. One-half-g amounts of this polymer (containing 0.125 g biomass) were mixed by shaking with five separate 25 mL amounts of a 5 ppm mixed metal solution containing Pb(II), Cu(II), Zn(II) and Cd(II) ions prepared in 0.05M KH_2PO_4 with pHs adjusted to 2, 3, 4, 5 and 6. After 1 hr the polymer and liquid phases were separated and the residual metal-ion content of the liquid determined by atomic absorption spectrophotometry.

Field Monitoring Studies with Polymer-Immobilized and Non-Immobilized Biomass

The strong metal-ion binding properties of certain type of biomass makes it possible to use immobilized biomass for monitoring dissolved metal-ion concentrations of natural waters. Samples of *Sphagnum* (peat) moss incorporated into polysulfone polymer (to give 25% by weight peat moss), a marine alga, *E. bicyclis*, and a lichen, *P. praesorediosum*, not incorporated into polymer, were used to study the dissolved metal content of a local polluted waterway. Ten sampling stations were selected, as shown in Figure 6.1. At each station a water sampler was immersed just below the surface. The water sampler was constructed of PVC pipe (30 cm length, 2.54 cm i.d). The samplers contained in separate nylon mesh bags 4 g each of peat moss/polysulfone polymer, dried *P. praesorediosum*, and dried *E. bicyclis*. After an exposure time of three weeks, the samplers were brought to the laboratory where the following extractions were performed. Metal ions bound to peat moss in the polymer were extracted by shaking the polymer with 25 mL of 1M HNO_3 for 1 hr. For the lichen and marine alga, 0.1 g amounts of each biomaterial, rinsed and dried, were heated in a high-pressure decomposition vessel (Parr 4745) with 2 mL of concentrated HNO_3 for 2 hours to give complete sample dissolution, followed by dilution to 25 mL with deionized water. Surface sediment samples were also collected at each station. Metals in sediment were extracted by the same procedure used for lichen and alga samples, with the exception that a 1-g sample was used. After heating, any undissolved sediment material was separated from the liquid phase by filtration before the solution was diluted to 25 mL. All analyses were performed by atomic absorption spectrophotometry.

Results and Discussion

The binding of metal ions by a wide variety of biomass types has been studied. In general binding is strongly affected by pH, as seen in Figure 6.2 where the percentages of Pb(II), Cu(II), Cd(II) and Zn(II) bound at the 5 ppm level by the marine algae *C. prolitera*, *P. pavonica* and *U. lactuca* are plotted vs. pH. The strong pH dependence is typical but pH-profile variations among organisms are great. In general, binding is greatest at pH 5-6, except for gold ions which are in the form of the $AuCl_4^-$ ion which because of the negative charge on the ion, binds most strongly at low pH values around 2.

It may be possible, and highly advantageous, to increase the degree of binding of metal ions by algae and other organisms by chemical and physical pretreatment of the dried biomass. This has been investigated in some detail. Many different

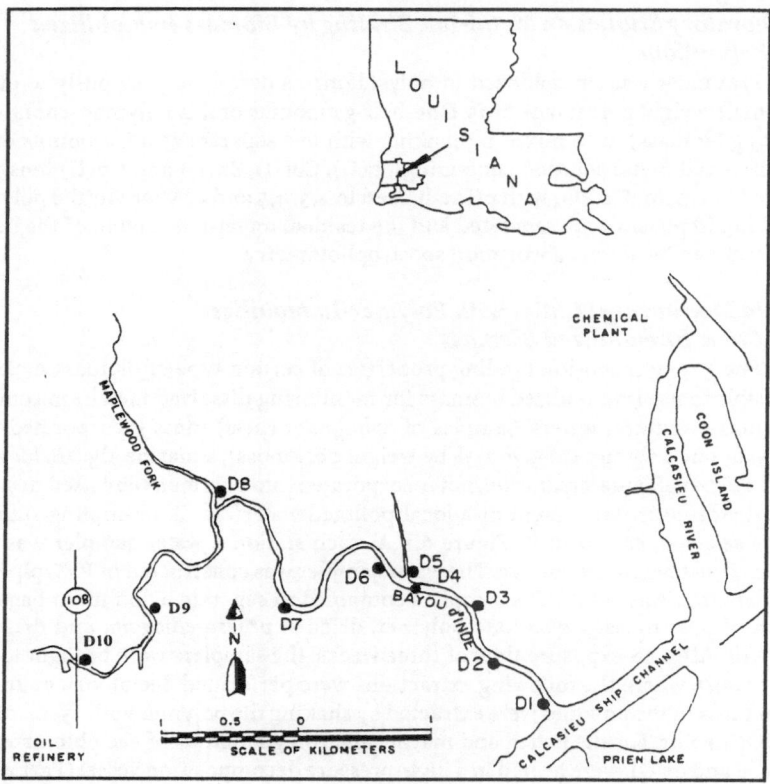

Fig. 6.1. Map of waterway in southwest Louisiana tested for dissolved metal concentrations. Sampling stations are indicated as D1 to D10.

treatments have been used, including sodium lauryl sulfate and Triton X-100 detergents, acetone, 0.1% EDTA, 0.1M HNO_3, 1 M HNO_3, 1 M HCl, 0.1 M NaOH, 1 M NaOH and warm (60°C) water.

The effect of treatment temperature and time for acidic and basic treatments on subsequent Pb(II), Cu(II), Cd(II) and Zn(II) binding by *S. fluitans*, *E. bicyclis* and *G. conferta* was studied.[25] As seen in Table 6.1 for *S. fluitans* and *E. bicyclis*, neither 1M NaOH treatment time nor temperature had much effect on percentage of metal ion bound. A 15-min time was slightly better than 30 or 60-min times. Essentially the same was observed for 1M HCl and 1M HNO_3 treatments of all three algae. As a result, all subsequent experiments were carried out using 15-min treatments at 25°C.

All treatments used showed some ability to increase the amount of metal ion bound but no single treatment was universally beneficial. In Figure 6.3 are shown a comparison of the effects of three different pretreatments, 1M HCl, 1M NaOH, and 60°C water, on subsequent Pb(II), Cu(II), Zn(II) and Cd(II) uptake by two marine algae, *C. prolitera* and *P. pavonica*, and the sea grass *Z. marina* at pH 5. The beneficial effects of all three treatments are apparent. Also, although acidic

Table 6.1. Effect of NaOH molarity, temperature and treatment time on percentage of metal subsequently removed from solution by nonliving biomass of E. bicyclis *and* S. fluitans

Biomass	NaOH molarity	Temperature (°C)	Treatment Time	% Metal Removed			
				Pb	Cu	Zn	Cd
E. bicyclis	0.1	25	15	92	88	42	79
	0.1	60	15	90	79	68	73
	1.0	60	15	89	70	56	69
	1.0	25	15	88	80	62	73
S. fluitans	0.1	25	15	97	91	84	88
	0.1	60	15	96	93	83	86
	1.0	25	15	94	91	83	87
	1.0	25	30	92	89	82	88
	1.0	25	60	88	62	73	82
	1.0	60	15	96	91	83	85
Untreated S. f.	-	-	-	92	77	74	78

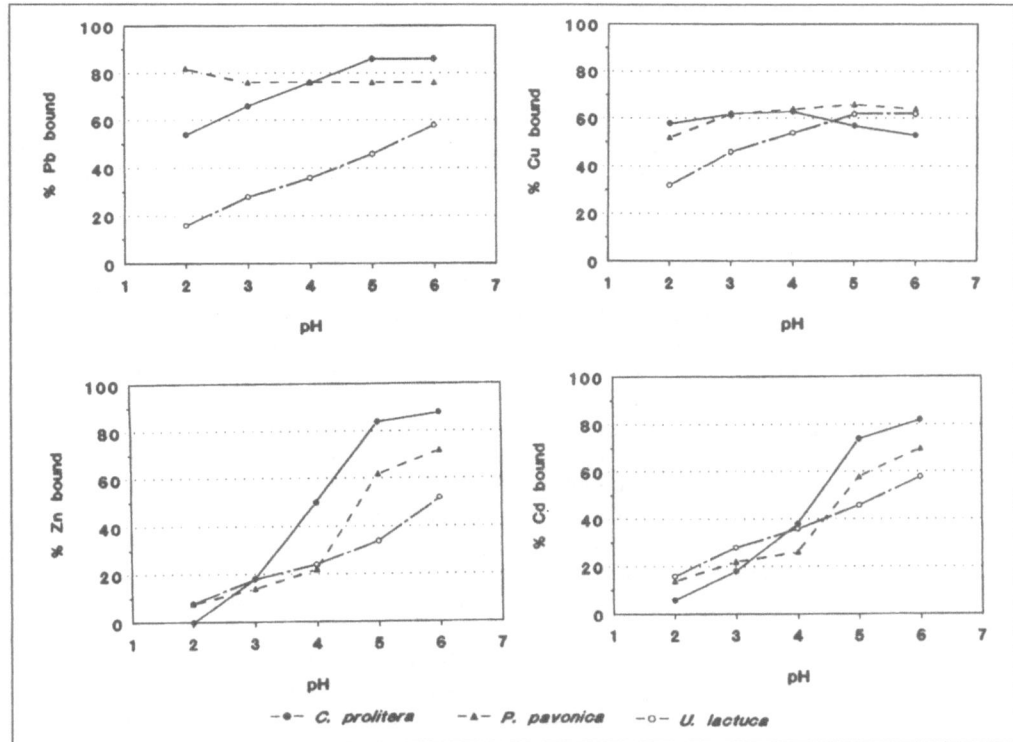

Fig. 6.2. Percentages of Pb(II), Cu(II), Zn(II) and Cd(II) ions bound by *C. prolitera, P. pavronica* and *U. lactuca* at pH 2-6.

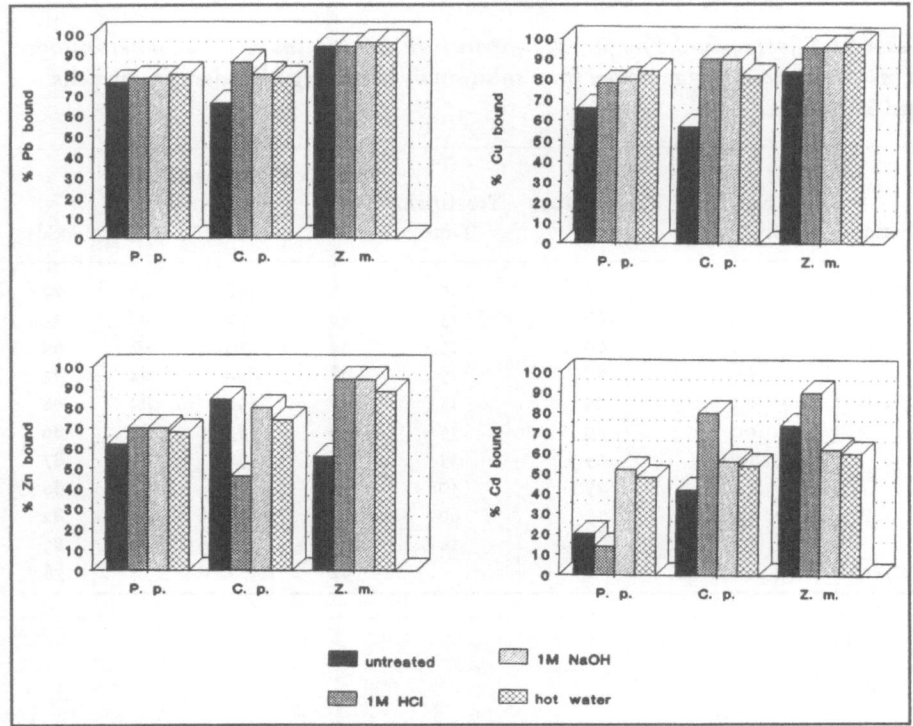

Fig. 6.3. Effect of three pretreatments of subsequent uptake of Pb(II), Cu(II), Zn(II) and Cd(II) ions by two marine algae, *P. pavronica* (P. p.), *C. prolitera* (C. p.) and the sea grass, *Z. marina* (*Z. m.*) at pH 5.

treatment increased Zn and Cd uptake considerably for *P. pavonica* and *Z. marina,* it and the other two treatment drastically decreased the uptake of these ions by *C. prolitera*. The latter effect is unique among all organisms studied so far and is unexplained. All treatments increased Cu uptake, and to a much lesser extent, Pb uptake.

The effects of some treatments on Au(III), Ag(I) and Hg(II) uptake by four marine algae, *S. fluitans, G. conferta, E. bicyclis* and *U. lactuca* at pH 5.5, are shown in Figure 6.4. Here, 1 M HNO$_3$ and 1 M HCl acid treatments were applied along with 0.1M NaOH and 1M NaOH. All four treatments increased metal-ion uptake. The treatments were most effective for organisms such as *E. bicyclis* which were quite ineffective metal-ion sorbers before treatment. After treatment all four marine algae were nearly equally effective in binding these metal ions and binding was in the 90-100% range.

The effect of the same four treatments on the uptake of Cr(III), Ni(II), Co(II), Mn(II) and Tl(I) ions by *S. fluitans, C. conferta, E. bicyclis* and *U. lactuca* was less pronounced, as seen in Figure 6.5. The two NaOH treatments were most beneficial but very little improvement was seen in thallium binding.

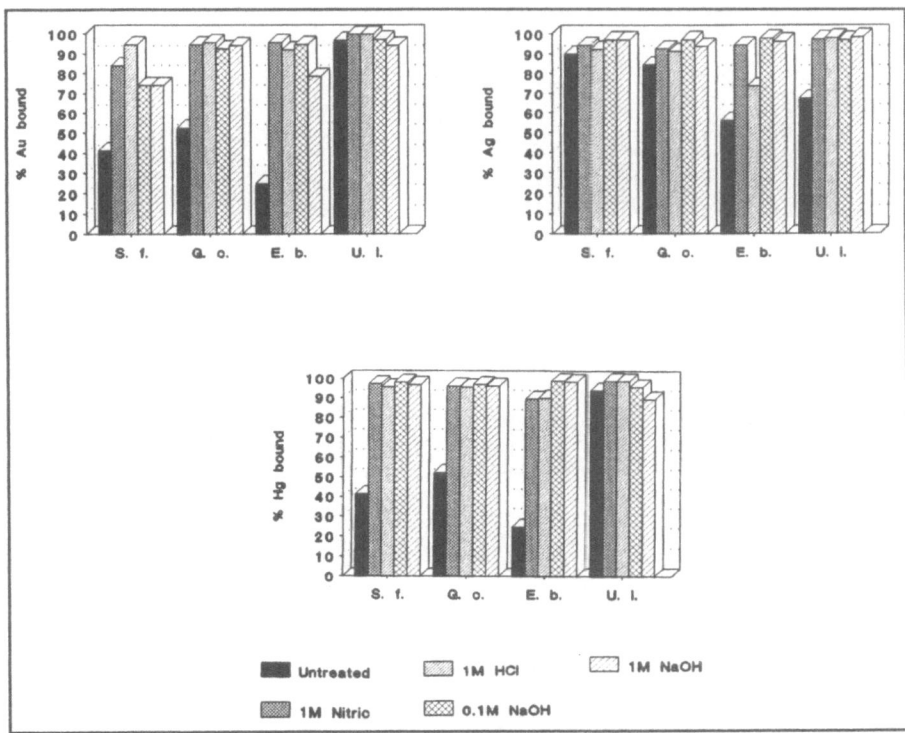

Fig. 6.4. Effect of four pretreatments on subsequent uptake of Au(III), Ag(I) and Hg(II) ions by four marine algae, *S. fluitans* (S. f.), *G. conferta* (G. c.), *E. bicyclis* (E. b.) and *U. lactuca* (U. l.) at pH 5.5.

These positive effects of pretreatments led to a more extensive study. Six different treatments were applied to dried biomass derived from four marine algae, *S. fluitans*, *G. conferta*, *E. bicyclis* and *U. lactuca*. Effects of the six treatments are shown in Figure 6.6 where metal uptake percentage at pH 5.5 (shown to be generally the optimum pH) are plotted for four elements. Again, the positive effects of pretreatments on subsequent metal-ion uptake are observed, particularly 0.1 M NaOH and 1M NaOH treatments.

Lichens, which consist of green or blue-green algal cells embedded in fungal cells coexisting in a symbiotic relationship, also bind metal ions in a pH-dependent manner.[12] Many different types of lichen have been studied. The strong effect of pH on the binding of metal ions by some of the lichens studied in this work is similar to that observed for the marine algae with maximum binding observed at pH 5-6. Lead and copper ions were found to bind to a higher degree, in general, for all seven lichen strains studied than the other metal ions. Metal-uptake ability was not increased greatly for any of the lichens studied by any of the treatments used. *P. praesorediosum* was most affected. This is shown in Figure 6.7 for copper, zinc,

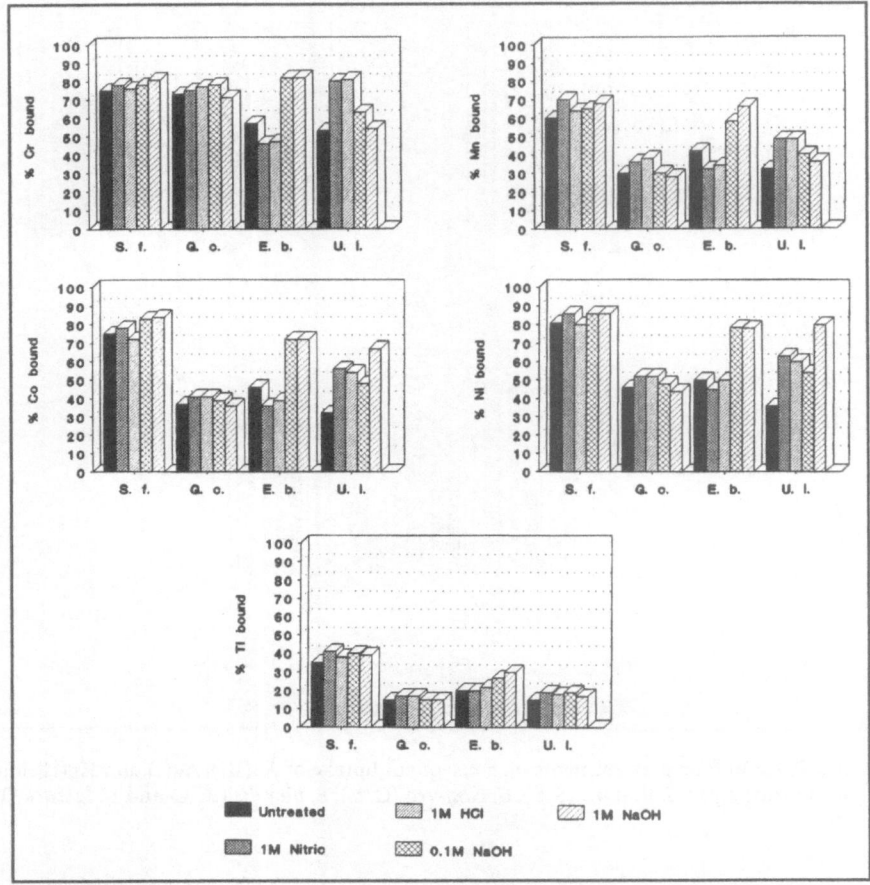

Fig. 6.5. Effect of four pretreatments on subsequent uptake of Cr(III), Mn(II), Co(II), Ni(II) and Tl(I) ions by four marine algae, *S. fluitans* (S. f.), *G. conferta* (G. c.), *E. bicyclis* (E. b.) and *U. lactuca* (U. l.) at pH 5.5.

cadmium and nickel in which the percentage of available metal remaining in solution after 15-min exposure to 5 ppm metal ion is plotted versus pH for each metal/organism combination under a variety of treatment conditions.

Dried, partially decomposed *Sphagnum* moss (peat moss) and water hyacinth root are exceptionally good sorbing materials for many metal ions. As was the case for lichen biomaterials, lead and copper ions are generally sorbed to a greater degree than the other metal ions studied, as shown in Figures 6.8 and 6.9. The optimum pH for lead sorption is 3 for both peat moss and water hyacinth root, whereas for copper the optimum pH is 4-6 for both. Lichens generally show little lead sorption at pH 3 with a maximum sorption in the pH range 4-6. Copper sorption by lichens is maximum near pH 6.

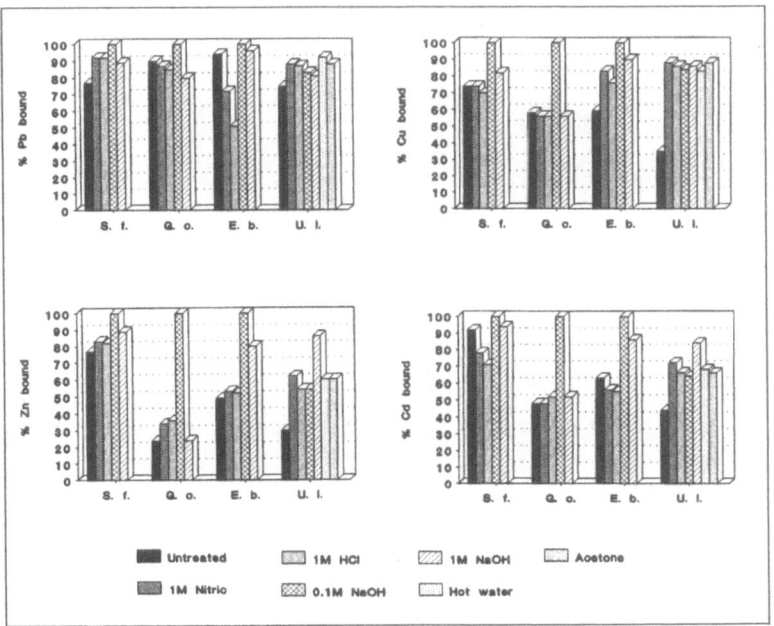

Fig. 6.6. Effect of six pretreatments on subsequent uptake of Pb(II), Cu(II), Zn(II) and Cd(II) ions by four marine algae, *S. fluitans* (S. f.), *G. conferta* (G. f.), *E. bicyclis* (E. b.) and *U. lactuca* (U. l.) at pH 5.5.

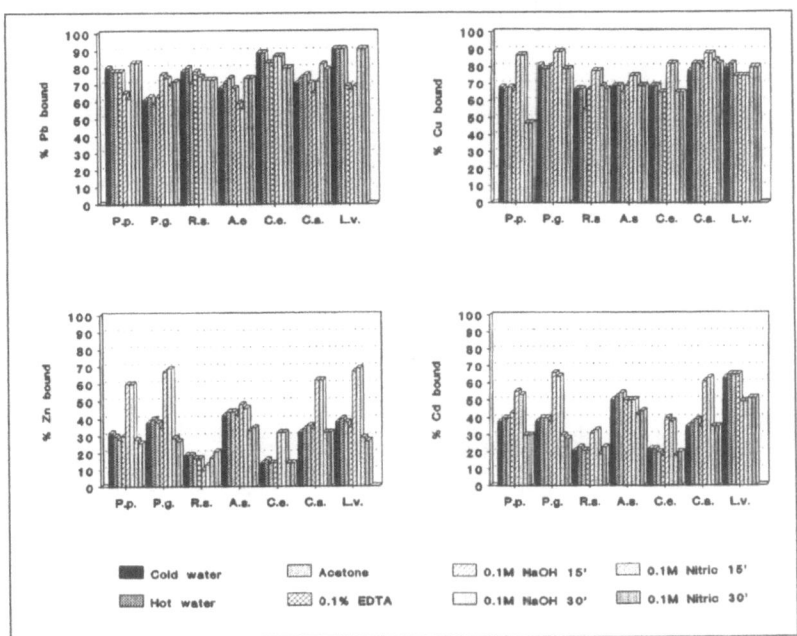

Fig. 6.7. Effect of five pretreatments on subsequent uptake of Pb(II), Cu(II), Zn(II) and Cd(II) ions by seven lichen strains, *P. praesorediosum* (P. p.), *P. glauca* (P. g.), *R. stenospora* (R. s.), *A. sarmentosa* (A. s.), *C. evansii* (C. e.), *C. abbreviatula* (C. a.) and *L. vulpina* (L. v.) at pH 5.

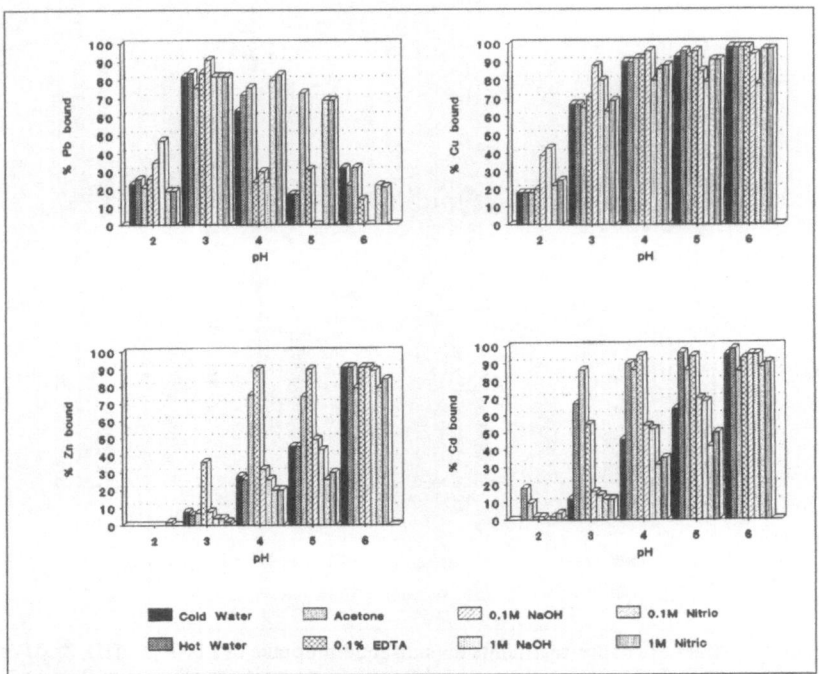

Fig. 6.8. Effect of seven pretreatments on subsequent uptake of Pb(II), Cu(II), Zn(II) and Cd(II) ions by *Sphagnum* (peat) moss at pH 2-6.

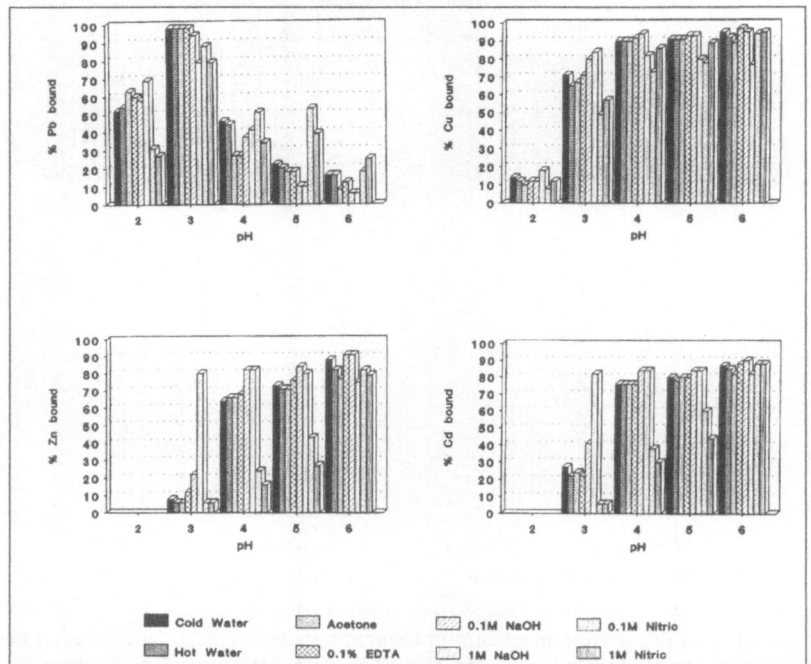

Fig. 6.9. Effect of seven pretreatments on subsequent uptake of Pb(II), Cu(II), Zn(II) and Cd(II) ions by water hyacinth (*E. crassipes*) root at pH 2-6.

Several of the treatments altered the pH sorption profiles for algae, lichens, peat moss and water hyacinth root. The effect of some treatments was to increase dramatically the degree of metal ion sorption at lower pHs in the range 3-4. Often the percentage of metal ion sorbed at the lower pH approached the amount sorbed for untreated biomass at higher pH. This is shown in Figure 6.10 for chromium, manganese, cobalt and nickel sorption by peat moss.

The sorption of gold and silver has received much attention due probably to precious metal reclamation applications. It is well known that these two elements are strongly sorbed by certain types of biomass.[13,17,24,25] The effect of treatments on subsequent uptake of gold and silver was studied for peat moss and water hyacinth root biomass. Each organism bound gold (III) ions to a high degree, as seen in Figure 6.11. None of the treatments increased the degree of binding to any significant degree. Binding decreased with increasing pH because the available gold in solution was in the form of the $AuCl_4^-$ ion. The negative charge results in a behavior opposite to positively-charged ions which compete with protons for binding sites. Silver was bound much more completely by water hyacinth root than peat moss, even if the peat moss was subjected to treatments. The two NaOH treatments increased the degree of Ag(I) binding by peat moss at pH 4-5, as shown in Figure 6.12.

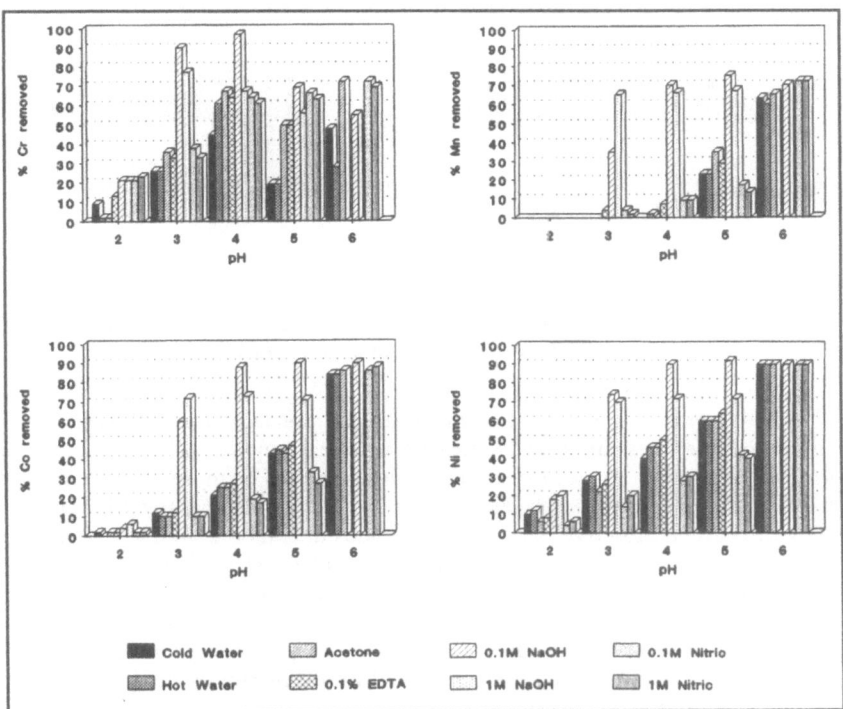

Fig. 6.10. Effect of seven pretreatments on subsequent uptake of Cr(III), Mn(II), Co(II) and Ni(II) ions by *Sphagnum* (peat) moss at pH 2-6.

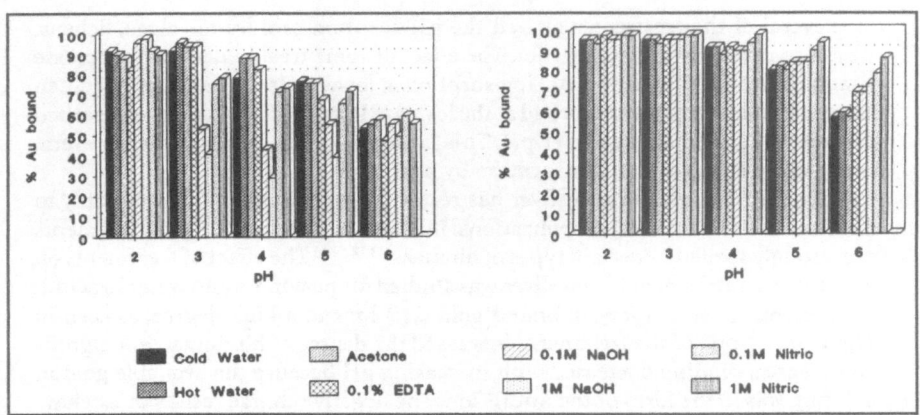

Fig. 6.11. Effect of seven pretreatments on subsequent uptake of Au(III) by *Sphagnum* (peat) moss (left) and water hyacinth (*E. crassipes*) root at pH 2-6 (right).

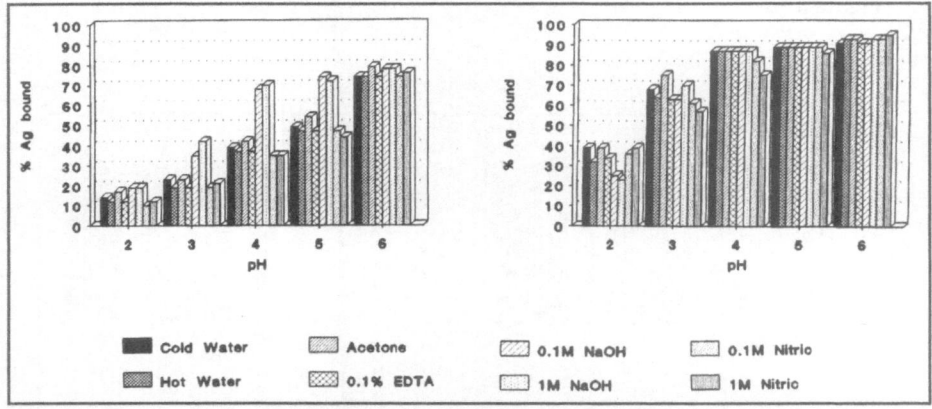

Fig. 6.12. Effect of seven pretreatments on subsequent uptake of Ag(II) by *Sphagnum* (peat) moss (left) and water hyacinth (*E. crassipes)* root at pH 2-6 (right).

All the treatments used showed some benefit in certain cases, although no one single treatment was universally effective. One-tenth molar NaOH appears to be the best treatment. A 15-min treatment time appears to be more effective than 30 minutes. Treatment temperatures of 25° and 60°C did not show any differences in subsequent metal-ion binding. Metal/organisms interactions which were already optimum in terms of degree of sorption were least affected by chemical or physical treatment. Interactions which produced little metal sorption, such as Zn(II), Cd(II), Co(II), Ni(II) and Mn(II) binding, led to increased metal binding after treatment. Peat moss and water hyacinth root were generally superior to all the lichens studied in terms of metal sorbing ability, especially after certain treatments.

The effect of pH on the uptake of Pb(II), Cu(II), Zn(II) and Cd(II) ions by peat moss immobilized in polysulfone at a biomass concentration of 25% was studied. The results shown in Figure 6.13 show that pH greatly affects the binding of all four

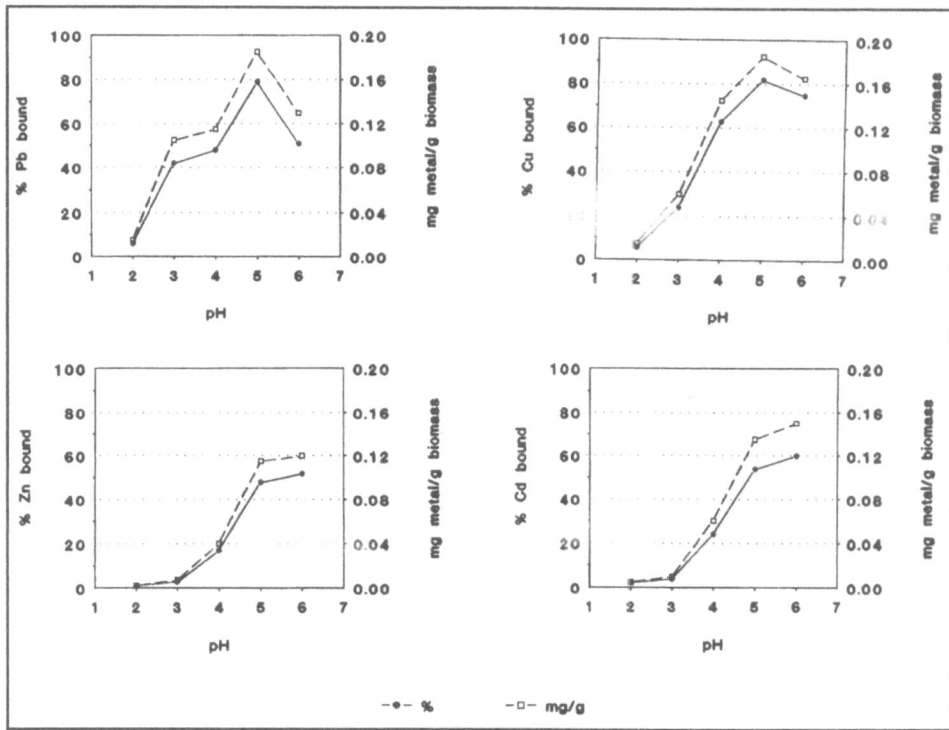

Fig. 6.13. Percentage of Pb(II), Cu(II), Zn(II) and Cd(II) ions bound and mg metal ion bound per g of biomass immobilized in polysulfone (25% biomass) at pH 2-6.

ions by peat moss when immobilized. The pH profiles are similar to those previously observed for algae and lichens with maximum binding seen at about pH 5. However, the pH profiles are somewhat different than those observed for non-immobilized peat moss, as can be seen by comparing Figure 6.13 with Figure 6.8. It is not known why the profiles are so different, expecially for lead.

The success of the lab experiments with peat moss immobilized in polysulfone prompted an investigation into the use of such polymers for monitoring dissolved metal concentrations in the environment. A local waterway known to be contaminated with certain metals[32] was studied. Samplers containing peat moss/peat moss polymer, together with a lichen, *P. praesorediosum*, and a marine brown alga, *E. bicyclis* were placed at 10 locations along Bayou d'Inde and left in place for three weeks. These were basically the same locations used in a previous study in which lichen biomass was immersed in the water to monitor dissolved metals.[33] Back in the lab any sorbed metal ions were stripped from the polymer with dilute nitric acid and analyzed. The results are shown in Figure 6.14 for Pb, Cu, Zn and Ni. Maximum values for each of these elements are seen at station 5. This station is located at the confluence of an industrial ditch with the bayou. A previous study using immersed lichen biomass indicated high levels of dissolved copper, zinc, chromium, nickel and mercury at this station.[33] The industrial plant no longer

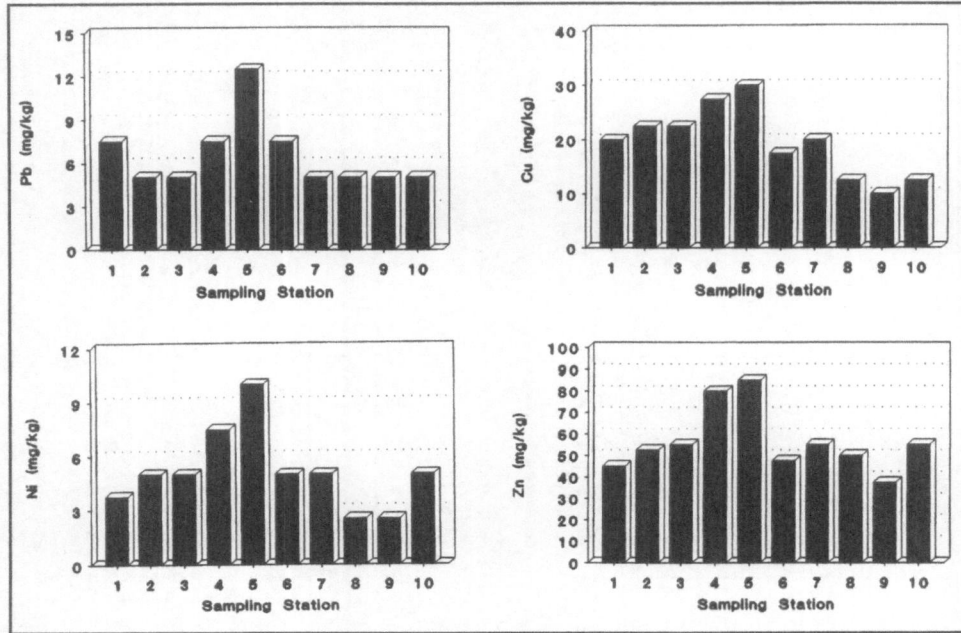

Fig. 6.14. Dissolved lead, copper, nickel and zinc (mg metal/kg biomass) at ten stations along a polluted waterway monitored by peat moss immobilized in polysulfone polymer (25% biomass) after a 3-week exposure.

releases any significant amount of these metals and a comparison of the data from 10 years ago shows that the metal ions in question are more evenly distributed along the bayou, but station 5 still has the highest concentration of Pb, Cu, Zn and Ni. Presumably metals deposited previously in the sediments[32] are still being mobilized into the water column. Other elements studied such as aluminum, iron and chromium were more evenly distributed along the waterway. This confirms the practicality of using biomass immobilized in polymers such as polysulfone for monitoring dissolved metal contents of natural waters.

Acknowledgment

The authors are grateful to Dr. Ulku Ramelow for her assistance in the preparation of the polymers.

References

1. Harris P, Ramelow G. Binding of metal ions by particulate biomass derived from *Chlorella vulgaris* and *Scenedesmus quadricauda*. Environ Sci Technol 1990; 24:220-228.
2. Beveridge TJ, Murray RGE. Uptake and retention of metals by cell walls of *Bacillus subtilis*. J Bacteriol 1976; 127: 1502-1518.
3. Horikoshi T, Nakajima A, Sakaguchi T. Uptake of uranium by *Chlorella regularis*. Agric Biol Chem 1979; 32:617-623.

4. Nakajima A, Horikoshi T, Sakaguchi T. Recovery of uranium by immobilized microorganisms. Europ J Appl Microbiol Biotech 1982; 16:88-91.

5. Tsezos M, Volesky B. Biosorption of uranium and thorium. Biotechnol Bioeng 1981; 23:583-604.

6. Tobin JM, Cooper DG, Neufeld RJ. Uptake of metal ions by *Rhizopus arrhizus* biomass. Appl Environ Microbiol 1984; 47:821-824.

7. Khummongkol D, Canterford GS, Fryer C. Accumulation of heavy metals in unicellular algae. Biotechnol Bioeng 1982; 24:2643-2660.

8. Les A, Walker RW. Toxicity and binding of copper, zinc, and cadmium by the blue-green alga *Chroococcus paris*. Water Air Soil Pollut 1984; 23:129-139.

9. Drake LR, Rayson GD. Plant-derived materials for metal ion-selective binding and preconcentration. Anal Chem 1996; 68:22A-27A.

10. Crist RH, Oberholser K, Schwartz D, Marzoff J, Ryder D, Crist DR. Interactions of metals and protons with algae. Env Sci Tech 1988; 22:755-760.

11. Huel-Yang D Ke, Rayson GD. Characterization of Cd binding sites on *Datura innoxia* using ^{113}Cd NMR spectrometry. Environ Sci Technol 1992; 26:1202-1205.

12. Ramelow GJ, Zhang Y, Liu L. Uptake of metallic ions from aqueous solution by dried lichen biomass. Microbios 1991; 66:95-105.

13. Darnall D, Greene B, Hosea M, McPherson M, Henzi M, Alexander MD. Recovery of heavy metals by immobilized algae. In: Thompson R, ed. Trace Metal Removal from Aqueous Solution, Special Publication No. 61, Royal Society of Chemistry:1-25.

14. Greene B, Hosea M, McPherson R, Henzi M, Alexander MD, Darnall D. Interaction of gold(I) and gold(II) complexes with algal biomass. Env Sci Technol 1986; 20:627-632.

15. Ramelow GJ, Liu L, Himel C, Fralick D, Zhao Y, Tong C. The analysis of dissolved metals in natural waters after preconcentration on biosorbents of immobilized lichen and seaweed biomass. Int J Environ Anal Chem 1993; 53: 219-232.

16. Trujillo M, Jeffers TH, Ferguson C, Stevenson HQ. Mathematically modeling the removal of heavy metals from a wastewater using immobilized biomass. Env Sci Technol 1991; 25:1559-1565.

17. Kuyucak N, Volesky B. Accumulation of cobalt by marine alga. Biotechnol Bioeng 1989; 33:809-814.

18. Kuyucak N, Volesky B. Accumulation of gold by algal biosorbent. Biorecovery 1989; 1:189-204.

19. Sawidis Th, Voulgaropoulos AN. Seasonal accumulation of iron, cobalt, and copper in marine algae from Thermaikos Gulf of the northern Aegean Sea, Greece, Mar Environ Res 1986; 19:39-47.

20. Ho YB. *Ulva lactuca* as bioindicator of metal contamination in intertidal waters in Hong Kong. Hydrobiologia, 1990; 203:73-81.

21. Richardson DHS, Kiang S, Ahmadjian V, Nieboer E. Lead and uranium uptake by lichens. In: Brown DH, ed. Lichen Physiology and Cell Biology New York: Plenum, 1985:227-246.

22. Ramelow GJ, Fralick D, Zhao Y. Factors affecting the uptake by aqueous metal ions by dried seaweed biomass. Microbios 1992; 72:81-93.

23. Tong C, Ramelow US, Ramelow GJ. Evaluation of polymeric supports for immobilizing biomass to prepare sorbent materials for metals. Intern J Environ Anal Chem 1994; 56:175-191.

24. Hao Y, Zhao Y, Ramelow GJ. Uptake of metal ions by nonliving biomass derived from marine organisms-effect of pH and chemical treatments. J Environ Sci Health 1994; A29:2235-2254.

25. Zhou Y, Hao Y, Ramelow GJ. Evaluation of treatment techniques for increasing the uptake of metal ions from solution by nonliving seaweed algal biomass. Environ Monitor Assess 1994; 33:61-70.
26. Richardson DHS, Nieboer E. Lichens and pollution monitoring. Endeavor 1981; 5:127-133.
27. Breuer K, Melzer A. Heavy metal accumulation (lead and cadmium) and ion exchange in three species of *Sphagnaceae*. I. Main principles of heavy metal accumulation in *Sphagnaceae*. Oecologia 1990; 82:461-467.
28. Heaton C, Frame J, Hardy JK. Lead uptake by *Eichhornia crassipes*. Toxicol Environ Chem 1986; 11:125-135.
29. Chigbo F, Smith RW, Shore FL Uptake of arsenic, cadmium, lead, and mercury from polluted waters by the water hyacinth *Eichhornia crassipes*. Environ Pollut Ser A 1982: 27:31-36.
30. Gonzalez H, Lodenius M, Otero M. Water hyacinth as indicator of heavy metal pollution in the tropics. Bull Environ Contam Toxicol 1989; 43:910-914.
31. Hale ME. How to Know the Lichens, 2nd ed. Dubuque, Iowa: Wm. C. Brown, 1979.
32. Mueller CS, Ramelow GJ, Beck JN. Spatial and temporal variations of heavy metals in sediment cores from the Calcasieu River/Lake Complex. Water, Air, Soil Pollut 1989; 43:213-230.
33. Beck JN, Ramelow GJ. The use of lichen biomass to monitor dissolved metals in natural waters. Bull Environ Contam Toxicol 1990; 44:302-308.

CHAPTER 7

Metal Resistance and Accumulation in Cyanobacteria

Marli F. Fiore, David H. Moon and Jack T. Trevors

Introduction

The indiscriminate discharge of metals into the environment generated from various industrial processes, modern agricultural practices, acid mine drainage and human wastes, has long been recognized as an important source of these pollutant. Metals constitute more than 75% of all known elements and occupy groups 1A to 6A (representative metals), groups 1B to 8B (transition metals) and the lanthanide and actinide metals.[1] The term "heavy metals" has been redefined over the years and although not completely satisfactory from a chemical point of view, the most widely used is those elements with a density greater than 5 g/cm^3. The definition problem has been thoroughly discussed by Gadd[2] and to avoid further confusion the term heavy metal will not be used here.

Microbial cells can accumulate metals essential for growth and metabolic processes[3] as well as other metals with no known biological functions.[4,5] Due to their well documented ability to bind metals and to their abundance in natural environments, a significant contribution to metal sorption has been attributed to microbial cells.[6-10]

For about 150 years cyanobacteria were considered a special group of algae, the blue-green algae.[11] However, the comparison of 16S and 5S rRNA sequences suggest that the cyanobacteria are a phylogenetically coherent group within the Gram-negative eubacteria.[12] According to Gibbons and Murray[13] the cyanobacteria are members of the Kingdom Procaryotae and are included within the division Gracilicutes (bacteria with a Gram-negative cell wall), class Photobacteria, subclass Oxyphotobacteriae, order Cyanobacteriales. A provisional classification was proposed by Rippka et al[14] where cyanobacteria can be subdivided into five major groups (sections) and this classification is included in the eighth edition of the Bergey's Manual of Determinative Bacteriology.[15]

Cyanobacteria are the largest, most diverse, and most widely distributed group of oxygenic photosynthetic microorganisms in freshwater, marine and terrestrial environments.[16] They also have the distinction of being the oldest known fossils,

Wastewater Treatment with Algae, edited by Yuk-Shan Wong and Nora F.Y. Tam.
© Springer - Verlag and Landes Bioscience 1998.

more than 3.5 billion years old, and morphologies within the group have remained much the same for all this time.[17] These microorganisms occur as unicellular, colonial, filamentous and branched filamentous forms and unlike most of the other prokaryotes, are usually seen and often recognizable (even at the species levels) in the environment and present in many cases as plankton blooms or as dense turfs or mats containing few other microbial species. They can exist as macrophytes forming branched thalloid plants or colonies several centimeters in length or diameter[18] and have ecologically important roles as initial colonizers of arid land, primary producers of organic matter and through their ability to fix atmospheric nitrogen.[19,20] Cyanobacteria are often encountered growing in metal-contaminated freshwater habitats[21-24] probably due to their ability to accumulate and effectively detoxify intracellular metals.[25-35]

Before entering into an explanation about uptake/accumulation and toxicity, it should be stated that some of these so-called toxic metals have metabolic roles, for example, Zn, Ni and Cu are essential micronutrients for some cyanobacteria[36] and only in increased concentrations do these metals become toxic. When considering a metal bioremoval role for cyanobacteria, one can assume the necessary mechanisms for this beneficial processes already exist in nature, for example, the case of *Oscillatoria*, which blooms in the presence of raised concentrations of some metals.[37]

It is the intent of this chapter to introduce cyanobacterial characteristics which make it possible to propose them as detoxifying agents of metal polluted environments and due to the abundance of information on the subject we will confine ourselves to the relatively more important metals, Hg, Cd, Ni, Zn, Cr, Pb and Cu.

Metal Uptake

Passive accumulation of various metals has been demonstrated in *Synechococcus* sp. with concentration factors ranging from zero to 10^6,[27] but normally the internalization of metals by cyanobacteria involve two distinct phases: binding of cations to the negatively charged groups on the cell surface (passive) and the subsequent, metabolism-dependent, uptake/internalization (active). The passive process is rapid, occurring on exposure of the cyanobacteria to the metal, whereas the metabolically dependent process is much slower. Metal adsorption followed by metabolism-dependent uptake/internalization has been described for Cd in *Anacystis nidulans* and *Chrococcus paris*, Cu in *Anacystis nidulans*, *Chrococcus paris*, *Nostoc calcicola* and *Nostoc muscorum*, Zn in *Anacystis nidulans* and *Chrococcus paris*, Ni in *Anabaena cylindrica*, Hg in *Nostoc calcicola*, and Cr in *Anabaena doliolum*.[38-46]

The uptake rate of metals by cyanobacteria is dependent on the physicochemical state of the cells as well as environmental variables such as pH, redox potential, salinity, temperature, available nutrients, metal concentration, extracellular metabolites, organic acids—all of these factors can alter metal toxicity.[47-49] One important factor which modifies the physicochemical state of the metal is pH.[50] An increase in toxicity under acidic conditions has been reported for Cd^{2+} in *Nostoc calcicola* and *Anacystis nidulans*, Cu^{2+} in *Anacystis nidulans* and *Nostoc muscorum* and was suggested to be due to the increase in free metal ions available in the suspending medium.[42,44,51] Under acidic conditions metals tend to exist in the free ionic form, whereas under alkaline conditions they may precipitate as insoluble complexes or in hydroxylated forms which can alter their uptake/toxicity.[47,52] However, acidic conditions can result in competition between free metal ions and H^+ for the same uptake sites which can reduce cellular metal uptake/toxicity.[50]

The presence of other cations can also affect metal uptake/toxicity in cyanobacteria. Fe, a particularly common environmental metal ion, has been noted to have a protective effect against Ni and Cu toxicity in *Anabaena doliolum*.[53] A decrease in toxicity of several metals has been described as a result of direct competition between the different cations for the same uptake/binding site (Table 7.1). Cations can also have a synergistic affect on metal uptake due to adsorption of one cation and facilitating metal uptake probably due to increased membrane permeability (Table 7.2). Ca and Mg salts also form complexes with toxic metals in hard and eutrophic waters which reduce their toxicity.[48,62] Organic ligands synthesized by the cyanobacteria or from other sources are also capable of binding metals, leaving less free metal available and consequently decreasing the uptake/toxicity of such metals.[63,64]

Mechanisms of Metal Tolerance

Numerous genera of cyanobacteria have been isolated from metal contaminated environments such as zinc-enriched water,[21,22] mine tailings containing high concentrations of zinc (22.8 mg/l), cobalt (0.44 mg/l), nickel (0.43 mg/l) and lead (0.28 mg/l)[24] and copper-rich soils.[23] Subsequently, many of these isolates have proven to be naturally tolerant to these metals[24,65] and laboratory studies have also demonstrated that metal-tolerance can be selected for.[43,66,67] Cyanobacteria which grow in the presence of metals have evolved several mechanisms of tolerance, such as extracellular binding or precipitation, impermeability/exclusion and internal detoxification.[49,68]

Extracellular Binding or Precipitation

Little is known about metal deposition within the cyanobacteria cell envelope, since most metal-binding studies have been carried out using whole cells. Sheaths isolated from culture of *Gloeothece* ATCC 27152, grown with and without $NaNO_3$, showed substantial adsorption of Cd,[69] with the $NaNO_3$ grown cells binding more

Table 7.1. Antagonistic interaction among metal cations in cyanobacteria

Metal	Antagonist	Species	Reference
Hg	Ca,Cd,Cu,Mg,Ni	*Nostoc calcicola*	54,55
		Anabaena inaequalis	33,40,56
Cd	Ca,Zn,Ni,Fe	*Anacystis nidulans*	23,57,58
	Ca,Cd,Hg		
		Anabaena inaequalis	33
Ni	Ca,Cd,Hg	*Nostoc muscorum*	59
		Anabaena inaequalis	33
Cr	Ca,Mg,Mn	*Anabaena doliolum*	60,61
Cu	Ca	*Nostoc muscorum*	42
Pb	Ca	*Nostoc muscorum*	42

Table 7.2. Synergistic interaction among metal cations in cyanobacteria

Metal	Synergist	Species	Reference
Hg	Cd,Ni,CH$_3$Hg	*Anabaena inaequalis*	33
		Anacystis nidulans	57,58
		Nostoc calciola	55,56
Cd	Pb	*Anacystis nidulans*	23
Cr	Ni,Co,Zn	*Anabaena doliolum*	60,61

Cd than N$_2$-fixing cultures, most probably due to the variation in the chemical composition of the sheath. Sheaths isolated from *Calothrix parietina* and *C. scopulorum* were also found to bind metals in the order Zn>Cu>Ni.[70] Pregrown *Nostoc muscorum* cells were treated with 1 mM Cr for 10 minutes and whole mounts of unstained cells studied using energy dispersive X-ray spectroscopy (EDS). Figure 7.1 shows that Cr precipitated in their mucilage.

Some information is available regarding the ability of cyanobacteria to produce extracellular substances to protect themselves from toxic metals, notably the production of polypeptides by *Anabaena cylindrica* which complexed cupric and zinc ions[71] and by *Anabaena flos-aquae*.[72] According to Murphy et al[73] when blooms of *Anabaena flos-aquae* were present in the water column, hydroxamates could be detected and although these compounds are induced by limiting Fe concentrations, their primary function seems to be chelation of metal ions and to give a selective advantage to the producer.

Metal Efflux and Exclusion

In *Nostoc calcicola*, an energy-dependent Cu^{2+} efflux system was demonstrated in a resistant mutant, which resulted in a net reduction in internal Cu^{2+}, although the uptake rate did not differ from that of the original parent.[46] A Cu tolerant strain of *Anabaena doliolum*[74] was isolated by repeatedly subculturing into modified Allen and Arnon's medium containing increasingly higher levels of copper (0.5 and 1.0 µg/ml) and analysis of the mutant showed a larger lipid fraction and insignificant loss of K and Na in the presence of Cu. These data were interpreted as indicating changes in plasma membrane properties probably associated with reduced permeability for Cu. Singh and Pandey[51] isolated a Cd resistant mutant of *Nostoc calcicola* which also demonstrated multiple resistance to metals (Zn and Hg) coupled to antibiotic resistance, suggesting plasmid-associated metal resistance.

Internal Detoxification

Internally accumulated metals, such as Cd, Cu, Hg, Ni and Zn, have been shown by in situ X-ray dispersive analysis, to be localized in polyphosphate bodies in various cyanobacteria including *Plectonema boryanum*,[29] *Anabaena cylindrica* and *Anacystis nidulans*[32,75] and *Anabaena flos-aquae*.[76] These studies suggest that polyphosphate bodies are a means of binding metallic cations in a non-toxic form within cells and which under permissive growth conditions may serve as storage

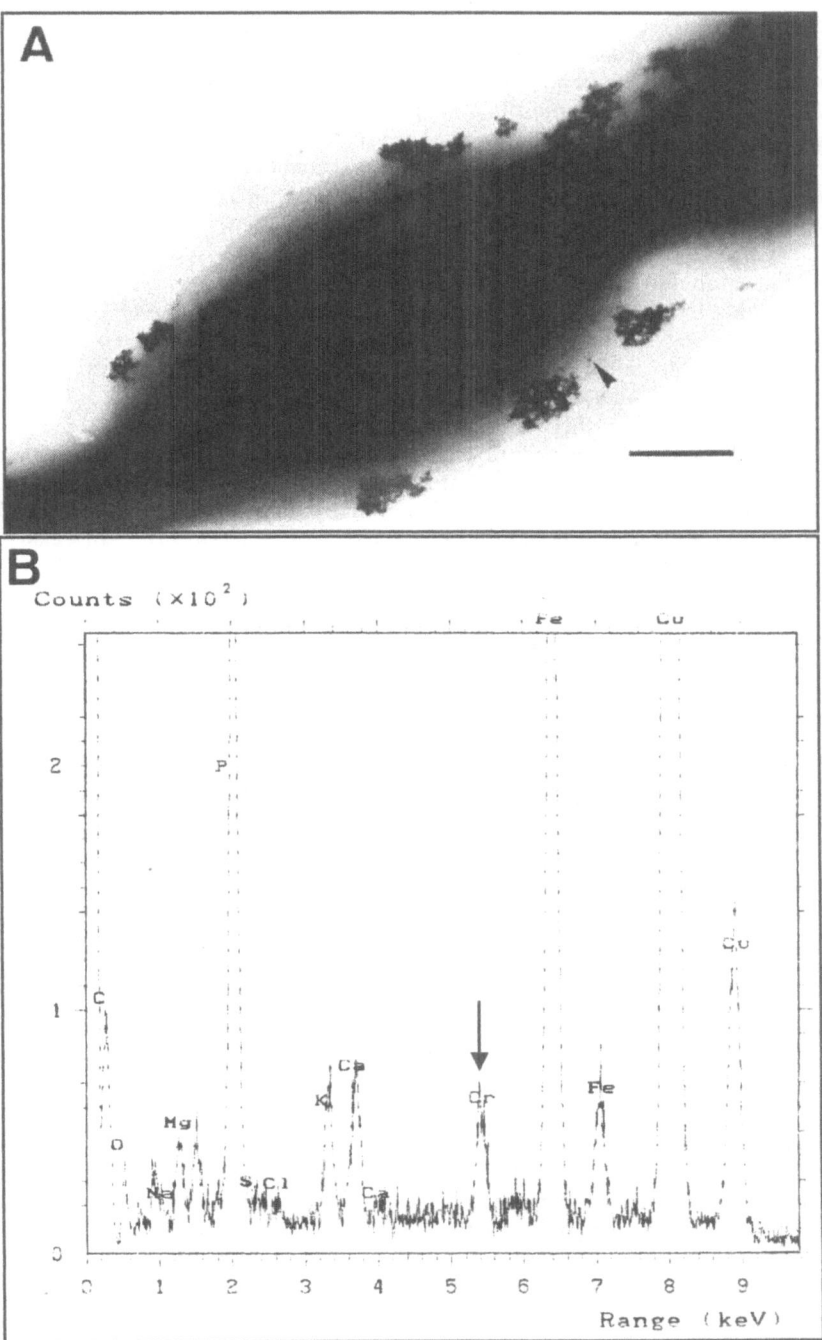

Fig. 7.1. Transmission electron microscope micrograph of unstained whole cell mount of *Nostoc muscorum* (A). The arrowhead on the micrograph indicates the precipitate analyzed by EDS shown in (B). The Cu peaks are from the copper grid, the Fe peaks are from the column of the electron microscope, the CI peak is from the plastic embedding resin and the other elements are from the cell components.

sites for essential metallic micronutrients. Other examples include *Synechococcus* sp. which synthesize large quantities of an intracellular polymer that binds nickel, leaving the cell interior highly granular.[66] An intracellular detoxification mechanism was also suggested to explain the presence of extra intracellular membrane whorls in *Plectonema boryanum* exposed to Ni, Zn, Hg, Cu and Cd.[77] *Nostoc muscorum* is Mo-requiring and sensitive to tungsten (W) and Cr, but analysis of tolerant mutants revealed that they became W- and Cr- requiring, suggesting a modification of the Mo uptake/metabolism.[78]

Metallothionein (MT)

Another aspect of internal compartmentalization is the synthesis of metal-binding macromolecules which may function in detoxification. This group of metal-binding proteins, called metallothioneins, are characterized as a group of small molecular weight proteins (<10 kD), which bind metals ions to clusters of thiolate bonds conferred by the high cysteine content and preponderance of -SH groups. Another distinctive characteristic is their amino acid composition; no aromatic forms or histidine are present.[79,80] The first metallothionein was isolated from the renal cortex of a Cd-intoxicated equine[81] and since this discovery, proteins termed 'MT-like' have been isolated and characterized from a large variety of eukaryotic[79,82-85] and a few prokaryotic organisms.[86-91] The true metabolic role of the MT is still unknown, but functions such as cadmium detoxification have been proposed, since MT-like proteins are classically cadmium inducible.[79] Other functions for MTs, including the translocation and availability of essential metals, principally copper and zinc,[92] and Zn homeostasis in the regulation of gene expression[93-95] have been suggested.

Production of an MT-like protein in prokaryotic organism was first described in the cyanobacterium *Anacystis nidulans* linked to Cd resistance.[86] Although the presence of MT- like proteins have been reported in several prokaryotic organisms, only in the cyanobacteria, *Synechococcus*, has the MT gene been isolated and sequenced.[91] The abundance of *smt*A transcripts in *Synechococcus* PCC 6301, evaluated by hybridization, increases following exposure to increased oncentrations of Cd, Zn and Cu with maximal transcription after exposure to 5 μM Cd, 10 μM Zn with Cu being a less potent inducer. A homologue of the *smt*A gene has also been identified in *Synechococcus* PCC 7942 by hybridization.[91] It should be noted that *Anacystis nidulans*, *Synechococcus* PCC 6301 and PCC 7942 have been considered to be the same species.[96]

To examine the metal-binding properties of its product, the *smt*A gene was expressed in *Escherichia coli* as a carboxyterminal extension of glutathione-S-transferase.[97] The expressed fusion protein was shown to have affinities for Cd, Cu, Hg and Zn comparable to those of equine MTs. Although the *E. coli* expressing this gene showed enhanced (ca. 3-fold) accumulation of Zn suggesting Zn-binding in vivo, no evidence of metalloregulation was reported.

Huckle et al[98] isolated and structurally characterized a genomic fragment carrying the MT gene from *Synechococcus* PCC 7942 and analyzed its regulation by metal ions. The MT locus, designated *smt*, includes *smt*A and a divergently transcribed gene, *smt*B. The abundance of *smt*A transcripts, evaluated by hybridization, increases in response to elevated concentrations of Cd, Co, Cr, Cu, Hg, Ni, Pb and Zn, but not after a heat shock. Of all metal ions tested at maximum permissive concentrations, the greatest expression was observed in response to Zn. Metal-ion-induced expression of the *smt*A gene is directed by an operator-promoter under the control of metal responsive factors with no detectable effect of metal (Cd) on *smt*A transcript stability being observed. Sequences upstream of *smt*A, fused to a promoterless *lacZ* gene, conferred metal-dependence on *lacZ* expression. In *smt*B⁻ mutants highly elevated expression of *lacZ* (driven by the *smt*A operator-promoter) was detected, even in the absence of metal ions. Re-introduction of *smt*B reduced *lacZ* expression and restored metal-dependency, indicating that *smt*B was a metal-responsive repressor of *smt*A transcription.

Further analysis of the function and regulation of *smt* showed that *Synechococcus* PCC 7942 mutants deficient in both *smt*A and *smt*B were sensitive (ca. 5-fold reduction in tolerance) to Zn, and showed some reduction in tolerance to Cd.[99] These cells retained normal Cu tolerance, indicating that Cu resistance was independent from *smt*-mediated metal tolerance. *Synechococcus* cells containing re-introduced *smt* have been successfully isolated from *smt*⁻ cells based upon restored tolerance to Zn and Cd. The finding that a Cd tolerant *Synechococcus* PCC 6301 mutant gained increased tolerance to Cd resulting from a deletion within the *smt*B gene[100] and that mutant cells lacking the repressor (*smt*B), *smt*A⁺/B⁻ were more tolerant to elevated Zn^{2+} and Cd^{2+} than cells containing an intact *smt*[101] gave strong evidence that reduced metal tolerance of *smt*⁻ mutants was a function of the loss of *smt*A and not the accompanying loss of *smt*B.[101] Furthermore, Gupta et al[102] demonstrated that when cyanobacteria cells were grown in increasingly high concentrations of Cd^{2+}, increased tolerance to Cd was correlated with the amplification of the *smt*A gene copy number, resulting in higher transcriptional levels producing high cellular levels of metallothionein. Increased accumulation of Zn in *Synechococcus* PCC 7942 *smt*⁺ when compared to *smt*⁻ mutants provided further evidence of Zn-binding in vivo.[101]

Based upon the known MT gene sequence of *Synechococcus* PCC 7942 published by Robinson et al[91] we designed primers to isolate the *smt*A and *smt*B gene fragments from the *Synechococcus* PCC 7942 which were cloned and sequenced. In a preliminary screening study, primers specific for the *smt*A were used in an attempt to amplify gene homologues from several genera and species of cyanobacteria. The PCR products were immobilized on a nylon membrane and probed with the *Synechococcus* PCC 7942 *smt*A gene fragment. *Phormidium autumnale* was shown to possess an *smt*A gene homologue (Fig. 7.2) by the strong hybridization signal with the *smt*A probe.

Use of Cyanobacteria for Metal Bioremoval

Bioremoval has been defined as the accumulation and concentration of pollutants from principally aqueous solutions by the use of biological material, traditionally microbial biomass, thus allowing the recovery and/or environmentally acceptable disposal of the pollutants.[103] The use of microalgae, as the biological biomass, has been considered an advantage because they utilize light as their energy source, allowing metabolic activity in the absence of organic carbon sources. Also, they utilize an active system to take up a variety of essential and non-essential metals and since it is not an equilibrium process, a simple batch culture system should result in low residual metal concentration.[103]

Becker,[104] using what are considered to be optimal conditions, measured the time taken to reduce the initial metal concentration by 50% and showed that between 14 and 19 days were needed depending on the model used. He concluded

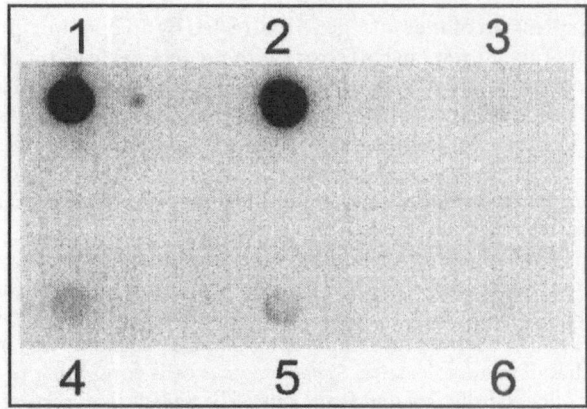

Fig. 7.2. Dot-blot analysis of the PCR products, amplified using specific primers for *smt A*, from several genera and/or species of cyanobacteria and probed with the *Synechococcus* PCC 7942 *smtA* gene fragment. Dot 1, *Synechococcus* PCC 7942; dot 2, *Phormidium autumnale*; dot 3, *Nostoc muscorum*; dot 4, *Cylindrospermum* sp; dot 5 *Nostoc* sp; dot 6, *Anabaena flos-aquae*.

that in practice, the use of phytoplankton algae is not feasible and that their use should be limited to only very low levels of metal contamination. In spite of this conclusion various early pilot studies have been set up to test the feasibility of using cyanobacteria to remove metals from contaminated water systems. Filip et al[67] tried an algal-intermittent sand filter system, demonstrating that a large percentage (approx. 70-90%) of the contaminating Cu, Cr and Cd can be removed using a small-scale laboratory based model system. On the other hand, Gale and Wixon[105] assessed the use of tailing ponds and artificial stream meanders to encourage the growth of algae, including cyanobacteria, in the treatment of mining and milling effluents. It was observed that the promotion of this algal growth effectively reduced the amount of particulate material and metals released into the receiving streams of the surroundings. A more recent review[103] considers the use of microalgae in the bioremoval of metals and suggest that with the advances in the technology, including selection of improved strains, improving biomass containment or immobilization of algal cells, will be of great importance in the future where low level residues of metal contamination are desired. Maquieira et al[106] used *Spirulina platensis* immobilized on controlled pore glass (CPG) to preconcentrate Cu, Zn, Cd, Pb and Fe from aqueous solution and showed that the retention of Cu, Zn and Cd occurred over a wide range of pH values whereas the retention of Fe and Pb was highly pH-dependent (pH 6-7). The column retained activity for approximately three months and acheived concentration factors of between 100-1000. The authors suggested that because of the limitations, these columns should be used for relatively low levels of contamination. Bender et al[107] investigated the use of microbial mats (MM), principally cyanobacteria mixed with sediment from the contaminated area, in the bioremoval of metals. Microbial mats are likely to play an important role in adsorbing, filtering and transforming various nutrients and contaminants, such as metals, which enter wetlands. When contaminated water was passed over MM immobilized on glass wool there was a rapid removal of the metals from the water. They also present evidence that when the MM was exposed to water contaminated with soluble metal salts, a transformation occurred, converting these metals into insoluble precipitates. They concluded that the MM system has the advantage of being flexible (use in both salt and fresh waters), durable and tolerant to high concentrations of metals and metalloids and the unique capacity of the mat to form associations with new microbial species should allow the design of microbial mats with defined combinations for specific problems.

References

1. Borovik AS. Characterization of metal ions in biological systems. In: Shaw AJ ed. Heavy Metal Tolerance in Plants: Evolutionary Aspects. Boca Raton: CRC Press, 1990:3-19.
2. Gadd GM. Metals and microorganisms: A problem of definition. FEMS Microbiol Lett 1992; 100:197-204.
3. Nies DH, Silver S. Ion efflux system involved in bacterial metal resistances. J Ind Microbiol 1995; 14:186-199.
4. Beveridge TJ, Hughes MN, Lee H et al. Metal-microbe interaction: Contemporary approaches. Adv Microbial Physiol 1997; 38:177-243.
5. Hughes MN, Poole RK, eds. Metals and Micro-organisms. London: Chapman & Hall, 1989.
6. Beveridge TJ. Mechanisms of the binding of metallic ions to bacterial walls and the possible impact on microbial ecology. In: Klug MT, Reddy CA eds. Current Perspectives in Microbial Ecology. Washington DC: American Society for Microbiology, 1984:601-607.
7. Beveridge TJ. The immobilization of soluble metals by bacterial wall. Biotechnol Bioeng Symp 1986; 16:127-139.
8. Greene B, Darnall DW. Microbial oxygenic photoautotrophs (cyanobacteria and algae) for metal ion binding. In: Ehrlich HL, Brierley CL eds. Microbial Mineral Recovery. New York: McGraw-Hill, 1990: 277-302.
9. MacHardy BM, George JJ. Bioaccumulation and toxicity of zinc in the green algae, *Cladophora glomerata*. Environ Poll 1990; 66:55-66.
10. Trevors JT, Stratton GW, Gadd GM. Cadmium transport, resistance and toxicity in bacteria, algae and fungi. Can J Microbiol 1986; 32:447-464.
11. Rippka R. Recognition and identification of cyanobacteria. Meth Enzymol 1988; 167:28-67.
12. Woese CR. Bacterial evolution. Microbiol Rev 1987; 51:221-271.
13. Gibbons NE, Murray RGE. Proposal concerning the higher taxa of bacteria. Int J Syst Bacteriol 1978; 28:1-6.
14. Rippka R, Deruelles J, Waterbury JB et al. Generic assignments, strains histories and properties of pure cultures of cyanobacteria. J Gen Microbiol 1979; 111:1-61.
15. Staley JT, Bryant MP, Pfenning N et al, eds. Systematic Bacteriology. Baltimore: Williams & Wilkins, 1989.
16. Fogg GE, Stewart WDP, Fay P et al, eds. The Blue-green Algae. London: Academic Press, 1973.
17. Lee RE, ed. Phycology. Cambridge: Cambridge University Press, 1989.
18. Castenholz RW. Culturing methods for cyanobacteria. Meth Enzymol 1988; 167: 69-93.
19. Carr NG, Whitton BA, eds. The Biology of Blue-green Algae. Berkeley: University California Press, 1973.
20. Fay P, ed. The Blue-greens. Studies in Biology no. 160. London: Edward Arnold, 1983.

21. Say PJ, Whitton BA. Change in flora down a stream showing a zinc gradient. Hydrobiologia 1980; 76:255-262.
22. Whitton BA. Zinc and plants in rivers and streams. In: Nriagu JO, ed. Zinc in the Environment. Part II. New York: John Wiley & Sons, 1980:364-400.
23. Whitton BA, Shehata FHA. Influence of cobalt, nickel, copper and cadmium on the blue green algae *Anacystis nidulans*. Environ Poll 1982; 27:275-281.
24. Whitton BA, Gale NL, Wixson BG. Chemistry and plant ecology of zinc-rich wastes contaminated by blue-green algae. Hydrobiologia 1981; 83:331-341.
25. Avery SV, Codd GA, Gadd GM. Caesium accumulation and interactions with other monovalent cations in the cyanobacterium *Synechocystis* PCC 6803. J Gen Microbiol 1991; 137:405-413.
26. Baxter M, Jensen T. Uptake of magnesium, strontium, barium and manganese by *Plectonema boryanum* (Cyanophyceae) with special reference to polyphosphate bodies. Protoplasma 1980; 104:81-89.
27. Fisher NS. Accumulation of metals by marine picoplankton. Mar Biol 1985; 87:137-142.
28. Horikoshi T, Nakajima A, Sakaguchi T. Uptake of uranium from sea water by *Synechococcus elongatus*. J Ferment Technol 1979; 57:191-194.
29. Jensen TE, Baxter M, Rachlin JW et al. Uptake of heavy metals by *Plectonema boryanum* (Cyanophyceae) into cellular components, especially polyphosphate bodies: an X-ray energy dispersive study. Environ Pollut Ser A 1982; 27:119-127.
30. Laube VM, McKenzie CN, Kushner DJ. Strategies of response to copper, cadmium, and lead by a blue-green and green algae. Can J Microbiol 1980; 26: 1300-1311.
31. Massalski A. Laube VM, Kushner DJ. Effects of cadmium and copper on the ultrastructure of *Ankistrodesmus braunii* and *Anabaena* 7120. Microb Ecol 1981; 7:183-193.
32. Pettersson A, Kunst L, Bergman B et al. Accumulation of aluminium by *Anabaena cylindrica* into polyphosphate granules and cell walls: an X-ray energy-dispersive microanalysis study. J Gen Microbiol 1985; 131:2545-2548.
33. Stratton GW, Corke CT. The effect of mercuric, cadmium, and nickel ion combinations on a blue-green alga. Chemosphere 1979; 10:731-740.
34. Verma SK, Singh SP. Factors regulating copper uptake in a cyanobacterium. Curr Microbiol 1990; 21:33-37.
35. Wang HK, Wood JM. Bioaccumulation of nickel by algae. Environ Sci Technol 1984; 18:106-109.
36. Rippka R. Isolation and purification of cyanobacteria. Meth Enzymol 1988; 167:3-27.
37. Ray S, White W. Selected aquatic plants as indicator species for heavy metal pollution. J Environ Sci Health 1976; A11: 717-725.
38. Campbell PM, Smith GD. Transport and accumulation of nickel ions in the cyanobacterium *Anabaena cylindrica*. Arch Biochem Biophys 1986; 244:470-477.
39. Les A, Walker RW. Toxicity and binding of copper, zinc, and cadmium by the blue-green alga, *Chroococcus paris*. Water Air Soil Pollut 1984; 23:129-139.
40. Pandey PK, Singh SP. Hg^{2+} uptake in a cyanobacterium. Curr Microbiol 1993; 26:155-159.
41. Rai LC, Dubey, SK, Mallick N. Influence of chromium on some physiological variables of *Anabaena doliolum*: interaction with metabolic inhibitors. BioMetals 1992; 5:13-16.
42. Schecher WD, Driscoll CT. Interactions of copper and lead with *Nostoc muscorum*. Water Air Soil Pollut 1985; 24: 85-101.
43. Shehata FHA, Whitton BA. Zinc tolerance in strains of the blue-green alga *Anacystis nidulans*. Br Phycol J 1982; 17:5-12.

44. Singh DP. Cu²⁺ transport in the unicellular cyanobacterium *Anacystis nidulans*. J Gen Appl Microbiol 1985; 31:277-284.
45. Singh SP, Yadava V. Cadmium uptake in *Anacystis nidulans*: effect of modifying factors. J Gen Microbiol 1985; 31:39-48.
46. Verma SK, Singh HN. Evidence for energy-dependent copper efflux as a mechanism of Cu²⁺ resistance in the cyanobacterium *Nostoc calcicola*. FEMS Microbiol Lett 1991; 84:291-294.
47. Gadd GM, Griffiths AJ. Microorganisms and heavy metals toxicity. Microb Ecol 1978; 4:303-317.
48. Rai LC, Gauer JP, Kumar HD. Phycology and heavy-metal pollution. Biol Rev Cambridge Philos Soc 1981; 56:99-151.
49. Reed RH, Gadd GM. Metal tolerance in eukaryotic and prokaryotic algae. In: Shaw AJ ed. Heavy Metal Tolerance in Plants: Evolutionary Aspects. Boca Raton: CRC Press, 1990:105-118.
50. Peterson HG, Healey FP, Wagemann R. Metal toxicity to algae: a highly pH-dependent process. Can J Fish Aquat Sci 1984; 41:974-979.
51. Singh SP, Pandey AK. Cadmium-mediated resistance to metals and antibiotics in a cyanobacterium. Mol Gen Genet 1982; 187:240-243.
52. Babich H, Stotzky G. Developing standards for environmental toxicants: The need to consider abiotic environmental factors and microbe-mediated ecologic processes. Environ Health Perspect 1983; 49:247-260.
53. Mallick N, Rai LC. Response of *Anabaena doliolum* to bimetallic combinations of Cu, Ni and Fe with special reference to sequential addition. J Appl Phycol 1989; 1:301-306.
54. Singh CB, Singh SP. Protective effects of Ca²⁺, Mg²⁻, Cu²⁺, and Ni²⁺ on mercury and methylmercury toxicity to a cyanobacterium. Ecotoxicol Environ Safety 1992; 23:1-10.
55. Singh CB, Singh SP. Effect of mercury on photosynthesis in *Nostoc calcicola*: role of ATP and interacting heavy metal ions. J Plant Physiol 1987; 129:49-58.
56. Singh CB, Verma SK, Singh SP. Impact of heavy metals on glutamine synthethase and nitrogenase activity in *Nostoc calcicola*. J Gen Appl Microbiol 1987; 33: 87-91.
57. Singh SP, Yadava V. Cadmium induced inhibition of nitrate uptake by *Anacystis nidulans*: interaction with other divalent cations. J Gen Appl Microbiol 1983; 29:297-304.
58. Singh SP, Yadava V. Cadmium induced inhibition of ammonium and phosphate uptake in *Anacystis nidulans*: interaction with other divalent cations. J Gen Appl Microbiol 1984; 30:79-86.
59. Rai LC, Raizada M. Effect of nickel and silver ions on survival, growth, carbon fixation and nitrogenase activity in *Nostoc muscorum*: regulation of toxicity by EDTA and calcium. J Gen Appl Microbiol 1985; 31:329-337.
60. Dubey SK, Rai LC. Toxicity of chromium and tin to *Anabaena doliolum*. Interaction with bivalent cations. Biol Metals 1990; 3:8-13.
61. Rai LC, Dubey SK. Impact of chromium and tin on a nitrogen-fixing cyanobacterium *Anabaena doliolum*: interaction with bivalent cations. Ecotoxicol Environ Safety 1989; 17:94-104.
62. Whitton BA. Toxicity of heavy metals to algae: A review. Phykos 1970; 9:116-125.
63. Clarke SE, Stuart J, Sanders-Loehr J. Induction of siderophore activity in *Anabaena* spp., and its moderation of copper toxicity. Appl Environ Microbiol 1987; 53:917-922.
64. Wurtsbaugh WA, Horne AJ. Effects of copper on nitrogen fixation and growth of blue-green algae in natural plankton associations. Can J Fish Aquat Sci 1982; 39:1636-1641.

65. Shehata FHA, Whitton BA. Field and laboratory studies on the blue-green algae from aquatic sites with high levels of zinc. Verh Int Ver Theor Angew Limnol 1981; 21:1466-1471.

66. Wood JM, Wang HK. Microbial resistance to heavy metal. Environ Sci Technol 1983; 17:582A-590A.

67. Filip DS, Peters T, Adams VD et al. Residual heavy metal removal by an algae-intermittent sand filtration system. Water Res 1979; 13:305-313.

68. Fiore MF, Trevors JT. Cell composition and metal tolerance in cyanobacteria. BioMetal 1994; 7:83-103.

69. Tease BE, Walker R. Comparative composition of the sheath of the cyanobacterium *Gloeothece* ATCC 27152 cultured with and without combined nitrogen. J Gen Microbiol 1987; 133: 3331-3339.

70. Weckesser J, Hofman K, Jurgens UJ et al. Isolation and chemical analysis of the sheaths of the filamentous cyanobacteria *Calothrix parietina* and *C. scopulorum*. J Gen Microbiol 1988; 134:629-634.

71. Fogg GE, Westlake DF. The importance of extracellular products of algae in fresh-water. Verh Int Ver Theor Angew Limnol 1955; 12:219-231.

72. Wang WS, Tischer RG. Studies of the extracellular polysaccharides produced by a blue- green alga *Anabaena flos-aquae* A-37. Arch Microbiol 1973; 91:77-81.

73. Murphy TP, Lean DRS, Nalewajko C. Blue-green algae: their excretion of iron selective chelators enables them to dominate other algae. Science 1976; 192:900-902.

74. Rai LC, Mallick N, Singh JB et al. Physiological and biochemical characteristics of a copper tolerant and wild type strain of *Anabaena doliolum* under copper stress. J Plant Physiol 1991; 138:68-74.

75. Crang RE, Jensen TE. Incorporation of titanium in polyphosphate bodies of *Anacystis nidulans*. J Cell Biology 1975; 67:80a.

76. Rachlin JW, Jensen TE, Warkentine B. The toxicological response of the alga *Anabaena flos-aquae* (Cyanophyceae) to cadmium. Arch Environ Contam Toxicol 1984; 13:143-151.

77. Rachlin JW, Jensen TE, Baxter M et al. Utilization of morphometric analysis in evaluating response of *Plectonema boryanum* (Cyanophyceae) to exposure to eight heavy metals. Arch Environ Contam Toxicol 1982; 11:323-333.

78. Singh HN, Vaishampayan A, Singh RK. Evidence for the involvement of a genetic determinant controlling functional specificity of group VI B elements in the metabolism of N_2 and NO_3^- in the blue-green alga *Nostoc muscorum*. Biochem Biophys Res Comm 1978; 81:67-74.

79. Hamer DH. Metallothionein. Annu Rev Biochem 1986; 55:913-951.

80. Fowler BA, Hieldebrand CE, Kojima Y et al. Nomenclature of metallothionein. Experientia, Suppl 52, 1987:19-22.

81. Margoshes M, Vallee BL. A cadmium protein from equine kidney cortex. J Am Chem Soc 1957; 79:4813-4814.

82. Nordberg M, Kojima Y. Metallothionein and other low molecular weight metal-binding proteins. In: Kagi JHR, Nordberg M, eds. Metallothionein Basel: Brikhauser Verlag, 1979:41-124.

83. Hartmann HJ, Li YJ, Weser U. Analogous copper (I) coordination in metal-lothionein from yeast domains of the mammalian protein. BioMetals 1992; 5:187-191.

84. Rauser WE. Phytochelatins. Annu Rev Biochem 1990; 59:61-86.

85. Robinson NJ, Tommey AM, Kuske C et al. Plant metallothioneins. Biochem J 1993; 295:1-10.

86. Maclean FI, Lucis OJ, Shakh ZA et al. The uptake and subcellular distribution of Cd and Zn in microorganisms. Fed Proc 1972; 31:699.

87. Khazaeli MB, Mitra RS. Cadmium-binding component in *Escherichia coli* during accommodation to low levels of this ion. Appl Environ Microbiol 1981; 41:46-50.

88. MacEntee JD, Woodrow JR, Quirk AV. Investigation of cadmium resistance in *Alcaligenes* sp. Appl Environ Microbiol 1986; 51:515-520.

89. Higham DP, Sadler PJ. Cadmium-resistant *Pseudomonas putida* synthesizes novel cadmium proteins. Science 1984; 225:1043-1046.

90. Olafson RW, McCubbin WD, Kay CR. Primary-and secondary-structural analysis of a unique prokaryotic metallothionein from *Synechococcus* sp. Cyanobacteria. Biochem J 1988; 251:691-699.

91. Robinson NJ, Gupta A, Fordham-Skelton AP et al. Prokaryotic metallothionein gene characterization and expression: chromosome crawling by ligation-mediated PCR. Proc R Soc Lond B 1990; 242:241-247.

92. Wolf WR, Irgolic KJ, Ludwicki KJ et al. Importance and determination of chemical species in biological systems. In Bernhard M, Brinckman FE, Sadler PJ, eds. The Importance of Chemical Speciation in Environmental Processes. Berlin: Springer-Verlag, 1986:17-25.

93. Vallee BL. Introduction to metallothionein. Meth Enzymol 1991; 205:3-7.

94. Zeng J, Heuchel R, Schaffner W et al. Thionein (apometallothionein) can modulate DNA binding and transcription activation by zinc finger containing factor. Spl. FEBS Lett 1991; 279:310-312.

95. Zeng J, Valle BL, Kägi JHR. Zinc transfer from transcription factor IIIA to thionein clusters. Proc Natl Acad Sci USA 1991; 88:9984-9988.

96. Wilmotte AMR, Stam WT. Genetic relationships among cyanobacterial strains originally designated as *Anacystis nidulans* and some other *Synechococcus* strains. J Gen Microbiol 1984; 130:2737-2740.

97. Shi J, Lindsay WP, Huckle JW et al. Cyanobacterial metallothionein gene expressed in *Escherichia coli*: Metal-binding properties of the expressed protein. FEBS Lett 1992; 303:159-163.

98. Huckle JM, Morby AP, Turner JS et al. Isolation of the *smt*A gene encoding a prokaryotic metallothionein. Mol Microbiol 1993; 7:177-187.

99. Turner JS, Morby AP, Whitton BA et al. Construction and characterization of Zn^{2+}/Cd^{2+} hypersensitive cyanobacterial mutants lacking a functional metallothionein locus. J Biol Chem 1993; 268:4494-4498.

100. Gupta A, Morby AP, Turner JS et al. Deletion within the metallothionein locus of cadmium-tolerant *Synechococcus* PCC 6301 involving a highly iterated palindrome (HIP1). Mol Microbiol 1993; 7:189-195.

101. Turner JS, Robinson NJ, Gupta A. Construction of Zn^{2+}/Cd^{2+}-tolerance cyanobacteria with a modified metallothionein divergon: Further analysis of the function and regulation of *smt*. J Ind Microbiol 1995; 14:259-264.

102. Gupta A, Whitton BA, Morby AP et al. Amplification and rearrangement of a prokaryotic metallothionein locus *smt* in *Synechococcus* PCC 6301 selected for tolerance to cadmium. Proc R Soc Lond B 1992; 248:273-281.

103. Wilde EW, Benemann JR. Bioremoval of heavy metals by the use of microalgae. Biotech Adv 1993; 11:781-812.

104. Becker EW. Limitations of heavy metal removal from waste water by means of algae. Water Res 1983; 17:459-466.

105. Gale NL, Wixson BG. Removal of heavy metals from industrial effluents by algae. Dev Ind Microbiol 1979; 20:259-273.

106. Maquieira A, Elmahadi HAM, Puchades R. Immobilized cyanobacteria for on-line trace metal enrichment by flow injection atomic adsorption spectrometry. Anal Chem 1994; 66:3632-3638.
107. Bender J, Lee RF, Phillips P. Uptake and transformation of metals and metalloids by microbial mats and their use in bioremediation. J Ind Microbiol 1995; 14:113-118.

CHAPTER 8

Modeling the Uptake of Metal Ions by Living Algal Cells

Ian G. Prince, Y.P. Ting and Frank Lawson

Introduction

The intrinsic capability of both living and dead microorganisms, including algae, to sequester and possibly accumulate high levels of metal ions from dilute aqueous solutions has attracted much attention over the years and a number of substantial reviews are available.[1-5] This interest is due to both the ecological implications, not the least of which includes the entry of potentially toxic heavy metals into the food chain, as well as the possible exploitation of this phenomenon to economically clean up waste and other waters.

The greatest progress towards industrial application of microbial metal uptake has occurred with nonliving biomass, usually immobilized in some manner, to decontaminate waste streams or for the recovery of metals of high value.[5] In these cases quantitative analysis and modelling has focused on understanding and enhancing the microbial (passive) adsorption process. By contrast, despite, a large number of qualitative, ecological studies on the uptake of various metal ions (e.g., copper, lead, cadmium and zinc) being carried out over the years,[6] very few studies have attempted to quantitatively model the uptake by the living biomass. This undoubtedly reflects the complexity of the phenomenon, as well as various perceived disadvantages in the process application of living cells.[4] The strong, ongoing interest in the environmental consequences of metal ion uptake, and the capability of mechanistic modelling to identify and focus on uncertainties and deficiencies in the knowledge makes such research of much importance, however.

In this chapter we review the current literature on existing physical and mathematical models, describe the uptake of metal ions by microorganisms, and a present specific model developed to address the uptake behavior of living algal cells. additionally, the complexities inherent in the quantitative descriptions are highlighted. While the treatment specifically relates to algal cultures, many of the observations are applicable to other microbial species. The treatment is also primarily concerned with dispersed micro algae, rather than the macro algae (seaweeds). A list of relevant nomenclature may be found at the end of this chapter.

Wastewater Treatment with Algae, edited by Yuk-Shan Wong and Nora F.Y. Tam.
© Springer - Verlag and Landes Bioscience 1998.

General Considerations in Modeling the Uptake Process

The interactions between metal ions and microorganisms, living or dead, are numerous and extremely complex, and understanding the interaction phenomena remains far from complete. Algal cells, being extremely small in size, present very large specific surface areas, and consequently, surface adsorption, which involves various mechanisms, constitutes an important aspect of any metal uptake by these cells. The first stage is a rapid uptake of the metal ion which does not involve any metabolic process or energy. Hence, dead cells exhibit this behavior, since the metal ion accumulates on the external cell surface, or where the cell structure is no longer maintained intact, on other of the cell structures.

It has long been postulated,[7] with much experimental evidence, that a second stage, present only for living cells, involves energy-consuming transport against a concentration gradient, and/or diffusion-controlled transport across the energy barrier presented by the cell membrane. Such a pathway for essential metal ions such as zinc, also appears to be followed by non-essential metal ions such as cadmium. There may also be a third stage where the metal ion is immobilized within the cell by being incorporated into an insoluble organo-metallic structure such as a metallothionin or a metal-containing protein. (This latter stage provides the cell a measure of protection from otherwise toxic species.) Thus, the active biological uptake of metal ions across the membrane and into the intracellular structure entails a series of energized events. These events are much slower than the initial surface adsorption.

Adsorption Processes

Studies on the physicochemical properties of the surfaces of algal cells have revealed their net negatively-charged nature. In general, the cell walls consist of a mosaic of various anionic and cationic exchange groups, including the hydroxyl, sulfydryl, carboxylic and amino groups. Thus, algal cell walls may be seen as providing ample surface adsorption sites for metal ions present in an aqueous system. However, it is unlikely that only a single adsorption mechanism prevails. Among the various possible interactions are physical adsorption, chemical adsorption (or chemisorption), ion-exchange, precipitation and complexation. Many factors are likely to be important in determining the extent of biosorption of metal ions on algal surfaces, including: the algal species; the charge and valence of the metal containing ion; hydration and hydrolysis reactions of the metal ion; preference for the coordination of metals by certain organic ligands; bio-availability of the metal ion in the aqueous environment; other chemical reactivities of the metal ion in solution; and kinetic controls which are pertinent to the metal ion binding and transport. An excellent comprehensive review[5] has recently appeared which addresses these and other microbial adsorption issues.

Understanding of the situation is made more complex by poorly understood metal ion speciation and interactions between metal ion and other species in the aqueous environment. It is readily apparent that the extent of metal uptake depends on the particular metal ion, the presence of other metal ions, and the specific biological system—both the microorganism and the environment in which it is placed. All of these factors make comparison of reported studies very difficult.[5]

Quantitative description of adsorption processes has classically employed adsorption isotherms developed from equilibrium batch sorption tests, or dynamic continuous-flow sorption studies. The two most commonly used adsorption isotherms are the Langmuir adsorption isotherm and the Freundlich adsorption isotherm. These adsorption isotherms are represented as follows:

Langmuir Isotherm:

$$C = \frac{K_L m}{a + m}$$ 1)

Freundlich Isotherm:

$$C = K_F m^{1/n}$$ 2)

where:
C = specific metal uptake (mM adsorbed/mg adsorbent)
m = equilibrium concentration of adsorbate (mM/L)
K_F, K_L, n, a = constants (temperature dependent)

The derivation of the Langmuir Adsorption Isotherm requires that the adsorption is confined to less than one complete monolayer on the surface, that all the surface is energetically uniform and that there is no interaction between the adsorbed species (i.e., there is no transmigration of the adsorbate on the surface). The Freundlich adsorption isotherm is an empirically derived relationship and can be reduced to the Langmuir isotherm with appropriate simplifications. These isotherms have been employed to characterize the uptake of heavy metals since they serve as useful models in evaluating the adsorption capacities of the adsorbents (in this case, the microorganisms), and in describing the equilibrium conditions for adsorption in different systems. Another advantage is in illustrating the differences between different types of species, and in particular, between the different morphological types of the same organism. However, major deficiencies are readily identified,[5] not the least of which include: the absence of explicit recognition of any external environmental factors; a lack of account of a possible irregular pattern due to the complex nature of both the sorbent and the cell's varied multiple active sites; and the inability to offer predictive capability even in single ion situations.

Hence, while these isotherms are commonly employed (in the absence of feasible alternatives), various limitations on their use must be realized.[8] For example, when applied to complex biological systems, these isotherms are undoubtedly too simplistic, and do not offer predictive capability. It is doubtful that a true monolayer is formed in such systems, and that there is no movement of the species on the surface. (For instance, electron microscopy has revealed that uranyl ions adsorbed onto the yeast cell, *Saccharomyces cerevisiae* form three or more layers.[9] The isotherms assume equilibrium conditions and provide no information on the kinetics of the uptake process. Since the adsorption models assume a fixed number of adsorption sites, they are strictly not applicable in the case of growing cells. The isotherms are strictly not applicable for use when transport processes occur.

Despite these drawbacks, the isotherms can be useful in characterizing adsorption systems, and many such systems have been found which are modelled adequately by either the Langmuir adsorption isotherm,[10,11] or the Freundlich adsorption isotherm.[12-14] In some cases, both isotherms appear to be equally applicable in describing the uptake behavior of the metals by microorganisms.[8,15,16] More recently some very interesting work has emerged for the description of biosorption in binary[17,18] and ternary[19] metal ion systems involving three-dimensional mapping of experimental results; this work has sought to overcome some of the difficulties highlighted above, but their complexity is readily apparent.

The use of adsorption isotherms needs to be complemented by kinetic studies if the results are to be used for process development. Over the years various surface adsorption models have been presented for various microbial species which appear to have broader applicability. An example is the early model proposed by Weidemann et al.[9] Termed the Intermediate-State Adsorption Model, it describes the rate of transfer of uranyl ions onto yeast cells. The model assumes a two-step reversible reaction, analogous to an enzyme-substrate scheme, as shown by Equation 3.

$$M + C \underset{k_2}{\overset{k_1}{\longleftrightarrow}} M - C \underset{k_4}{\overset{k_3}{\longleftrightarrow}} MC \qquad\qquad 3)$$

[M] is the concentration of the uranyl ion available for adsorption, C represents the adsorption sites at the cell walls, M-C is an intermediate complex formed before the bonding to the cell wall, and MC is the tightly-bound complex at the cell external surface. As with almost all other similar models, Weidemann and co-workers[9] assumed that the cells were not growing, and thus the number of binding sites remained constant.

Tsezos et al[20] presented a mass transfer kinetic model for the biosorption of uranium by immobilized biomass in a batch reactor. The adsorbent (i.e., the immobilized biomass) is assumed to be made of a number of spherical particles, consisting of a uniform non-biomass porous outer layer enclosing a uniform porous inner core. Biosorption is assumed to take place within the inner core, which consists of an intricate network of microbial cells. The approach adopted by these authors is similar to the modelling of the intra-particle adsorption kinetics of organic solutes in activated carbon particles.[21] In a recent paper, Tsezos and Deutschmann,[22] showed that the previously developed mass transfer kinetic model generated curves which compared favorably with experimental results. A sensitivity analysis performed on the model also suggested that the model results were affected by the immobilized biomass particle size, overall mass transfer coefficient and the effective solute diffusivity in the immobilized biomass particle core. The model appears to be able to describe the metal biosorption phenomenon of an immobilized biomass. In all these developments the (passive) adsorption process remains central.

Modeling Metal Uptake by Living Microorganisms

As indicated earlier, the quantitative analysis of metal uptake by living cells remains a complex issue, with far fewer contributions appearing than for the description of passive adsorption processes. Consideration of the growth and active metabolism of the cells must be added to all of the complexities highlighted above for the adsorption process.

Davies[23] presented an early model for the uptake of zinc by the diatom *Phaeodactylum tricornutum*, as illustrated in Figure 8.1. Davies[23] proposed that initially zinc is rapidly adsorbed on the extracellular surfaces of the diatom where

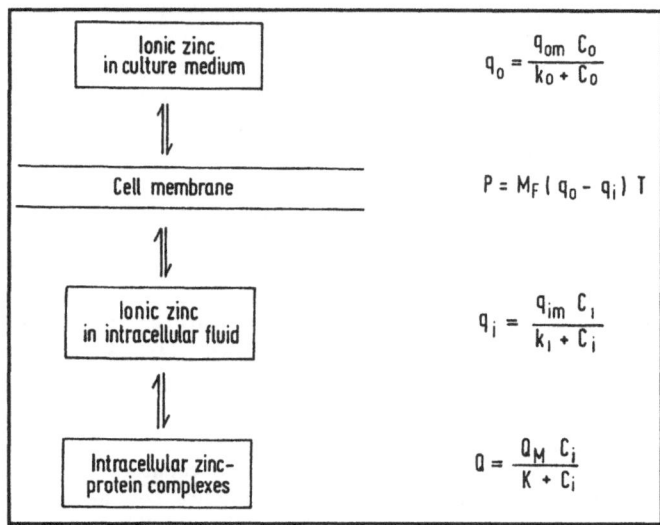

Fig. 8.1. A kinetic model for the uptake of zinc by the diatom *Phaeodactylum tricornutum* (after Davies[23]).

the surface bound concentration of zinc, q_o, is related to the concentration in the medium, C_o, by a Langmuir adsorption isotherm. Subsequent transport across the diffusion barrier presented by the membrane proceeds at a rate, which is proportional to the concentration gradient caused by the difference in the quantity of surface bound zinc and the amount at the inner surface of the cell membrane, q_i. In turn, q_i is related to the concentration of ionic zinc in solution in the intracellular fluid, C_i, again by a Langmuir adsorption isotherm. Finally, he hypothesized that most of the intracellular zinc is bound to a protein, the maximum binding capacity of which (Q_m) regulates the values of C_o, which in turn regulates q_i, thereby controlling the uptake of zinc.

In order to determine the amount of zinc bound onto the extracellular surfaces of the cell, desorption experiments were performed, and a Langmuir adsorption isotherm was obtained over a range of concentrations. Having derived the constants q_{om} and K_L, the amount of externally adsorbed zinc q_o corresponding to the values of C_o may be calculated. In turn, the intracellular zinc content may be evaluated. Davies[23] found that by subtracting the surface-bound zinc from the total cellular zinc content, the uptake of the metal by *Phaeodactylum tricornutum* could be explained in terms of a rapidly equilibrating adsorption of the ions on the cellular surfaces, followed by rate controlling diffusion of the ions across the cell membrane. However, there was no direct experimental evidence given by Davies[23] for the formation of zinc-protein complexes as he proposed and as shown in Figure 8.1.

The model relies on the implicit assumption that in the separate desorption experiments where q_o and C_o were determined, the Langmuir isotherm could be used to describe the actual uptake data. Also, central to the model is the assumption that the only process controlling the rate of transport of zinc into (or out of) the cells is diffusion. While this mechanism could explain the observed uptake of zinc ions by the diatom *Phaeodactylum tricornutum*, the model is certainly not universally applicable, since many microorganisms have been observed to be capable of accumulating the metals to a very high degree against a concentration gradient. Diffusion (a form of passive uptake) as the sole operating mechanism cannot explain the commonly reported high concentration factors.

The cell wall structure of microorganisms is usually porous, and therefore allows easy entry of the small metal ions even when hydrated. However, free access of the metal ions into the cell is restricted due to the hydrophilic charged nature of the most simple species available to the cell from the environment, and the fact that cell membranes are hydrophobic. Thus, the metal ion cannot traverse an intact membrane without the involvement of carrier ligands. A general model for the uptake and transport of a metal ion in a microbial cell (adapted from the early work of Williams[24] and Wood[25] is shown in Figure 8.2. [M] denotes the free external metal ion concentration, $L1$ represents a variety of bases or ligands (e.g., humic acids) in the aqueous solution which complex with the metal ion external to the cell, and $L2$ to $L4$ are ligands synthesized by the cell.

As shown in Figure 8.2, the ligands $L1$ and $L2$ are capable of coordinating with the metals external to the cell, with $ML2$ being complexed on the cell outer surface. The metal ion M can be transported into the cell through the agency of the carrier ligand $L3$. Ligand $L4$ complexes with the metal ion within the cell. The extent of the metal M uptake would then depend on:

1) The concentration of metal M which diffuses freely between the cell wall and the cell membrane;
2) The rate of formation and dissociation of the carrier complex $ML3$ which diffuses through the membrane and releases M into the cytoplasm of the cell; and
3) The ligand $L4$ which specifically binds to the metal ion.

Thus this model postulates that transport through the cell membrane and the concentration of the metal ions within the cytoplasm is an active process, requiring metabolic energy for the synthesis of the ligands, as well as for the maintenance of the pH gradient and the redox potential difference. It is clear, however, that this more general model remains a greatly simplified view of the actual uptake process. It is acknowledged that there is a lack of real understanding of the selectivity principles involved in the metal ion transport and of the mechanisms which energize the intracellular transport.[24]

Neither of these models incorporate considerations of cell growth, and hence increase in cell surface area in batch culture. These issues are addressed in the following detailed treatment of a model developed and experimentally tested in our laboratories.

Adsorption and Membrane Transport Model

The model now presented in some detail had its origins with the work of Khummongkol,[7] with refinements and extension by Ting et al[26-28] and most recently by Khoshmanesh et al.[29,30] The model has been used to describe the uptake of various metal ions, both singly and in combination, by living algal cells. Termed the Adsorption and Membrane Transport Model, it assumes that metal uptake by the microorganism is a two-stage process: a rapid passive uptake on the surface of the cells, followed by a slow active transport into the intracellular structure. Through this mechanism, the metal ion is transported across the membrane and into the

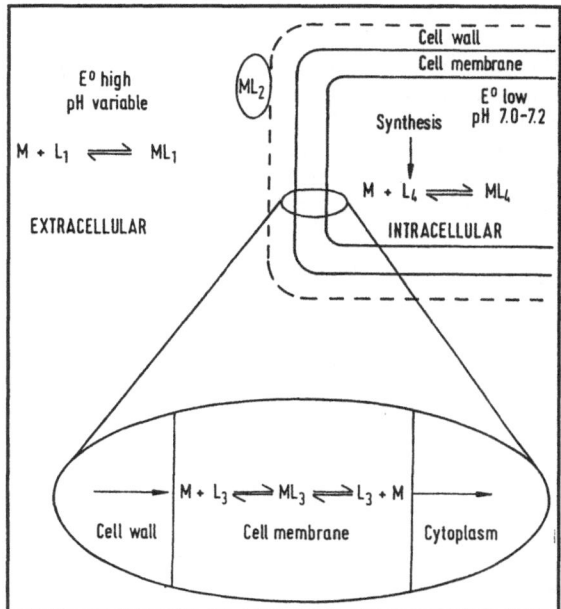

Fig. 8.2. A generalized conceptual model of metal uptake by microorganisms.

cytoplasm, resulting in an intracellular concentration of the metal ions which is substantially higher than in the external environment. Growth of the cells is explicitly accommodated. It is clearly recognized, however, that many of the complexities highlighted earlier in this chapter remain to be addressed.

A brief treatment of the development of the model is provided below to assist in the subsequent discussion. In essence, the following assumptions are made:

1) The metal in solution at the cell surface, m (mM metal/L), is in equilibrium with the metal adsorbed on the cell, C_1 (mM metal/mg cell dry wt). A linear relationship between these two is assumed, viz.

$$m = KC_1 \qquad\qquad 4)$$

where

K = adsorption constant (mg cell dry wt/L)

2) The metal ion adsorbed on the cell surface traverses the cell membrane through the agency of carrier molecules present on both sides of the membrane. The carrier-mediated mechanism is analogous to an enzyme-substrate coupling scheme:

$$\overset{k_1}{\underset{k_{-1}}{}} \quad \overset{k_2}{\underset{k_{-2}}{}}$$
$$C_1 + E \leftrightarrow CE \leftrightarrow C_2 + E \qquad\qquad 5)$$

where:

C_1 = extracellular metal
C_2 = intracellular metal

E = carrier

CE = metal-carrier complex

k_1, k_{-2} = chemical reaction rate constants (mg/mM.h)

k_{-1}, k_2 = chemical reaction rate constants (h^{-1})

3) The diffusion of the metal-carrier complex CE through the cell membrane is very rapid and the concentrations of E and CE are small compared with $[C_1]$ and $[C_2]$, the metal ion concentrations outside and inside the cells, respectively.

4) Pseudo steady-state is assumed; there is no net accumulation of the metal-carrier complex and:

$$\frac{d(CE)}{dt} = 0$$

5) The carrier content of each cell is essentially constant, and so the total carrier concentration in the system, E_t, is given by

$$E_t = [E] + [CE] \qquad\qquad 6)$$

6) One mole of carrier reacts with one mole of metal ion to produce one mole of metal-carrier complex.

It can be shown that

$$\frac{d(x[C_2])}{dt} = xR_1\{[C_1] - R_2[C_2]\} \qquad\qquad 7)$$

where the left-hand side denotes the rate of intracellular transport;

x = the biomass concentration

$R_1 = p(Z_1)$, the carrier rate constant

(which is a measure of the forward carrier reaction, p being the carrier concentration in the cell) and the ratio of rate constants (which is a measure of the relative predominance of the reverse carrier reaction to the forward carrier reaction); and

$$Z_1 = \frac{k_1 k_2}{k_{-1} + k_2} \qquad\qquad 8)$$

$$Z_2 = \frac{k_{-1} k_{-2}}{k_{-1} + k_2} \qquad\qquad 9)$$

The metal mass balance gives

$$A = m + x\left\{[C_1] + [C_2]\right\} \qquad\qquad 10)$$

where A is the total metal concentration in a unit volume of the system and m is the concentration of the metal ion in solution.

Substituting Equations 4 and 10 into Equation 7, one obtains

$$\frac{d(x[C_2])}{dt} = \frac{xR_1}{K+x}\{A - [C_2](x + R_2(K+x))\} \qquad\qquad 11)$$

This final expression provides a description of the model from which an analytical solution may be obtained. Next, various cell growth scenarios are considered.

Uptake by viable cells without growth

In the simplest case where the cells are not growing, and $[C_2] = 0$ and $x = x_0$ at $t = 0$, the solution becomes

$$[C2] = \frac{U}{V}\{1 - \exp(-Vt)\} \qquad\qquad 12)$$

$$[C1] = \frac{A}{K + x_0}\left\{1 - \frac{Ux_0}{VA}[1 - \exp(-Vt)]\right\} \qquad\qquad 13)$$

and

$$m = \frac{KA}{K + x_0}\left\{1 - \frac{Ux_0}{VA}[1 - \exp(-Vt)]\right\} \qquad\qquad 14)$$

where

$$U = \frac{R_1 A}{K + x} \qquad\qquad 15)$$

$$V = \frac{R_1}{K + x}\{x + R_2(K + x)\} \qquad\qquad 16)$$

Uptake by cells with linear growth

In this case $\quad \dfrac{dx}{dt} = L \qquad\qquad 17)$

where:

L = linear growth rate (mg dry wt/L.h)

By defining and substituting Equation 17 into Equation 11, one obtains

$$\frac{d(x[C_2])}{dx} + S_1\left(\frac{x}{K + X} + R_2\right)x[C_2] = \frac{S_1 A x}{K + X} \qquad\qquad 18)$$

The integrating factor for Equation 18 is:

$$I(x) = exp\{S_1[x(1 + R_2) - K\ln(K + x)]\} \qquad\qquad 19)$$

and hence Equation 18 can be transformed to

$$\frac{d(I(x)x[C_s])}{dx} = \frac{S_1 A x}{K + x}I(x) \qquad\qquad 20)$$

If $[C_2] = 0$ at $t = 0$, the solution to Equation 20 becomes:

$$[C_2] + \frac{S_1 A}{xI(x)}\int_{x_0}^{x}\frac{yI(y)}{k + y}dy \qquad\qquad 21)$$

Uptake by cells with exponential growth

Under this condition:

$$\frac{dx}{dt} + \mu x \qquad\qquad 22)$$

where: μ = specific growth rate (h^{-1})

By defining $S_2 = \dfrac{R_1}{\mu}$ and substituting Equation 22 into Equation 11, one obtains

$$\frac{d(x[C_2])}{dx} + S_2\left(\frac{1}{K+x} + \frac{R_2}{x}\right)x[C_2] = \frac{S_2 A}{K+x} \qquad\qquad 23)$$

The integrating factor for Equation 23 is

$$I(x) = \exp\{(K+x)^{S_2} x^{S_2 \, R_2}\} \qquad\qquad 24)$$

and hence Equation 23 can be transformed to

$$\frac{d(I(x)x[C_2])}{dx} = \frac{S_2 A}{K+x} I(x) \qquad\qquad 25)$$

Assuming the same initial conditions as were used in the previous cases, the solution where exponential growth occurs becomes

$$[C_2] = \frac{S_2 A}{xI(x)} \int_{x_0}^{x} \frac{I(y)}{K+y}\,dy \qquad\qquad 26)$$

Neither Equation 21 nor 26 can be solved analytically. However, a numerical solution can be obtained by solving Equation 11, together with the appropriate growth equation (i.e., Equations 17 or 22), using a first-order differential equation solver on the IMSL routine (DVERK). With known values of the model parameters K, R_1 and R_2, the solution yields a prediction of the uptake of metals by the combined effect of surface adsorption and membrane transport for growing cells.

Estimation of model parameters

At the initial stage of the metal uptake, only surface adsorption is significant so that $[C_2]$ may be assumed to be zero at time $t = 0$, $x = x_0$. Hence, from Equations 4 and 10

$$K = \left(\frac{m}{A-m}\right)x_0 \qquad\qquad 27)$$

The adsorption constant K can therefore be estimated using results obtained at the beginning of each experiment.

From Equations 4 and 10, the intracellular metal concentration $x[C_2]$ may be written as

$$x[C_2] = \left(\frac{KA - m(k+x)}{K}\right) \qquad\qquad 28)$$

Substituting Equations 4 and 28 into Equation 7 yields

$$\frac{d\{(KA - m(K+x))/K\}}{dt} = \frac{xmR_1}{K} - \left\{\frac{KA - m(K+x)}{K}\right\}R_1 R_2 \qquad\qquad 29)$$

Defining two variables, a and b as

$$\alpha = \frac{KA - m(K+x)}{K} \text{ and } \beta = \frac{xm}{K}$$

Equation 29 becomes

$$\frac{d\alpha}{dt} = \beta R_1 - \alpha R_1 R_2 \qquad\qquad 30)$$

From experimental data, α can be expressed as a function of t. Likewise, β can be calculated using values of x and m at different times. By substituting $\frac{d\alpha}{dt}$, and α and β evaluated at different times into Equation 30, the carrier rate constant (R_1), and the ratio of rate constants (R_2) can be estimated using linear least-squares analysis routine. Using estimates of the three constants (K, R_1 and R_2) obtained from experimental data, Equation 14 can be solved by numerical means.

Analysis of data obtained from experimental studies on the uptake of metal ions by a range of simple green freshwater algae has shown that the model describes the uptake behavior quite well.[26,28,30,31]

Accommodating Cell Surface Area

The sizes of the algal cells of different species and even the sizes of cells from the same species can vary substantially, and hence the sorption areas per unit mass available for the passive or active uptake of metal ions can also vary. A better comparison of the uptake behavior of different algal species can be made by taking the specific surface areas into account. This approach has been described by Khoshmanesh.[29]

Surface Adsorption Isotherm

Since the first stage of any active uptake depends on the surfaces provided by the algal cells, the specific surface area should be an important parameter in any adsorption equation. The adsorption constant K is inversely proportional to the specific surface area of the cells [Equation 4] and so the values of the adsorption constants for small algal cells are low since the specific surface areas are large and vice versa. One may define new surface adsorption constant, K_s as

$$K_s = KS_p \qquad \qquad 31)$$

and hence Equation 4 becomes

$$m = K_s C_{ls} \qquad \qquad 32)$$

where:
K = adsorption constant (mg/L)
K_s = surface adsorption constant (m²/L)
m = residual metal concentration in solution (mM/L)
S_p = specific surface area of the cells (m²/mg of dried cells)
C_{ls} = specific surface metal uptake on cell (mM/m² cell surface area)
   and

$$C_{ls} = \frac{C_l}{S_p} \qquad \qquad 33)$$

where:
C_l = specific metal uptake on the cell (mM/mg cells)

Equation 32 suggests that for any single algal species one should obtain a fixed value for the isothermal surface adsorption constant independent of the algal cell size and its specific surface area. This has been verified experimentally.[30,31]

The Membrane Transport Equation

The membrane transport equation, Equation 11, takes into account the intra-cellular uptake. This same equation may also be used to describe the intracellular uptake C_2 for the different size distributions with reasonable accuracy. The values of model parameters K, R_1, R_2, and L may be estimated from experimental data as before.

The inclusion of the specific surface area as a parameter in Equation 11 and estimating the intracellular uptake *per unit surface area* does not make physical sense since the intracellular uptake C_2 depends on the values of R_1 and R_2--the two parameters in the membrane transport equations--and these in turn depend on the rates of the forward and back carrier reactions through the membrane of the cell walls are not directly dependent on the external surface areas of the cells.

From a theoretical point of view, the final metal uptake per unit surface area given by:

$$\text{total metal uptake} = C_{ls} + \frac{C_2}{S_p}$$

is expected to be constant for a single algal species with any initial cell size distribution under fixed environmental conditions, and this too has been verified experimentally.[30,31]

Simulation and Sensitivity Analysis

Numerical simulations may be performed using estimates of the model pa-rameters K_s, R_1, R_2 and L to investigate the behavior of the system under different conditions. This often yields valuable insights into the operating mechanisms in the accumulation process. A study on the sensitivity analysis of each parameter is useful in that it shows the response of the model to changes in the specific param-eters keeping the remaining variables constant.

As an illustration, in a typical experiment on the uptake of cadmium ions at 21°C by C. *vulgaris* under constant illumination with an initial metal concentra-tion of 1.78×10^{-2} mM/L yields the following data pertinent to the modelling of the accumulation process.

Initial cell dry weight $X_0 = 180$ mg/L
Linear growth rate $L = 7.3$ mg/L
Carrier rate constant $R_1 = 0.1 \text{ h}^{-1}$
Ratio of rate constants $R_2 = 1.3$ (dimensionless)
Surface adsorption constant $K_s = 7.0 \text{ m}^2/\text{L}$

These data will be used as the base data for the sensitivity analysis described below.

Variation in the Adsorption Constant, K_s

The effect of a variation in the adsorption constant, K_s is shown in Figures 8.3 and 8.4 where K_s was varied over the range from 0.7 to 700 m²/L. The residual metal concentration is strongly dependent on K_s ; the lower the value of K_s the lower the value of the metal remaining in solution. The total metal uptake (mM/m² of cells) is not very sensitive to the value of K_s, particularly as the time increases. It should be noted that there is a reduction in the total metal uptake with time since new areas are continually being developed in this living system.

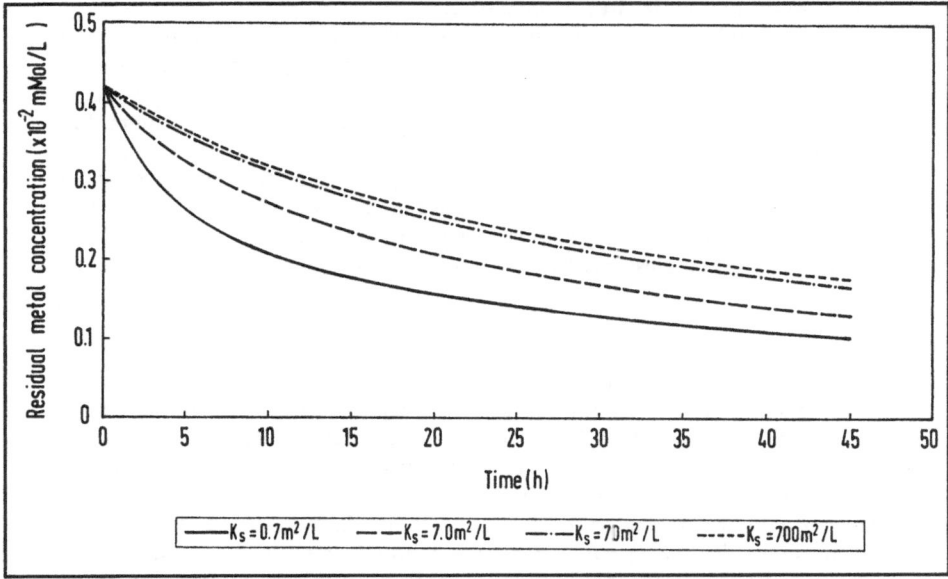

Fig. 8.3. Effect of K_s on the residual metal concentration (K_s from 0.7 to 700 m²/L).

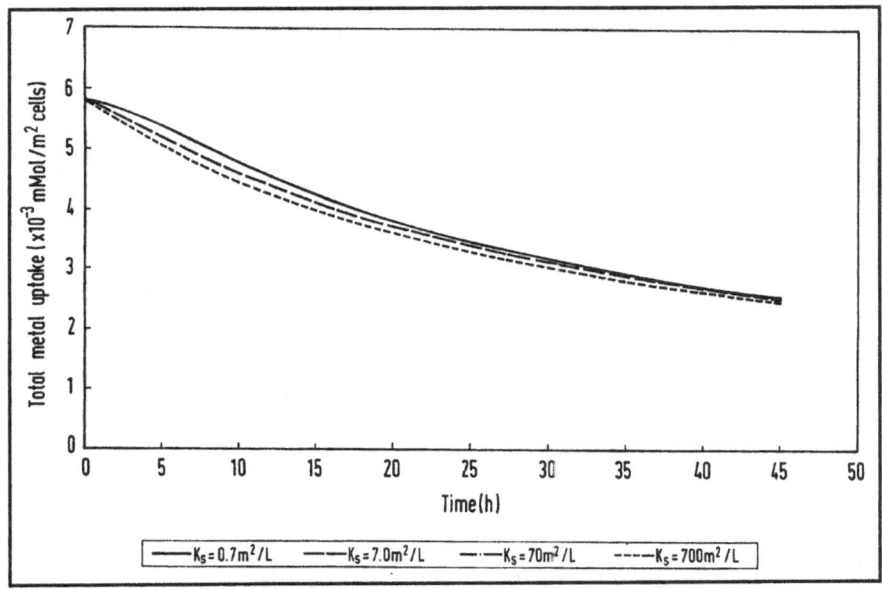

Fig. 8.4. Effect of K_s on the total metal uptake (K_s from 0.7 to 700 m²/L).

Variation in the Carrier Rate Constant, R_1

The sensitivity of the model to changes in the values of the carrier rate constant, R_1 is illustrated in Figures 8.5 and 8.6. Recalling that R_1 is defined as $p(Z1)$, the magnitude of R_1 is determined by the extent of the intracellular transport across the cell membrane, higher values of R_1 imply greater membrane transport. As is evident in Figure 8.5, the short-term drop in the metal concentration in the bulk solution depends greatly on R_1,; the higher the value of R_1, the greater the short-term metal uptake in the cells and the greater the change in the metal concentration remaining in solution. It is interesting to note that the long time results (25 hours and longer) are essentially the same provided R_1 is greater than about 0.12 h^{-1}.

Variation in the Ratio of Rate Constants, R_2

R_2, defined as the ratio of the rate constants, Z_2/Z_1, is a relative measure of the reverse and forward reaction rates for the transport processes. With a fixed forward reaction rate constant, Z_1, a decrease in R_2 implies a decrease in the back reaction rate and hence an increase in the metal ions being transported across the cell membrane. Thus, lower values of R_2 result in greater metal uptake, as shown in Figures 8.7 and 8.8. Initially, the effect of the variation in R_2 on the residual metal concentration, the total specific metal uptake, and the intracellular uptake are not significant other than for very high values of R_2.

Effect of Cell Productivity, L

The effect of a variation in the cell productivity L on the residual metal concentration, and the total specific metal uptake is shown in Figures 8.9 and 8.10. When the other model parameters are held constant, the extent of the growth of the cells

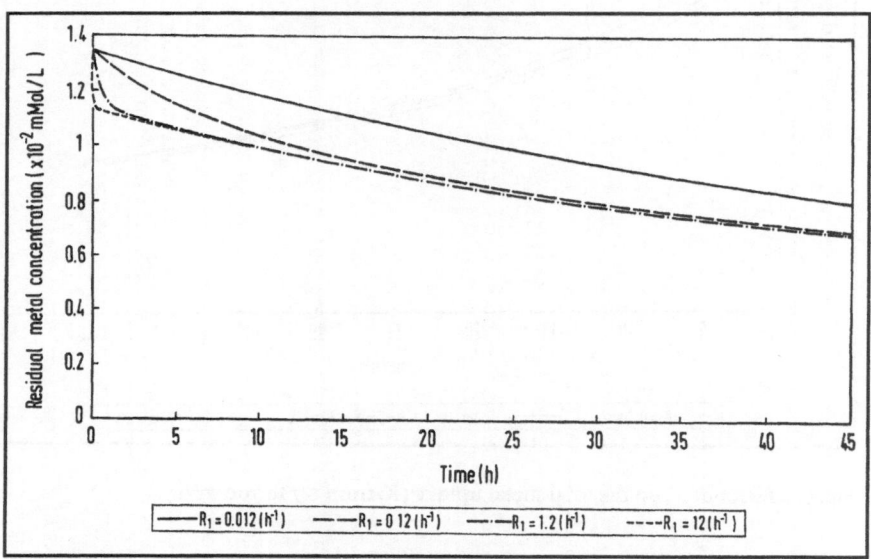

Fig. 8.5. Effect of R_1 on the residual metal concentration (R_1 from 0.012 to 12 h^{-1}).

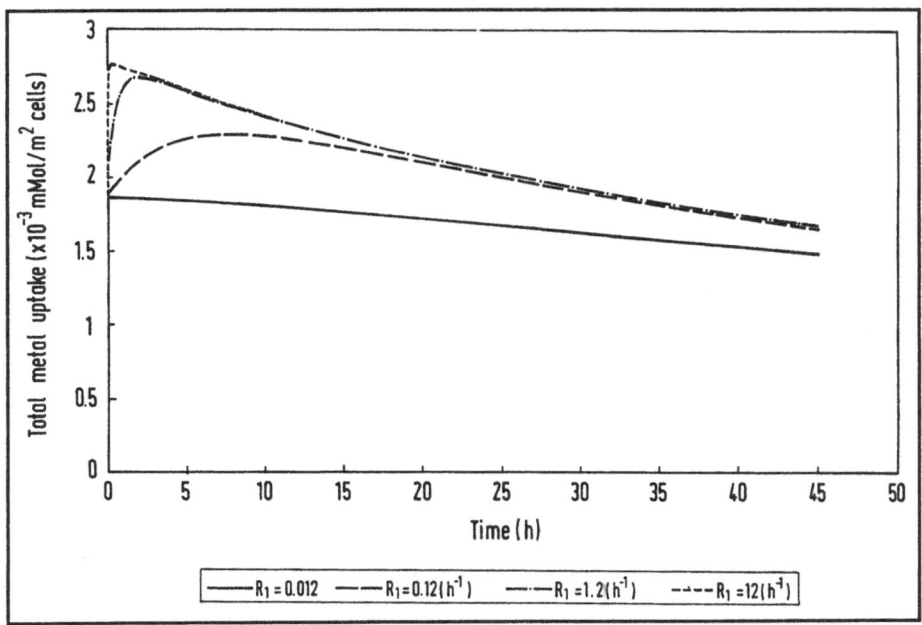

Fig. 8.6. Effect of R_1 on the total metal uptake (R_1 from 0.012 to 12 h^{-1}).

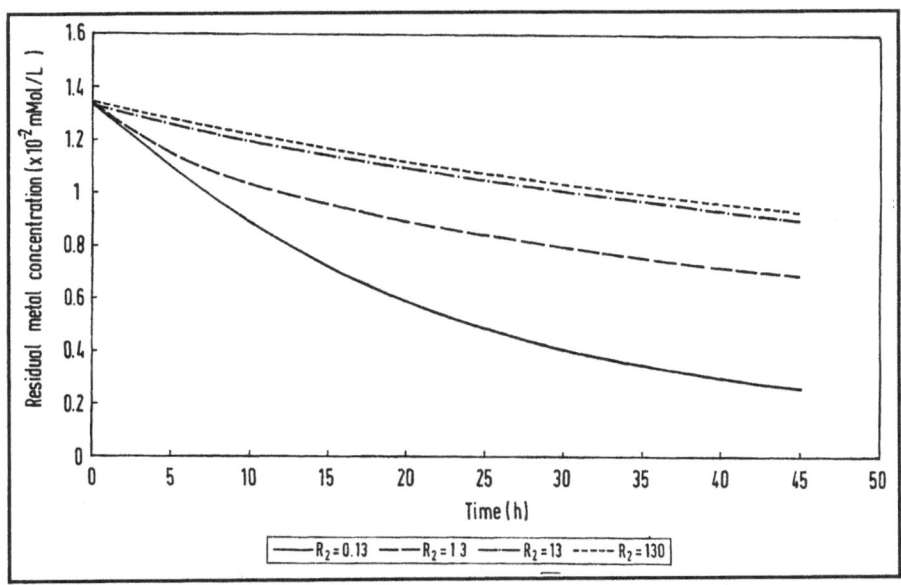

Fig. 8.7. Effect of R_2 on the residual metal concentration (R_2 from 0.13 to 130).

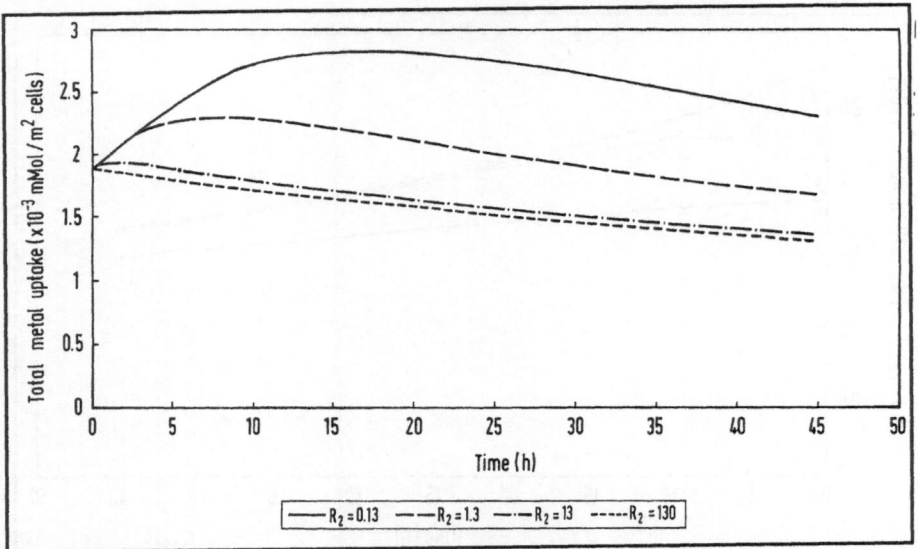

Fig. 8.8. Effect of R_2 on the total metal uptake (R_2 from 0.13 to 130).

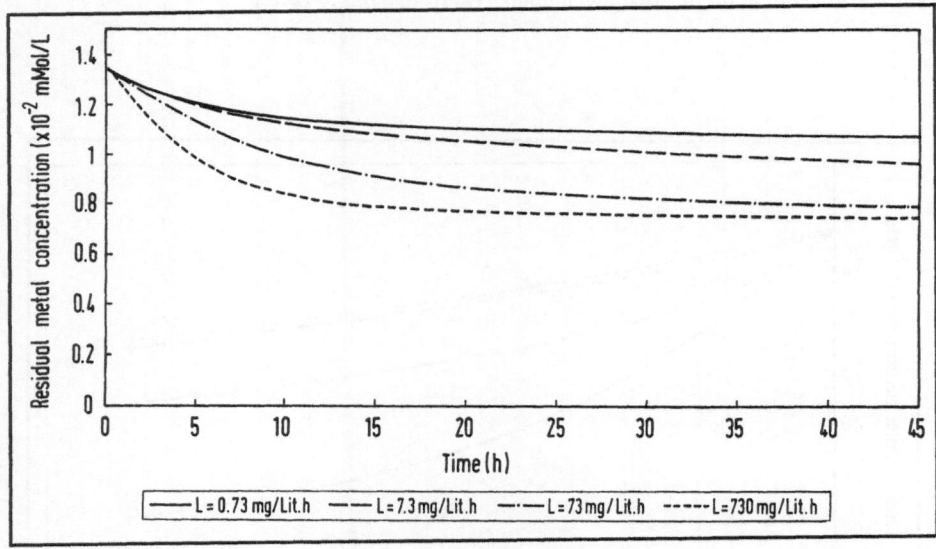

Fig. 8.9. Effect of L on the residual metal concentration (L from 0.73 to 730 mg/L.h).

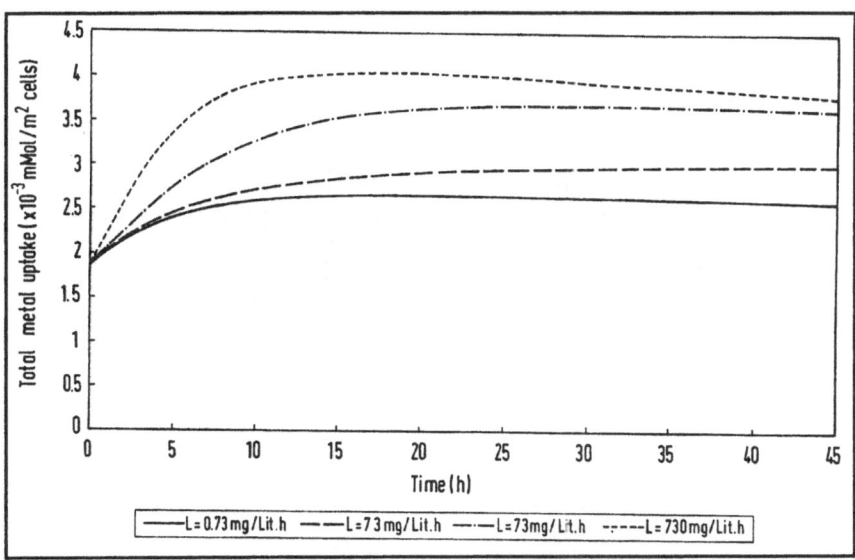

Fig. 8.10. Effect of L on the total metal uptake (L from 0.73 to 730 mg/L.h).

influences the total active surface area available for adsorption over a period of time and hence, high cell productivities give rise to large metal uptakes and low residual metal concentrations remaining in solution.

Conclusions

Algal cells are capable of sequestering metal ions from aqueous solution. The first step in any sorption process is the adsorption or absorption onto the cell external surface. If the cells are dead, this is all that occurs and the level of metal uptake can be generally described in terms of either the Langmuir or Freundlich isotherm expressions, though inherent weaknesses in such simple isotherms are recognized. More complex three-dimensional mapping representations may offer advantages. If, however, the cells are living, there are at least two additional effects that must be taken into account: the generation of new cell surfaces by cell growth and the transfer of metal ions across the cell wall into the cell cytoplasm. There may also be a third process whereby the metal ions may be "removed" by the living cell by precipitation within the cell as an organo-metallic compound, possibly a metal-containing protein.

Relatively few quantitative models of metal uptake by living algal cells have been presented. The model originally developed by Khummongkol et al,[7] and modified first by Ting et al,[26] and then by Khoshmanesh et al,[29] appears to be generally applicable and has been tested with various simple freshwater algal species. In the revised model, the growing algal surfaces adsorb metal species and then subsequently transfer them into the interior of the cells. The model parameters are as follows:

K_s = the surface adsorption constant (m²/L);
L = the linear growth rate (mg/L. h);

R_1 = the carrier rate constant (h^{-1});
R_2 = the ratio of the rate constants.

An "ideal" algal species as far as metal uptake from aqueous solution is concerned, should have a high linear growth rate (L), a low surface adsorption constant (K_s), a high carrier rate constant (R_1) and a low value for R_2 (which is a measure of the active metal transfer back across the cell wall).

Each of these parameters can be estimated from experimental data. While the model in its present form is able to accommodate the experimental systems upon which it is based, the complexities are easily recognized, but accommodated only with great difficulty. This is particularly the case for multi-ion situations. Hence the predictive capabilities are limited. As always, much remains to be investigated.

Nomenclature

A	metal concentration in the system (mg/L)[b]
a	constant (mM/L)
C	specific metal uptake (mM/mg)
	concentration of adsorption sites on cell wall (g/kg)[b]
C_b	solute concentration in bulk solution (mg/L)[c]
C_o	metal concentration in solution (µg/L)[a]
	concentration of adsorption sites on cell wall at time t = 0 (g/kg)[b]
C_1	metal concentration adsorbed on cell (mM/mg)
C_2	metal concentration within the cell (mM/mg)
Ci	metal concentration in intracellular fluid (µg/L)[a]
C_{ls}	specific surface metal uptake on cell (mM/m^2)
CE	metal-carrier complex concentration (mM/L)
E	free carrier concentration (mM/L)
E_t	total carrier concentration (mM/L)
I	integrating factor (-)
K	adsorption constant (mg/L)
	half saturation constant (ng.metal/10^6 cells)[a]
K_F	constant (L/mg)
K_L	constant (mM/mg)
K_M	Michaelis constant (mM/L)
Ks	surface adsorption constant (m^2/L)
k_1	rate constant (mg/mM.h)
k_{-1}	rate constant (h^{-1})
k_2	rate constant (h^{-1})
k_{-2}	rate constant (mg/mM.h)
L	cell productivity (mg/L.h)
M	concentration of uranyl ion available for adsorption (g/kg)[b]
M-C	loosely-bound intermediate complex (g/kg)[b]
MC	tightly-bound complex (g/kg)[b]
m	metal concentration in the solution (mM/L)
n	constant (-)
P	= $qi + Q$, total intracellular metal content of cells (ng metal/10^6 cells)[a]
p	carrier concentration in the cell (mM/mg)
Q	weight of metal bound to intracellular proteins (ng metal/10^6 cells)[a]
Q_M	maximum binding capacity of metal bound to intracellular proteins (ng metal/10^6 cells)[a]

qi metal adsorbed on inner surface of cell membrane (ng metal/10^6 cells)[a]

qim maximum capacity of inner surface of cell membrane for metal absorption (ng metal/10^6 cells)[a]

qom maximum capacity of inner surface of cell membrane for metal adsorption (ng metal/10^6 cells)[a]

R_1 carrier rate constant (h^{-1})

R_2 ratio of rate constants (-)

S substrate concentration (mM/L)

S_1 $= R_1/L$ (L/mg)

S_2 $= R_1/\mu$ (-)

S_p specific surface area of cells (m^2/mg)

T time (h)[a]

t time (h)

U constant (mM/mg.h)

V constant (h^{-1})

V_{max} reaction velocity (membrane transport) (mM/L.h)

v reaction velocity (membrane transport) (mM/L.h)

x cell concentration (mg/L)

         thickness of layer of metal adsorbed on inner and outer surfaces of membrane (cm)[a]

x_0 initial cell concentration (mg/L)

Z_1 constant (mg.mM.h)

Z_2 constant (mg/mM.h)

Greek symbols

α variable (mM/L)

β variable (mM/L)

μ specific growth rate (h^{-1})

a, b and c: Units as used in Davies,[23] Weidemann et al[9] and Tsezos et al,[20] respectively.

References

1. Hughes M, Poole R. Metals and Micro-organisms. New York: Chapman & Hall, 1989.
2. Volesky B, Biosorption of Heavy Metals. Boca Raton: CRC Press, 1990.
3. Ehlrich H, Brierley C. Microbial Mineral recovery, New York: McGraw-Hill, 1990.
4. Madgwick J. Biological sorption and uptake of toxic metal ions from wastewaters. Aust J Biotechnol 1994; 4(5): 292-297.
5. Volesky B, Holan Z. Biosorption of heavy metals. Biotechnol Prog 1995; 11: 235-250.
6. Wilde E, Benemann J. Bioremoval of heavy metals by the use of microalgae. Biotechnol Adv 1993; 11(4):781-812.
7. Khummongkol D, Canterford G, Fryer C. Accumulation of heavy metals in unicellular algae. Biotechnol Bioeng 1982; 24:2643-2660.
8. de Rome J, Gadd G. Copper adsorption by *Rhizopus arrhizus, Cladosporium resinae* and *Penicillium italicum*. Appl Microbiol Biotechnol 1987; 26:84-90.
9. Weidemann D, Tanner R, Strandberg G. Shumate II S, Modelling the rate of transfer of uranyl ions onto microbial cells. Enzyme Microb Technol 1981; 3:33-40.
10. Paton W, Budd K. Zinc uptake in *Neocosmospora vasinfecta*. J Gen Microbiol 1972; 72:173-184.

11. Byerley J, Scharer J, Charles A. Uranium (VI) biosorption from process solution. Chem Eng J 1987; 36:B49-B59.
12. Gutknecht J. Mechanism of radioactive zinc uptake by *Ulva lactuca*, Limnol Oceanogr 1965; 10:58-66.
13. Les A, Walker R. Toxicity and binding of copper, zinc, and cadmium by the blue-green alga *Chroococus paris*. Water, Air, Soil Pollut 1984; 23:129-139.
14. Bell J, Tsezos M. Removal of hazardous organic pollutants by biomass adsorption. J Water Pollutn Control Fed 1987; 59:191-198.
15. Geisweid H, Urbach W. Sorption of cadmium by the green microalgae *Chlorella vulgaris, Ankistrodesmus braunii* and *Eremosphaera*. Pflanzenphysiol 1983; 109:127-141.
16. Sag Y, Kutsal T. Application of adsorption isotherms to chromium adsorption on *Z. ramigera*. Biotechnol Lett 1989; 11:141-144.
17. de Carvalho R, Chong K, Volesky B. Evaluation of the Cd, Cu and Zn biosorption in two-metal systems using an algal biosorbent. Biotechnol Prog 1995; 11:39-44.
18. Chong K, Volesky B. Description of two-metal biosorption equilibria by Langmuir-type models. Biotechnol Bioeng 1995; 47:451-460.
19. Chong K, Volesky B. Metal biosorption equilibria in a ternary system. Biotechnol Bioeng 1996; 49:629-638.
20. Tsezos M, Noh S, Baird M. A batch reactor mass transfer kinetic model for immobilized biomass biosorption. Biotechnol Bioeng 1988; 32:545-553.
21. Peel R, Benedek A. Dual rate kinetic model of a fixed bed adsorber. J Environ Eng Div (Am Soc Civ Eng) 1980; EE4:797-813.
22. Tsezos M, Deutschmann A. The use of a mathematical model for the study of the important parameters in immobilized biomass biosorption. J Chem Tech Biotechnol 1992; 53:1-12.
23. Davies A. The kinetics of and a preliminary model for the uptake of radioactive zinc by *Phaeodactylum tricornutum* in culture. In: Krippner M, ed. Symposium on Radioactive Contamination of the Marine Environment, Seattle, USA 10-14 July, 1972: 403-420.
24. Williams R. Physicochemical aspects of inorganic element transfer through membrane. Phil Trans R Soc 1981; Lond, B294:57-74.
25. Wood J. In: Sigel H, ed. Metal Ions in Biological Systems. Circulation of Metals in the Environment. Marcel Dekker 1984; 18:223-237.
26. Ting Y, Lawson F, Prince I. Uptake of cadmium and zinc by the alga *Chlorella vulgaris*: Part 1. Individual ion species. Biotechnol Bioeng 1989; 34:990-999.
27. Ting Y, Lawson F, Prince I. Uptake of heavy metal ions by algae. Aust J Biotechnol 1990; 4(3):192-204.
28. Ting Y, Lawson F, Prince I. Uptake of cadmium and zinc by the alga *Chlorella vulgaris*: Part 2. Multi-ion solution. Biotechnol Bioeng 1991; 37:445-455.
29. Khoshmanesh A. Modelling the algal uptake of metals. M.Eng.Sc. thesis 1995; Monash University, Australia. 1995.
30. Khoshmanesh A, Lawson F, Prince I. Cadmium uptake by unicellular green microalgae. Chem Eng J 1996; 62:81-88.
31. Khoshmanesh A, Lawson F, Prince I. Cell surface area as a major parameter in the uptake of cadmium by unicellular green microalgae. Chem Eng J 1997; 65:13-19.

CHAPTER 9

Carrageenan as a Matrix for Immobilizing Microalgal Cells for Wastewater Nutrients Removal

P.S. Lau, Nora F.Y. Tam and Yuk-Shan Wong

Introduction

Microalgae are a group of single celled photoautotrophs and have been suggested as an alternative biological system in treating wastewater to the conventional activated sludge process.[1] Soluble contaminants in domestic or municipal wastewater comprise principally organics which are generally referred to as biochemical oxygen demand (BOD) and inorganic nutrients. Domestic wastewater contains basically all the essential elements for algal growth. Growth of the microalgal biomass assimilates inorganic nutrients directly from wastewater while the production of photosynthetic oxygen stabilizes organic matters in the wastewater. This conception has long been conceived in the moats of the medieval castles and realized in waste stabilization ponds.[2-5] The full potential in optimizing the photosynthetic oxygen generation for wastewater treatment is credited to William J. Oswald in the conception of the high rate algal pond systems which optimizes algal growth for wastewater treatment.[1,6] The algal wastewater treatment system then draws great attention not only as a means of purifying wastewater, but also for generating valuable algal biomass for other usage or extraction of valuable chemicals.[7-9] Earlier works were concerned more so with BOD reduction,[1,6] whereas attention gradually shifted to nutrients stripping from wastewater as eutrophication of receiving water became more serious.[10,11]

One of the major and practical limitations in algal treatment systems is the harvesting or separation of the algal biomass from the treated water discharge. Direct discharge of the algal suspension will result in a loss of energy or chemical-rich algal biomass, and lead to a "green bloom" instead of red-tide as in the case of eutrophication. Conventional techniques for algal separation range from simple sand filtration to energy intensive centrifugation.[12] Chemical flocculation in promoting algal aggregation, followed by sedimentation, has also been suggested.[13-16]

Wastewater Treatment with Algae, edited by Yuk-Shan Wong and Nora F.Y. Tam.
© Springer - Verlag and Landes Bioscience 1998.

Immobilization of algal cells for wastewater treatment was also proposed for circumventing the harvest problem as well as retaining the high value algal biomass for further processing.[9] Application of immobilization technology to algal wastewater treatment also provides more flexibility in the reactor design when compared with conventional suspension systems.

Cell immobilization is a process by which cells are localized onto a support or entrapped within a three-dimensional matrix with the preservation of some enzymatic activities.[17] Immobilized cells are therefore prevented from free movement within a liquid phase.[18] The most common matrices used for immobilization of microalgae for wastewater treatment are polysaccharide types such as alginate[19-23] and carrageenan[24,25] that result in high cell viability. Chitosan, a chemical flocculant, has also been utilized as a matrix for algal immobilization for wastewater nutrient removal.[26]

Conventional algal wastewater treatment systems can be viewed as a two-component system consisting of the algae and the wastewater. The incorporation of immobilization technology into algal wastewater treatment systems thus introduces a third component, the gel matrix, into the system. With this additional component, two more interactions can be identified in the immobilized system, namely the interaction between the gel and the algae, and that between the gel and the wastewater (Fig. 9.1). Conventional algal wastewater treatment in suspension culture represents the sum of interactions between the algal and wastewater components (I_1) such as: species selection,[27] effect of algal density,[10,28] prestarvation of the algae,[10] acclimation,[29] and the nature of the wastewater.[30,31] This also includes the effect of physical factors such as light intensity,[1,32] temperature,[32] pH[2,33-35] and CO_2.[36,37] The interaction between the gel matrix and the algae (I_2) can be addressed as "How would the matrix and the procedures of immobilization affect the algal cells per se?" Parameters commonly involved with this aspect include morphological changes,[38] growth characteristics,[39-41] and metabolic activities.[38,41,42] The possibility of physicochemical interactions between the gel matrix and the organic or inorganic chemical species, or the biological interaction between the matrix and the indigenous microfauna of the wastewater characterize the interaction I_3. One typical example is the dissociation of the alginate gel in PO_4^{3-}-rich environments.[39] However, this interaction (I_3) is seldomly reported in the literature. Before a scaling up of immobilization technology in wastewater treatment can be made, a better understanding of these individual interactions and the overall performance of the immobilized algal system are required to assure good quality control.

In this chapter, carrageenan as an immobilization matrix for the unicellular algal cells *Chlorella vulgaris* for wastewater nutrients removal will be used as a model system for studying the various interactions depicted in Figure 9.1. Emphasis is placed on aspects not previously reported in the literature. Carrageenan is a potential matrix for algal immobilization alternative to alginate. It has an advantage over alginate of not being susceptible to dissociation in PO_4^{3-}-rich environments. Its employment for immobilizing algae for wastewater treatment has received less attention than alginate. Hence, its potential deserves more in-depth study.

Carrageenan by itself is a polyester with alternating 1,3-linked β-D-galactose sulfate and 1,4-linked β-D-galactose.[43-45] The presence of sulfate groups accounts for the polyanionicity of the carrageenan gel. Although the setting of the carrageenan gel is primarily thermal in nature, the presence of K^+ ions will cure the gel

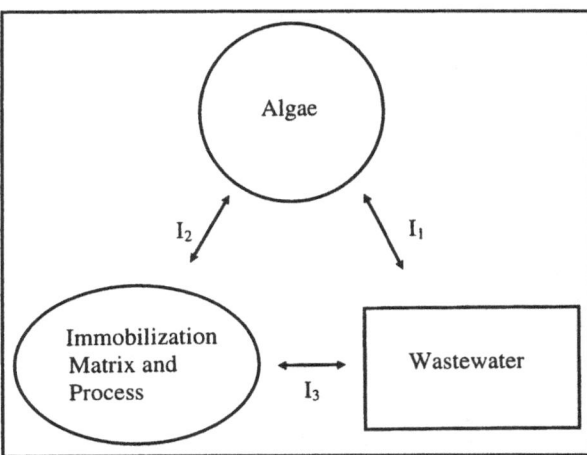

Fig. 9.1. Schematic diagram showing the possible interactions between the three components in the immobilized algal treatment system. I_1: Conventional algal treatment system in suspension culture. I_2: Physiological effect of the immobilization matrix and process on the algae. I_3: Possible physicochemical and/or biological interaction between the matrix and wastewater.

with better mechanical strength.[43,45] Ammonium has also been demonstrated to induce gelling of carrageenan with good mechanical strength.[44,45] As ammonium is the key inorganic nitrogen species in wastewater, the interaction between the ammonium ion and the carrageenan gel, and its relation to the algal uptake within the gel matrix becomes a crucial topic for investigation.

Here we review the immobilization process for carrageenan. The algal cells have to be kept at a temperature of 38-40°C for about 30 minutes, which may impose a heating stress to the algal cells concerned. However, no report has directly addressed this factor in relation to algal immobilization with carrageenan. Hence, the heating effect associated with the immobilization process in relation to algal growth and nutrient uptake from wastewater has been addressed here.

Finally, the nutrient removal efficiency from wastewater by carrageenan immobilized *C. vulgaris* was investigated, and the overall performance of the carrageenan immobilized algal system with respect to cell stocking density was assessed as well.

Materials and Methods

Routine Cultivation
Chlorella vulgaris from Carolina Biological Supply was cultivated in Bristol medium under axenic conditions and maintained by routine subcultures. Cells at exponential phase were double-counted with the improved Neubauer hemacytometer, then harvested by centrifugation at 3000 x g for 10 minutes, followed by washing twice with sterilized deionized water. The cells were then ready for inoculation.

Immobilization
K-carrageenan, Type III extracted from *Eucheuma cottonii* (Sigma, St. Louis, MO) was used for all immobilization studies. Carrageenan beads were formed according to the method of Kierstan and Coughlan.[46] A 20 ml of 5% (w/v) carrageenan solution was prepared and autoclaved at 121°C for 15 minutes. The pre-counted cells were resuspended in 20 ml sterilized deionized water. The resuspended culture

was then mixed thoroughly with 20 ml carrageenan solution to give a final gel concentration of 2.5% (w/v) and kept at $38 \pm 1°C$ in a temperature-controlled water bath. The algal-carrageenan solution was then poured into a sterile disposable 5 ml syringe and 3.0 mm diameter beads were formed by extruding the mixture through a 25 G gauge hypodermic needle into 2% KCl solution for curing. For blank beads (i.e., without algal cells), 20 ml deionized water was used instead of the algal suspension. The beads were cured in the 2% KCl solution overnight, then washed twice with sterilized deionized water for experimentation.

Study of Ionic Interaction With the Carrageenan Gel

Batches of 40 ml blank carrageenan gel beads were incubated with 120 ml solutions of 15 or 30 mg l^{-1} NH_4^+-N, 9 mg l^{-1} NO_3^--N or 6 mg l^{-1} PO_4^{3-}-P prepared from analytical grade salts of NH_4Cl, KNO_3 and K_2HPO_4, respectively. These values were typically found in domestic wastewater. The volume ratio of gel bead: ionic solution was kept at 1:3 (v/v). This volume ratio was chosen for the wastewater nutrient removal studies by the carrageenan immobilized microalgal cells as well. Preliminary results showed the adsorption of NH_4^+ to carrageenan gel, and the effects of pH and gel:solution volume ratio on NH_4^+ adsorption were investigated. The pH effect on NH_4^+ adsorption was conducted in the gel: ionic solution ratio of 1:3 (v/v) and the pH of NH_4Cl stock solution was adjusted with 0.1 M NaOH or HCl to cover the pH range from 4 to 10. The effect of volume factor was conducted by extending the gel bead: ionic solution ratio from 2:1 to 1:5 (v/v). In these adsorption studies, we constantly shook the mixture of beads and ionic solution to ensure homogenous exposure of the bead surface to the ions in the solution. Samples of the solution were collected at 1-minute intervals in the first 5 minutes and subsequently, at 5 to 15 minute intervals. Ammonium retained by the beads was assayed by redissolving five beads in 2 ml 0.9% NaCl solution.

Study of Heating Effect on C. vulgaris

In simulating the heating effect during the carrageenan immobilization process, equal volume of free suspended *C. vulgaris* cells was heated in the same water bath with constant shaking in parallel with the immobilization process. The heat-treated cells are referred to as "Heated" cells while *C. vulgaris* collected from the stock culture are termed "Fresh" cells hereafter.

The Fresh and Heated cells were then inoculated with an initial cell density of 1.0×10^6 cells ml^{-1} in 300 ml of the primary settled wastewater collected from the Hong Kong Shatin Sewage Treatment Work. The primary settled wastewater was characterized with 37.5 mg l^{-1} NH_4^+-N, 3.08 mg l^{-1} PO_4^{3-} -P, 138 mg l^{-1} BOD and a pH of 7.33. The algal cells were cultivated with a light intensity of 100 μE $m^{-2}s^{-1}$ from cool white fluorescent tubes at a light-dark cycle of 16-8 hours. The cultures were aerated with filtered air through mechanical pumps and temperature was kept at $25° \pm 1°C$. Blank wastewater was treated similarly to serve as control for the study. Daily samples were collected for analysis.

Study of Nutrient Removal by Carrageenan Immobilized C. vulgaris

Immobilized algal cells were prepared in batches of 100 ml carrageenan gel with a final concentration of 2.5% w/v. Two cell stocking beads were prepared. In the low cell stocking beads, the same total number of algal cells as used for the inoculation in the Fresh and Heated system were immobilized in 100 ml carrageenan gel. This resulted in beads with algal density of 1.9×10^5 bead^{-1}. The high cell stocking beads were prepared with cells 10 times that of the low cell stocking giving beads with a cell density of 18×10^5 bead^{-1}. The 100 ml gel beads were then inoculated in 300 ml of the primary settled wastewater, resulting in a gel : wastewater ratio of 1:3 (v/v). Apart from the blank wastewater control, a blank bead control of 100 ml blank carrageenan gel beads (without any algal cells) in 300 ml wastewater was also set up for comparison. The culture and sampling conditions were identical to those of the Fresh and Heated cell system.

Cell Count and Chlorophyll Content
Routine cell count for the Fresh, Heated and immobilized cells followed the methods described earlier.[28] Chlorophyll was extracted with methanol:chloroform mixture (2:1 v/v), diluted with acetone and assayed according to the method of Yung and Mudd.[47]

Chemical Analysis
Concentration of nitrate and phosphate were determined using the Lachat Quickchem™ automated ion analyzer. For the adsorption studies, ammonium level was determined with the Nesslerization method according to APHA et al.[48] For the wastewater, ammonium was determined with the Tecator Kjeltec 1030 auto analyzer.

Results and Discussion

Interaction Between Nutrient Ions and Carrageenan Gel
The interaction between the common nutrient ions found in wastewater (NH_4^+, NO_3^- and PO_4^{3-}) with the carrageenan gel beads in the first 30 minutes is shown in Figure 9.2. According to Guiseley,[43] diffusion was uninhibited for molecules of low molecular weight (<200 daltons) in carrageenan gel. Ions such as NO_3^-, PO_4^{3-} and NH_4^+ would therefore be able to diffuse freely in and out of the gel matrix. If size was the only factor concerned, a similar extent of reduction from the solution as a result of free diffusion into the gel beads was expected for all ions. This was true for anions of NO_3^- and PO_4^{3-} where a 17% and 16% reduction was recorded from the solution, respectively (Fig. 9.2). Theoretically, the volume factor of the gel will reduce the ion concentration by 25% as a result of free diffusion in the gel: ionic solution volume ratio of 1:3. Less than the theoretical 25% reduction recorded for NO_3^- and PO_4^{3-} ions was probably due to the additional ionic interaction apart from size. Repulsive interaction between the anions and the poly–anionic gel matrix interferred with the entrance and free movement of these anions in the gel matrix. The amount of the anions residing in the gel matrix would therefore be lower than the theoretical free and neutral molecules. Hence, the reduction in the ion concentration was less than expected.

On the contrary, over 40% NH_4^+ ions were reduced in the first 5 minutes in both high and low initial NH_4^+ input (Fig. 9.2). The significant reduction of NH_4^+ from the solution cannot be justified by size of the ions alone. The depletion profiles of NH_4^+ exhibited similar hyperbolic kinetics with a sharp initial drop in the first 5 minutes, which then levelled off in a state of equilibrium or saturation. This indicated that there is a strong ionic interaction and most probably, adsorption between the polyanionic carrageenan gel matrix and the small sized cationic NH_4^+

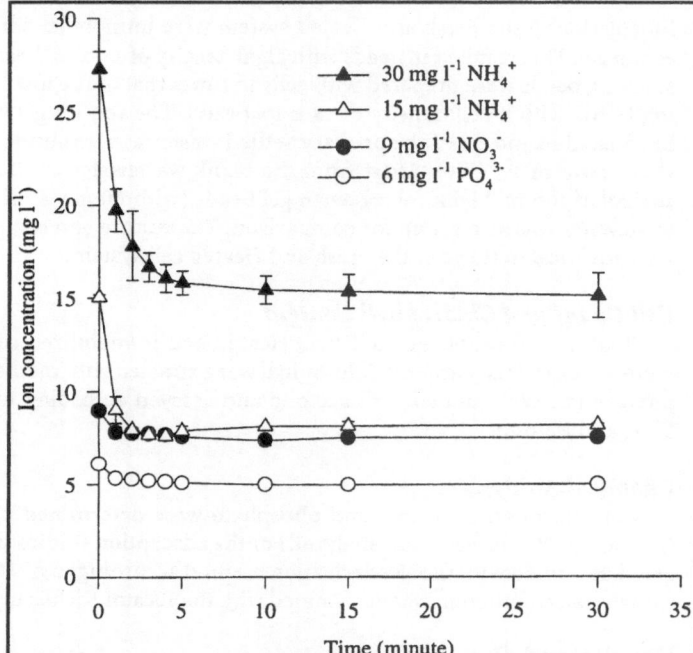

Fig. 9.2. The adsorption profiles of NO_3^-, PO_4^{3-} and NH_4^+ by blank carrageenan beads. Beads were incubated in the ionic solution with a volume ratio of 1:3 (v/v) and at a pH of 7. Each data point represented mean ± s.d. of triplicates.

ions. The adsorption nature depends on the availability of binding sites relative to the substrate level. The higher the binding site to substrate ratio, the greater extent will be the adsorption. The dependency of NH_4^+ adsorption on the volume ratio of the gel bead : ionic solution as shown in Table 9.1 supported such argument. With the volume ratio increased from 1:5 to 2:1, the percentage of NH_4^+ reduction increased from 37.3% to 88.3%, respectively. Furthermore, the mass balance analysis of NH_4^+ associated with the gel beads and its residual amount in solution at equilibrium showed that the sum of the two fractions was in good agreement to the total amount of NH_4^+ loaded at the beginning (Table 9.2). Reduction of NH_4^+ in the solution was actually redistributed and localized in the gel matrix. In other words, the carrageenan gel is capable of storing the NH_4^+ ions.

The reduction of NH_4^+ was not affected much by the ambient pH (Table 9.1). About 44% of the initial level was reduced from the solution in the pH range of 4 to 9. At pH 10, a 32.4% reduction was recorded. This slightly lower reduction was probably related to the equilibrium shift of NH_4^+ to NH_3 molecules, and a reduction in the effective NH_4^+ concentration or its availability to the gel matrix. The carrageenan gel as an adsorbent for NH_4^+ was therefore physically and chemically stable with respect to the acidity or alkalinity of the ambient environment.

Heating Effect on C. vulgaris
The heating effect at 38 ± 1°C for 30 minutes did not pose any adverse effect on *C. vulgaris* cells. There was no significant difference in the lag and early log phase as revealed from the growth profiles between the Fresh and Heated *C. vulgaris* cells (Fig. 9.3). The lag phase of the Heated cells was one day and the same as that

Table 9.1. Summary of the effect of (A) volume ratio of gel bead : solution (v/v) and (B) pH on the NH_4^+ reduction as percentage (mean of triplicates) of the initial concentration (30 mg N l^{-1}) due to physicochemical adsorption

A. Volume ratio effect

Gel bead : solution ratio

	2:1	1:1	1:2	1:3	1:4	1:5	
% reduction	88.3	75.8	61.0	52.2	48.4	37.3	

B. pH effect

pH	4	5	6	7	8	9	10
% reduction	44.7	45.3	45.7	43.6	45.5	43.4	32.4

Table 9.2. Results of mass balance analysis of NH_4^+ in the adsorption studies. Batches of blank carrageenan gel beads were mixed with NH_4^+ solutions with different initial concentrations and in different volume ratios of 1:1, 1:2 and 1:3 for 30 minutes

Gel bead: solution ratio	(a) NH_4^+ in solution	(b) NH_4^+ in gel bead	Sum of (a) + (b)	Amount added	deviation
(v/v)	(mg)	(mg)	(mg)	(mg)	%
1:1	0.550	0.981	1.531	1.426	7.36
1:1	1.242	1.606	2.848	2.857	-0.30
1:2	1.505	1.537	3.042	2.892	5.18
1:2	3.042	1.365	4.406	4.398	0.20
1:3	0.183	0.236	0.419	0.419	-0.04
1:3	2.322	1.764	4.086	4.158	-1.73

of the Fresh cells. This indicated that the heating process had not induced any effect on the preparatory stage for cell division. Both the Fresh and Heated cells entered the log phase after the lag period. Kinetic analysis gave the average growth constant of 0.342 and 0.382 day^{-1} for the Fresh and Heated system, respectively. The reproductive potential in terms of cell division was therefore not affected since no significant difference was found in the growth constant between the Fresh and Heated cells. From day 6 onwards, both systems entered the late log phase. The Heated cells even outgrew in number and were about 2-fold higher than the Fresh cells in the late log phase. While the Heated cells entered the stationary phase from day 8, there was a slight decline in the Fresh system. The multiplication in this late log period was likely at the expense of the vegetative development, as reported previously.[29] This speculation was supported by the total chlorophyll content which showed no difference between the Fresh and Heated system in the same period (Fig. 9.4). In other words, though the number of cells of the Heated system doubled that of the Fresh system, their cellular chlorophyll content was just half that of the Fresh cells (0.143 vs 0.297 µg 10^{-6} cells on day 10).

Fig. 9.3. The growth curves of the Fresh and Heated *C. vulgaris* in primary settled wastewater in the course of study. Each data point represented mean ± s.d. of triplicates.

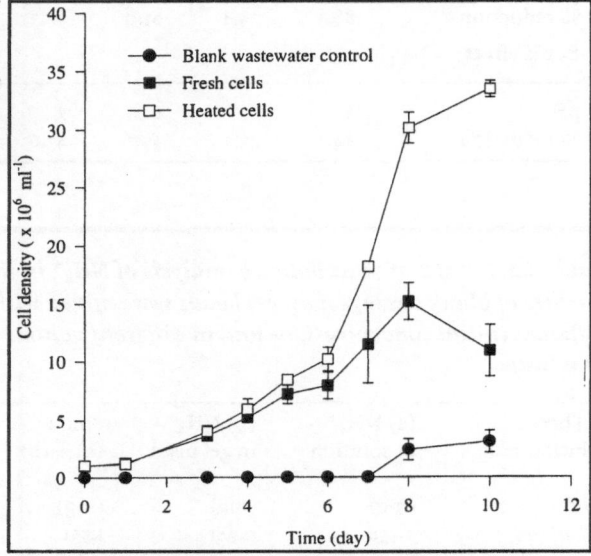

Fig. 9.4. The total chlorophyll content (mean ± s.d., n=3) of the Fresh and Heated cells cultured in primary settled wastewater in the course of study. Paired t-test showed no significant difference between the means (p < 0.05) except on day 6.

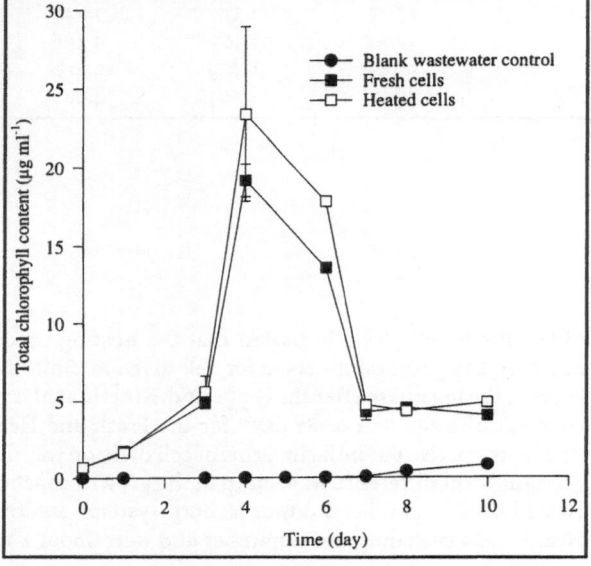

Photosynthetic ability of the Heated cells was retained, as revealed from the total chlorophyll content, a common photosynthetic indicator. There was virtually no difference between the Fresh and Heated systems in terms of kinetics and the order of magnitude of the photosynthetic pigment (Fig. 9.4). The chlorophyll increased gradually from the initial 0.7 μg ml^{-1} to about 5 μg ml^{-1} (about 7-fold) in the first 3 days. This was followed by a sharp increase on day 4, with the maximal value of 19 and 23 μg ml^{-1} recorded for the Fresh and Heated system, respectively. The chlorophyll content then dropped gradually back to 5 μg ml^{-1} on day 7 and maintained this level until the end of the experiment. Paired t-test analyses on the means showed that the total chlorophyll content was not significantly different between the Fresh and Heated system except on day 6. This suggested that the apparatus and machinery for the chlorophyll synthesis was well conserved in algal cells after the heating process.

Both the Fresh and Heated systems showed a significantly greater reduction in NH$_4^+$-N than the blank wastewater control. The net amount of NH$_4^+$-N uptake by the algal cells in the first 5 days was determined from the difference between the control and the Fresh or Heated system, and removal percentages are summarized in Table 9.3. The net uptake rates of NH$_4^+$-N were comparable between the two systems in the first 5 days, with the Heated cells showing a slightly higher uptake. The average uptake rate as determined by regression analysis gave the values of 3.47 and 4.13 mg l^{-1} day^{-1} for the Fresh and Heated system, respectively. The highest uptake rate was recorded in the first day in both systems with the corresponding initial specific uptake rates of 6.87 and 7.02 mg l^{-1} 10^{-6} cells day^{-1}, respectively (Table 9.3). Although the Heated cells had slightly higher rates than the Fresh cells, the difference was not statistically significant.

Table 9.3. Summary of the nutrient uptake (mean of triplicates) and percentage removal (in parentheses) by Fresh and Heated C. vulgaris *from wastewater*

Mean Uptake	NH$_4^+$-N		PO$_4^{3-}$-P	
	mg l^{-1} (% removal)		mg l^{-1} (% removal)	
Time (day)	Fresh	Heated	Fresh	Heated
1	6.8 (18.13)	7.3 (19.47)	1.18 (38.19)	1.13 (36.60)
2	9.5 (25.33)	13.6 (36.27)	1.20 (38.83)	1.39 (44.98)
3	12.6 (33.60)	16.4 (43.73)	1.21 (39.16)	1.40 (45.31)
4	17.0 (45.33)	21.4 (57.07)	1.34 (43.37)	1.63 (52.75)
5	20.4 (54.4)	24.5 (65.33)	1.54 (49.84)	2.19 (70.87)
Average uptake rate* (mg l^{-1} day^{-1})	3.47	4.13	0.162	0.268
Initial specific uptake rate# (mg l^{-1} 10^{-6} cells day^{-1})	6.87	7.02	1.19	1.08

*Average uptake rate was determined as the slope of the regression line of the net amount of nutrient uptake by algal cells vs time over the first 5 days.
#Initial specific uptake rate was defined as the net amount of nutrients removed by 10^6 cells on the first day.

The Fresh and Heated systems also showed a significant PO_4^{3-}-P depletion in the wastewater when compared to the control. This was especially noted on the first day when the drop was from initial 3.09 to 1.1 mg l^{-1} in both systems. The net PO_4^{3-}-P uptake and removal percentage by the algae are summarized in Table 9.3. Again, there was no significant difference in the mean uptake, average uptake rate and the initial specific uptake rate between the Fresh and Heated systems.

Results of nutrient uptake kinetics in terms of the depletion of NH_4^+-N and PO_4^{3-}-P from the wastewater and their average rate of uptake by the Heated cells were comparable with the Fresh cells. The highest specific uptake rates were recorded on the first day and agreed with reports in the literature.[21,49] Virtually no difference could be identified in terms of this maximal initial uptake rate for the nutrients. These implied that the physiology of nutrient uptake of the algal cells was also not affected by the heating process. In summary, various lines of evidence strongly indicated that the heating process associated with the carrageenan immobilization process imposed no adverse effect on the algal cells with respect to cell growth and multiplication, photosynthetic potential and nutrient uptake physiology.

Growth and Nutrient Removal by Carrageenan Immobilized C. vulgaris

The growth of *C. vulgaris* in the low cell stocking carrageenan beads followed a sigmoidal pattern, a lag followed by the log phase. Late log phase was observed on day 4, followed by a stationary phase from day 5 onward. The growth constant was 0.362 day^{-1}. This was in good agreement with the 0.347 day^{-1} of the Fresh cells and 0.382 day^{-1} of the Heated cells which started with the same initial total cell numbers.

For the high cell stocking beads, there was not much change in terms of cell number though the initial total number was 10-fold of the low cell stocking beads. The cell density varied just slightly at the initial density of 18×10^5 cells bead^{-1} throughout the experiment (Fig. 9.5). Growth in terms of cell multiplication was not perceived. The growth curve in Figure 9.5 suggested that the cells were in maintenance phase for most of the time. A similar effect was observed in alginate immobilized *C. emersonii* with similar cell stocking.[39] The growth of the algae at this high density might probably be light-limited due to self-shading.[50] Our previous study had demonstrated that free cultures of superconcentrated *C. vulgaris*, with an initial cell number equivalent to that of the high stocking beads in this study, could grow in a linear fashion for the first few days.[28] The problem of self-shading was partly compensated by the air mixing provided. In the present case of high cell stocking beads, the self-shading problem seemed unavoidable because the cells were packed in a smaller effective volume of 100 ml gel instead of 300 ml wastewater in the previous study.[28] This actually increased the relative density of the algal cells in the confined volume and thus intensified the shading problem. The situation was worsened by the fact that mixing in the present study could not alleviate the shading problem since the cells confined within the gel matrix and their movement was delimited within the matrix.

Despite the lack of increase in cell number, the total chlorophyll content of the immobilized *C. vulgaris* in the high cell stocking beads increased with a similar extent as the low stocking one (Fig. 9.6). Both the low and high stocking systems had a linear increase in the first 3 days at the corresponding rates of 0.0237 and

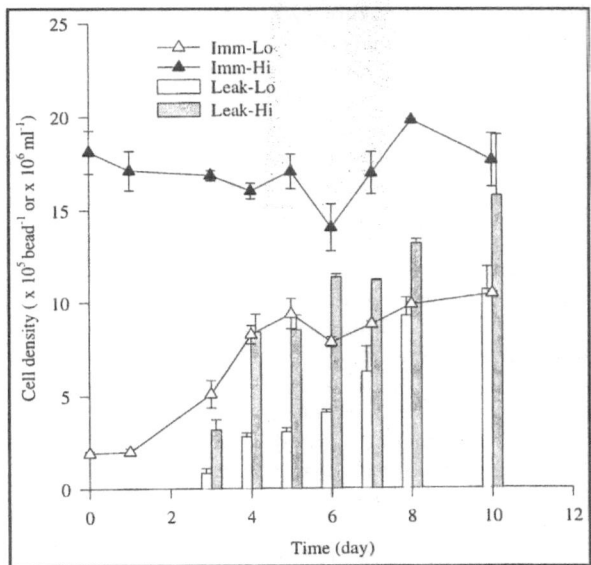

Fig. 9.5. The growth curves of carrageenan immobilized *C. vulgaris* with low (Imm-Lo) and high (Imm-Hi) initial cell stocking in the bead and of the leaked cells associated with beads of low (Leak-Lo) and high (Leak-Hi) cell stocking in the wastewater in the course of study. Each data point represented mean — s.d. of triplicates.

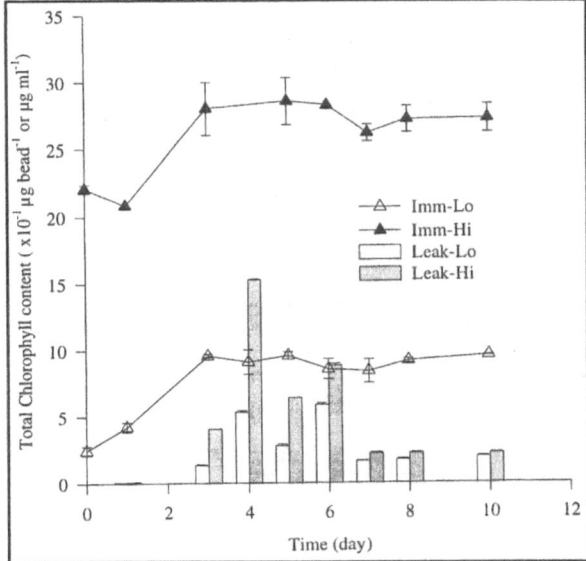

Fig. 9.6. The total chlorophyll content (mean ± s.d., n=3) of carrageenan immobilized *C. vulgaris* with low (Imm-Lo) and high (Imm-Hi) initial cell stocking in the bead and of the leaked cells associated with beads of low (Leak-Lo) and high (Leak-Hi) cell stocking in the wastewater in the course of study.

0.0196 µg bead^{-1} day^{-1}, respectively. The chlorophyll content was then maintained from day 3 onward at the level of 0.9 and 2.7 µg bead^{-1} in the low and high cell stocking beads, respectively. Apart from the active growing phase from day 1 to 3, the cellular chlorophyll in the high cell stocking beads was persistently higher than that in the low cell stocking of beads (average of 1.546 vs 1.005 µg 10^{-6} cells). This provided additional evidence that the self-shading effect was more serious in the high cell stocking beads and the algal cells compensated for it by a higher chlorophyll content.[51]

The depletion of NH_4^+-N is summarized in Figure 9.7. The NH_4^+-N in the blank wastewater control showed a reduction from 37.5 to 26 mg l^{-1} on day 5, accounting for 33.3% of the total. From day 5 onward, the NH_4^+-N in the control dropped sharply to zero on day 8. This was, however, matched by a coincident sharp increase in the NO_3^--N level from day 5 to day 8 with a maximum value of 22.1 mg l^{-1} (Fig. 9.8). The NO_3^--N level then dropped to 18.8 mg l^{-1} at the end of the experiment. The depletion of NH_4^+-N and the appearance of NO_3^--N showed that nitrification had taken place by the indigenous nitrifiers in the wastewater. The total inorganic nitrogen in the blank wastewater remained at about 20 mg N l^{-1}. Nitrification was not observed in all the immobilized systems, as no trace of NO_3^- was recorded (Fig. 9.8) and the inorganic nitrogen species was mainly NH_4^+-N. A rapid depletion of 12.5 mg l^{-1} (33%) NH_4^+-N was recorded in the first 5 minutes in all immobilized systems with or without cells. This was equivalent to a removal rate of 2.5 mg l^{-1} min^{-1} (or 178.6 M min^{-1}). This value was much higher than the 3.55 M min^{-1} as reported by Chevalier and de la Noüe.[52] Compared with the 43% being adsorbed by the gel beads, as reported earlier in the ionic interaction study, the initial reduction was mainly due to the NH_4^+-N adsorption within the gel bead rather than algal uptake. Adsorption took place initially at the surface layers of the gel beads. Prolonged incubation allowed NH_4^+-N ions to diffuse to the bead interior and ultimately saturated all the available adsorption sites of the beads. This accounted for about 65% NH_4^+-N reduction by the blank beads on day 2 (Table 9.4) and the relatively constant residual level of about 10 mg l^{-1} NH_4^+-N in the wastewater until the end of the study.

The algal uptake became prominent as time proceeded. Over 90% NH_4^+-N was removed from wastewater on day 2 by carrageenan beads stocked with algal cells. The high cell stocking beads had a slightly better removal efficiencies (Table 9.4). The algal uptake accounted for 30% of NH_4^+-N reduction in the first 2 days after reducing the portion due to bead adsorption. This was in good agreement with

Fig. 9.7. The residual NH_4^+-N (mean ± s.d., n=3) in the wastewater treated by carrageenan immobilized *C. vulgaris* with low (Imm-Lo) and high (Imm-Hi) cell stocking, blank wastewater and blank bead control.

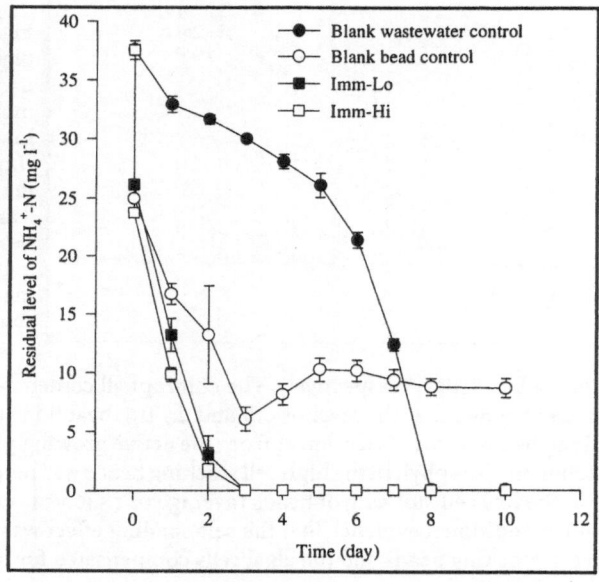

that of the Fresh or Heated cells (Table 9.3). Complete removal of NH_4^+-N from the wastewater was recorded from day 3 onward. It is noteworthy that despite the depletion of NH_4^+-N in the wastewater, cells in the beads kept growing in terms of number (Fig. 9.5) and chlorophyll content (Fig. 9.6). In other words, adsorbed or stored NH_4^+-N in the gel became available to algal cells. The carrageenan gel can thus serve as shuttle agents in preconcentrating NH_4^+-N initially and releasing it for algal growth later. In this way, the ionic property of the carrageenan gel enhanced NH_4^+-N removal from wastewater when compared with the Fresh cells, which took 6 days to attain 90% removal.

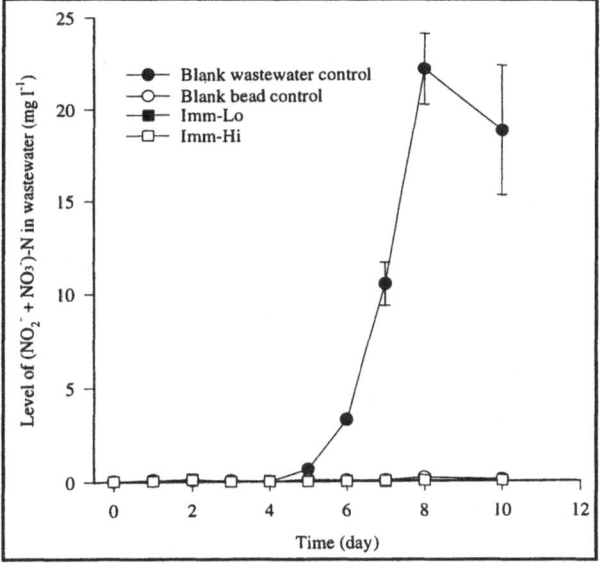

Fig. 9.8. The level of NO_x-N (mean ± s.d., n=3) in the wastewater treated by carrageenan immobilized *C. vulgaris* with low (Imm-Lo) and high (Imm-Hi) cell stocking, blank wastewater and blank bead control.

Table 9.4. Summary of the nutrient removal efficiencies (mean of triplicates) from wastewater by carrageenan immobilized C. vulgaris with low cell stocking (Imm-Lo), high cell stocking (Imm-Hi) and the Blank Bead Control

	Nutrient removal		
	Blank Bead	Imm-Lo	Imm-Hi
NH_4^+-N	(%)	(%)	(%)
5 minutes	33.3	33.3	33.3
Day 1	55.4	64.8	73.9
Day 2	64.8	92.0	95.2
PO_4^{3-}-P			
Day 1	40.4	89.3	82.3
Day 2	64.7	100	100

PO$_4$$^{3-}$-P removal by the immobilized systems was highly efficient when compared with the blank wastewater control (Fig. 9.9). The depletion profiles of all immobilized systems followed a hyperbolic pattern. Beads stocked with cells showed a sharper initial drop from the starting 3.09 to below 0.5 mg l^{-1} PO$_4$$^{3-}$-P. This accounted for over 80% PO$_4$$^{3-}$-P removal. Complete removal was recorded in day 2 by beads stocked with algal cells (Table 9.3). This was 6 days less than the Fresh or Heated system for a similar removal efficiency. The P removal was, however, independent of the cell stocking density of the beads. Particularly notable was the PO$_4$$^{3-}$-P reduction in the blank bead control system. A 40.4% PO$_4$$^{3-}$-P reduction was recorded in the first day. This increased to 64.8% on day 2 and a residual of 0.5 mg l^{-1} PO$_4$$^{3-}$-P was maintained in the blank bead system. This phenomenon suggested that some other kind of interaction between the bead and the PO$_4$$^{3-}$-P had been taking place in the context of wastewater. Further study was deemed necessary in order to define the possible interactions in this regard.

Concern and Limitation

Diffusion Limitation

Diffusion limitation of nutrients was believed to limit the cell growth at the periphery of the beads.[39] The results from this study on the ionic interaction suggested that the nutrient molecules are small enough to diffuse freely into the interior of the carrageenan gel beads. As a result, algal growth cells was not confined to the bead periphery. This was especially evident from the scanning electron micrographs. Cells were equally distributed on the equatorial plane which traversed the thickest layer of the low cell stocking bead (Fig 9.10a). This suggested that algal cells were capable of free division or multiplication in the bead interior. On the bead periphery, growth of the algal cells pushed some daughter cells out to the medium and freed them from the lattice of the gel matrix (Fig. 9.10b), leading to cell leaking from the beads to liquid medium. Thus, carrageenan as an immobilization matrix would not pose a growth limitation due to diffusion barrier of the gel. Instead, active growth on the periphery leads to the problem of cell leakage.

Cell Leakage

The leakage problem was one of the key concerns in cell immobilization since it obviates the primary purpose of delimiting viable cells in a confined matrix. Algal cells were observed in wastewater from day 3 onward in both immobilized systems of high and low cell stockings with a density of 1 x 10^6 and 3 x 10^6 cells ml^{-1}, respectively (Fig. 9.5). Growth of indigenous microalgal cells in blank wastewater has been reported after prolonged incubation[29] and observed from day 8 onwards in this study with a density of 1 x 10^6 cells ml^{-1} (Fig. 9.5). In terms of timing, it was unlikely that growth of indigenous algal cells in the wastewater accounted for the occurrence of cells in the immobilized systems. Rather, it was the consequence of the active growth at the periphery of the carrageenan beads. Robinson et al[39] reported that leakage of *C. emersonii* from alginate beads was most serious in the first few days. The extent of leakage was similar, regardless of cell loading, and the leaked cells had higher specific growth rates than the free cells.[39] The extent of cell leakage associated with the high cell stocking in this study agreed with the findings of Robinson et al.[39] The growth constant (0.978 day^{-1}) of the leaked cells in the high cell stocking was significantly higher than the Fresh cells (0.342 day^{-1}). For

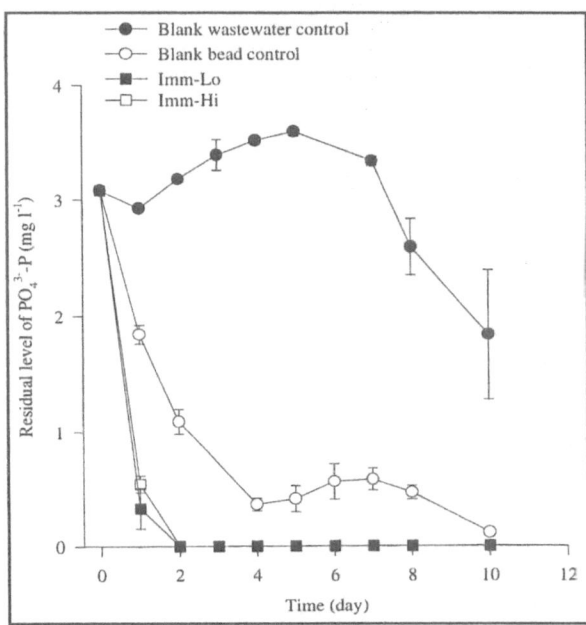

Fig. 9.9. The residual PO_4^{3-}-P (mean ± s.d., n=3) in the wastewater treated by carrageenan immobilized *C. vulgaris* with low (Imm-Lo) and high (Imm-Hi) cell stocking, blank waste-water and blank bead control.

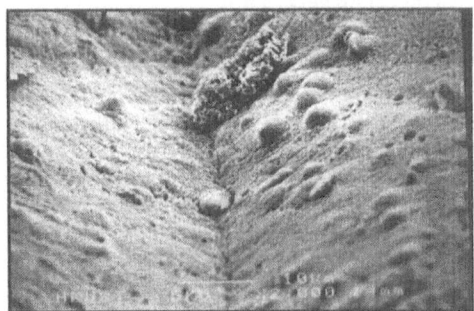

a.

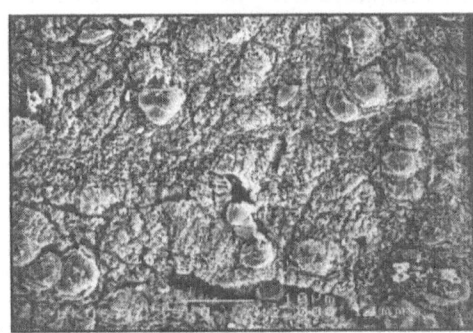

b.

Fig. 9.10. Scanning electron micrographs of *C. vulgaris* immobilized in the (a) equatorial plane and (b) periphery of the carrageenan bead.

the low cell stocking beads, cell leakage was also observed on day 3 (Fig. 9.5), but to a much less extent than the high cell stocking beads. The growth constant of the leaked cells from the low cell stocking beads was 0.337 day⁻¹ which was just comparable to that of the Fresh cells. Leakage in the high cell stocking beads was expected as more cells will be distributed on the bead periphery. The greater number of daughter cells upon multiplication have the chance of being freed from the gel matrix. Continuous leakage and new growth of the leaked ones in the medium intensified the leakage problem. Results of this study suggest that low cell stocking carrageenan beads is preferred to high cell stocking beads so as to: 1) minimize the cell leakage from the bead; 2) provide more space for algal growth; and 3) save material and energy since the efficiencies of the low cell stocking beads were not much different from the high cell stocking beads in terms of N and P removal. However, algal growth will eventually saturate the bead volume regardless of the initial cell density. Leakage is therefore unavoidable as time increases and new batch of algal beads has to replace the growth-saturated ones. Future studies should try to determine: 1) the critical cell density within the bead at which cell leakage is tolerable; and 2) the means of utilizing the algal beads.

Conclusion

This study provides clear evidence on the physicochemical adsorption of the NH_4^+ with the carrageenan gel beads. The adsorption process was characterized with a dependent pattern on the volume ratio of the gel bead : ionic solution, and was saturable and independent of the pH. The carrageenan bead could therefore act as a shuttle agent in temporarily accumulated NH_4^+ and release the adsorbed NH_4^+ when its ambient conditions are depleted of NH_4^+. The heating effect due to the immobilization process of the carrageenan gel did not pose any adverse effect on *C. vulgaris* in terms of growth, photosynthetic capability or the nutrient uptake kinetics from the wastewater. Carrageenan-immobilized *C. vulgaris* demonstrated efficient simultaneous removal of N and P from wastewater. The removal was independent of the cell density stocked in the beads. Hence, low cell stocking beads is the preferred method to minimize the leakage problem. Scanning electron micrographs provided evidence that growth of *C. vulgaris* was not limited within the gel bead and hence, diffusion would not be a limiting factor for growth in the carrageenan bead.

Acknowledgments

The study was supported by grants awarded from the Hong Kong Croucher Foundation and the Industrial Technology Development Council of Hong Kong.

References

1. Oswald WJ, Gotaas HB, Golueke CG et al. Algae in waste treatment. Sewage and Industrial Wastes 1957; 29(4):437-457.
2. Pearson HW. Algae associated with sewage treatment. In: Dasilva EJ, Dommergues YR, Nyns EJ, Ratledge C, eds. Microbial Technology in the Developing World. Oxford: Oxford University Press, 1987:260-288.
3. Fallowfield HJ, Garrett MK. The treatment of wastes by algal culture. Journal of Applied Bacteriology Symposium Supplement 1985; 187S-205S.
4. Oswald WJ. Ponds in the twenty-first century. Wat Sci Tech 1995; 31(12):1-8.
5. Mara DD. Waste stabilization ponds: effluent quality requirements and implications for process design. Wat Sci Tech 1996; 33(7):23-32.

6. Oswald WJ, Gotaas HB. Photosynthesis in sewage treatment. American Society of Civil Engineers Transactions 1955; 73-105.

7. Shelef G, Soeder CJ. Algae Biomass: Production and Use. Amsterdam: Elsevier/North Holland Biomedical Press, 1980: 852.

8. Borowitzka MA, Borowitzka LJ. Microalgal Biotechnology. Cambridge: Cambridge University Press, 1988:447.

9. De la Noüe J, De Pauw N. The potential of microalgal biotechnology: a review of production and uses of microalgae. Biotech Adv 1988; 6:725-770.

10. Lavoie A, de la Noüe J. Hyperconcentrated cultures of *Scenedesmus obliquus*: A new approach for wastewater biological tertiary treatment. Wat Res 1985; 19(11):1437-1442.

11. Tam NFY, Wong YS. Wastewater nutrient removal by *Chlorella pyrenoidosa* and *Scenedesmus* sp. Environ Pollut 1989; 58:19-34.

12. Middlebrooks EJ, Porcella DB, Gearheart TRA et al. Techniques for algae removal from wastewater stabilization ponds. Journal WPCF 1974; 46(2):2676-2695.

13. Al-Shayji YA, Puskas K, Al-Daher R. Production and separation of algae in a high-rate ponds system. Environment International 1994; 20(4):541-550.

14. Bustos Aragon A, Borja Padilla R, Fiestas Ros de Ursinos JA. Experimental study of the recovery of algae cultured in effluents from the anaerobic biological treatment of urban wastewaters. Resources, Conservation and Recycling 1992; 6:293-302.

15. Lavoie A, de la Noüe J. Harvesting microalgae with chitosan. J World Maricul Soc 1983; 14:685-694.

16. Nigam BP, Ramanathan PK, Venkataraman LV. Application of chitosan as a flocculant for the cultures of green alga: *Scenedesmus acutus*. Arch Hydrobiol 1980; 88(3):378-387.

17. Yang PY, Cai T, Wang ML. Immobilized mixed microbial cells for wastewater treatment. Biological Wastes 1988; 23:295-312.

18. Bucke C. Immobilized cells. Philosophical Transactions of the Royal Society, London, Series B 1983; 300:369-389.

19. Kaya VM, Picard G. The viability of *Scenedesmus biceelularis* cells immobilized on alginate screens following nutrient starvation in air at 100% relative humidity. Biotechnol Bioeng 1995; 46(5):459-464.

20. Tam NFY, Lau PS, Wong YS. Wastewater inorganic N and P removal by immobilized *Chlorella vulgaris*. Wat Sci Tech 1994; 30:369-374.

21. Mallick N, Rai LC. Influence of culture density, pH, organic acids and divalent cations on the removal of nutrients and metals by immobilized *Anabaena doliolum* and *Chlorella vulgaris*. World Journal of Microbiology and Biotechnology 1993; 9:196-201.

22. Travieso T, Benitez F, Dupeiron R. Sewage treatment using immobilized microalgae. Bioresource Technol 1992; 40:183-187.

23. Megharaj M, Pearson HW, Venkateswarlu K. Removal of nitrogen and phosphorus by immobilized cells of *Chlorella vulgaris* and *Scenedesmus bijugatus* isolated from soil. Enzyme Microb Technol 1992; 14:656-658.

24. Cañizares RO, Rivas L, Montes C et al. Aerated swine-wastewater treatment with k-carrageenan-immobilized *Spirulina maxima*. Bioresource Technol 1994; 47:89-91.

25. Chevalier P, de la Noüe J. Wastewater nutrient removal with microalgae immobilized in carrageenan. Enzyme Microb Technol 1985; 7:621-624.

26. Kaya VM, Picard G. Stability of chitosan gel as entrapment matrix of viable *Scenedesmus bicellularis* cells immobilized on screens for tertiary treatment of wastewater. Bioresource Technol 1996; 56:147-155.

27. Tam NFY, Wong YS, Liong E. Algal growth and nutrient removal in Hong Kong domestic wastewater. In: Hills P, Keen R, Lam KC, Leung CT, Oswell MA, Stokes M, Turner E, eds. Pollution in the Urban Environment. POLMET 88, vol. 2. Hong Kong: Vincent Blue Copy Co. Ltd., 1988:621-628.

28. Lau PS, Tam NFY, Wong YS. Effect of algal density on nutrient removal from primary settled wastewater. Environ Pollut 1995; 89:59-66.

29. Lau PS, Tam NFY, Wong YS. Wastewater nutrients removal by *Chlorella vulgaris*: Optimization through Acclimation. Environ Technol 1996; 17:183-189.

30. Lau PS, Tam NFY, Wong YS. Influence of organic-N sources on an algal wastewater treatment system. Resources, Conservation and Recycling 1994; 11:197-208.

31. Tam NFY, Wong YS. Effect of ammonia concentrations on growth of *Chlorella vulgaris* and nitrogen removal from media. Bioresource Technol 1996; 57:45-50.

32. Azov Y, Shelef G. Operation of high-rate oxidation ponds: theory and experiments. Wat Res 1982; 19:1153-1160.

33. Azov Y, Shelef G. The effect of pH on the performance of high rate oxidation ponds. Wat Sci Tech 1987; 19(12):381-383.

34. Goldman JC, Jenkins D, Oswald WJ. The kinetics of inorganic carbon limited algal growth. Journal WPCF 1974; 46(12):2785-2787.

35. Azov Y. Effect of pH on inorganic carbon uptake in algal cultures. Applied and Environ Microbiol 1982; 43(6):1300-1306.

36. Olaizola M, Duerr EO, Freeman DF. Effect of CO_2 enhancement in outdoor production system using *Tetraselmis*. J Appl Phycol 1991; 3:363-366.

37. Talbot P, Lencki RW, de la Noüe J. Carbon dioxide absorption characterization of a bioreactor for biomass production of *Phormidium bohneri*: comparative study of three types of diffuser. J Appl Phycol 1990; 2:341-350.

38. Jeanfils J. Immobilization of whole cells of green algae or cyanobacteria in insoluble matrices; morphological observation and nitrite reductase activities of immobilized cells. Arch Biol 1986; 97:209-222.

39. Robinson PK, Dainty AL, Goulding KH et al. Physiology of alginate-immobilized *Chlorella*. Enzyme Microb Technol 1985; 7:212-216.

40. Robinson PK, Goulding KH, Mak AL et al. Factors affecting the growth characteristics of alginate-entrapped *Chlorella*. Enzyme Microb Technol 1986; 8:729-733.

41. Lau PS. Wastewater nitrogen and phosphorus removal by free and immobilized microalgal systems. Ph.D. Thesis. Hong Kong: Hong Kong University of Science and Technology, 1995: 283.

42. Lukavsky J, Komárek J, Lukavská A et al. Metabolic activity and cell structure of immobilized algal cells (*Chlorella, Scenedesmus*). Arch Hydrobiol Suppl 1986; 73(2):261-279.

43. Guiseley KB. Chemical and physical properties of algal polysaccharides used for cell immobilization. Enzyme Microb Technol 1989; 11:706-716.

44. Tosa T, Sato T, Mori T et al. Immobilization of enzymes and microbial cells using carrageenan as matrix. Biotechnol Bioeng 1979; 21:1697-1709.

45. Marine Colloids Inc. Carrageenan Monograph No.1, Springfield, New Jersey, 1977.

46. Kierstan MPJ, Coughlan MP. Immobilization of cells and enzymes by gel entrapment. In: Woodward J, ed. Immobilized Cells and Enzymes, a Practical Approach. Oxford: IRL Press, 1985:39-48.

47. Yung KH, Mudd JB. Lipid synthesis in the presence of nitrogenous compounds in *Chlorella pyrenoidosa*. Plant Physiol 1966; 41:506-509.

48. APHA (American Public Health Association), American Water Works Association and Water Pollution Control Federation. Standard Methods for the Examination of Water and Wastewater. 17th ed. Washington: American Public Health Association, 1989:1524.

49. Jeanfils J, Canisius MF, Burlion N. Effect of high nitrate concentrations on growth and nitrate uptake by free-living and immobilized *Chlorella vulgaris* cells. J Appl Phycol 1993; 5:369-374.
50. Fogg HJ. Algal Cultures and Phytoplankton Ecology. 2nd ed. Wisconsin: The University of Wisconsin Press, 1975:175.
51. Darley WM. Algal Biology : A Physiological Approach. Basic Microbiology Vol. 9. Oxford: Blackwell Scientific Publications, 1982:168.
52. Chevalier P, de la Noiie J. Efficiency of immobilized hyperconcentrated algae for ammonium and phosphorus removal from wastewaters. Biotech Lett 1985; 7(6):395-400.

CHAPTER 10

Dynamics of Picoplankton and Microplankton Flora in the Experimental Wastewater Stabilization Ponds in the Arid Region of Marrakech, Morocco and Cyanobacteria Effect on *Escherichia coli* and *Vibrio cholerae* Survival

Nour-Eddine Mezrioui and Brahim Oudra

Introduction

Several treatment systems are available to reduce the load of organic matter and bacterial pollution in sewage water. Among them, the technique of stabilization pond treatment appears to be a good choice due to its simple management and its satisfactory level of treatment.[1]

The assimilative capacity of stabilization pond systems depends on the interaction between the ecosystem's various biotic and abiotic factors.[1-2] Among the biotic factors, algae plays a key role in the self-purification process. There is great interest in acquiring more information about properties and laws governing biological sewage purification using stabilization ponds, and through dynamic studies of algae in this type of system. In fact, pond micro-algae dynamics and biotic-abiotic components relationship studies have lead to a greater understanding of wastewater stabilization pond operation. This research topic has occupied much of the hydrobiological literature.[3-7]

Wastewater Treatment with Algae, edited by Yuk-Shan Wong and Nora F.Y. Tam.
© Springer - Verlag and Landes Bioscience 1998.

Aside from their importance in supplying oxygen for bacterial oxidation of organic matter, algae affect treatment processes in other ways, including heavy metals removal,[8] direct assimilation of organic matter[8-9] and nutrient removal.[5,10-11] Regarding the latter, several recent studies using microalgae for wastewater tertiary treatment have brought about satisfactory results.[12-14]

Seasonal fluctuations in bacterial abundances evaluated during stabilization pond treatment have been attributed to algae dynamics. Numerous researchers have analyzed the die-off of enteric bacteria.[63-64] Additionally, some information exists on the specific interaction between algae and pathogenic bacteria in stabilization ponds which contain appreciable concentrations of algae.[15]

In the experimental stabilization ponds of Marrakech, Morocco which function under arid climate, dynamics of non-O1 *Vibrio cholerae* (Vc) and fecal coliforms (FCs) were studied.[16] The results showed that the dynamics of Vc in this system followed a seasonal pattern that was the inverse of that displayed by FCs. Variation in Vc numbers, with a summer maximum and a winter minimum, was considered in relation to biotic and abiotic factors. An understanding of the die-off and aftergrowth characteristics of Vc and FCs in the presence of various algal species is an important design consideration.[16]

This investigation describes the dynamics of some algae species in the experimental stabilization ponds of Marrakech, Morocco, which function under arid climate. It also aims to verify whether the inverse relationship observed between Vc and FCs abundance depends upon blooms of picocyanobacteria which occur during summer periods.

Micro-Algae in Wastewater Stabilization Ponds (A Review)

Composition of Algal Flora Ponds

In both facultative and high rate oxidation ponds, algal development is clearly the second step in sewage self-purification, after bacterial establishment. However, in habitats where the water quality is unfavorable to great algal diversity, we have noted the extraordinary pollution tolerance of few species which are representative of this micro-algal population.

The first report on phytoplankton dynamic studies in sewage ponds was in 1961.[17] By now it is quite evident that algal composition should change principally with climatic condition and pond effluent characteristics. According to many reports in the literature (Table 10.1), the Euglenophyta and Chlorophyta (Volvocales and Chlorococcales), represent the greatest number of algal species. However, filamentous Cyanobacteria and Diatoms are not considered as phytoplanktonic pond species,[6-18] but rather, they characterize the phytobenthos pond flora.[5] Although the diversity of wastewater effluent quality and climatic conditions may vary, some algal genera have been found to be common to wastewater stabilization ponds: *Ankistrodesmus, Scenedesmus, Pandorina, Chlorella, Chlorogonium* (Chlorococcales), *Chlamydomonas* (Volvocales), *Euglena, Phacus, Trachelomonas* (Euglenophyta) and *Nitzschia* (Diatoms). In comparison with eutrophic freshwater natural ecosystems, sewage pond microflora have been found to be composed of less taxa number. It is interesting to note that Cyanobacteria blooms have been reported in high rate oxidation ponds by Lincoln and Hill.[46] Their recent study on facultative sewage pond Cyanobacteria will be one of the first reports in this area.

Table 10.1.

	1	2	3	4	5	6	7	8	9	10	11	Ref. No	Authors and locality
Cyanophyta													
Oscillatoria	+	+	+				+		+			1	Rasche, 1970 St.Ames (USA)
Anabeana													
Phormidium			+		+								
Dactylococcus					+		+					2	Kalisz, 1973 Kielce (Poland)
Synechococcus				+			+						
Synechocystis				+									
Euglènophyta												3	Kalisz, 1973 Czetochwa (Poland)
Euglena	+		+	+	+	+	+	+	+	+	+		
Phacus			+			+	+	+		+	+		
Trachelomonas			+				+	+		+	+	4	Prat, 1982 Meze (France)
Cryptophycea													
Cryptomonas				+			+				+		
Chlorophyta												5	D. Delfosse, 1976 Pres Duhm (France)
Chlorococcus						+							
Ankistrodesmus	+	+	+	+	+	+	+	+		+	+		
Scenedesmus	+			+	+		+	+	+	+	+	6	Patil, 1975 Dhawar (India)
Tetradesmus			+										
Micractinuim	+			+	+				+				
Chlorella	+			+		+	+		+		+	7	Delannoy, 1973 Warvin (France)
Coelastrum	+	+											
Oocystis	+	+							+				
Pyrobotrys						+						8	Steiner, 1984 Rauchaux (France)
Selanastrum													
Pandorina		+	+	+			+	+	+		+		
Chlamydomonas	+		+	+	+	+	+	+	+	+		9	Steiner, 1984 Borey (France)
Chlorogonium	+		+				+			+	+		
Eudorina	+			+						+			
Tetraderon				+						+		10	Steiner, 1984 Etival (France)
Spondylomorum		+	+							+			
Westella		+	+										
Diatoms												11	Cauderellier, 1985 (France)
Nitzschia	+	+	+										
Navicula	+	+				+							
Pinnularia						+							
Hantzshia	+												
Synedra		+	+			+	+						
Gomphonema	+	+											
Cyclotella													
Achnanthes						+							
Stephanodiscus					+								

Role of Micro-Algae on Biological Sewage Treatment

Sewage self-purification in algal treatment systems has been discussed by many authors;[19-21] the process is summarized as follows: after heterotrophic bacterial mineralization, some components such as carbon dioxide (CO_2) and mineral salts are produced. These compounds are used by algae for their own development and growth. The algal photosynthesis supplies the oxygen required for the decomposition of sewage organic waste matter by bacterial oxidation.[22-24] Oxygen production, breakdown of heavy metals and plant nutrients removal are now generally recognized applications of algae,[8,10,21,25] but algal contributions in disinfection and toxic organic removal are less developed.[21] Complete wastewater treatment may be defined as the degree of treatment necessary to restore the physical, chemical and biological purity of the water originally used. More efficient treatment would involve removal of not only all floating and settling solids and all biodegradable organics, but also safe, dependable disinfection of the water prior to reuse. Disinfection would induce the removal of disease-causing agents.

Factors Influencing Algal Growth in Wastewater Stabilization Ponds

For self-purification operations, it is first necessary to determine if the waste effluent quality (organic load) can, under normal conditions of time and temperature, support algal growth and photosynthetic activity sufficient to supply enough oxygen to meet the biochemical oxygen demand (BOD). Numerous studies indicate that the ratio of oxygen synthesized is variable, and depends mainly on the composition of algae, which changes with climatic conditions, waste effluent quality, stabilization pond operating modes (deep ponds), algal cell size and density, and the light and water mixing relationship.[5,21,26-30] For example, Roques[30] developed a relationship between temporal variation of oxygen concentration in water pond and solar radiation and deep pond.

$dC/dt=$ øS.nP / 3.68 h
$dC/dt=$ Temporal variation of oxygen concentration
øS= Solar radiation (cal / cm^2 / s)
nP= Photochemical balance (total energy not used by algal photosynthetic activity)
h= Average pond depth (m)
3.68= Constant energy ratio (cal) : (Free oxygen/light).

In wastewater stabilization ponds, toxic substances and zooplankton grazing are the major factors that influence algal growth and distribution in stabilization ponds.

Zooplankton grazing can sharply influence algal population "tunovers" in sewage ponds.[28,31] This phytoplankton dynamic control, essentially by Rotifers and Cladoceran species, has been described by many authors as harmful.[32-34] However, this action leads to rapid nutrient recycling and stimulates primary production.[35-36] Only some algae taxons undergo preferential grazing pressure, however.[28-37]

Domestic sewage usually contains a sufficient amount of nutrients required for algal growth, but micro-algae are sensitive to numerous toxic substances conveyed by domestic and industrial wastewater, such as heavy metals, pesticides, phenols and others. Other substances are toxic at higher levels such as flouride, nitrate, ammonia and sulphide hydrogen. Algal toxic effects of ammonia (NH_3) and ammonium (NH_4) at higher concentrations have been studied by many authors.[7,28,84-86] Such acute toxicity is closely related to water temperature and pH.

The increasing free-NH_3 concentration was correlated with higher temperature and pH.[87-88] Using algae species growing in wastewater stabilization ponds, the ammonium toxicity test showed that *Chlorella* was less sensitive than *Euglena*.[86] Pearson et al[7] tested the sensitivity of the major microalgae sewage pond species to ammonia and ammonium concentrations, resulting in the following tolerance sequence: *Chlorella, Scenedesmus, Euglena* and *Chlamydomonas* (Table 10.2). These toxic components in wastewater stabilization ponds influence algal photosynthesis directly and consequently, influence the sewage self-purification process.

Among the toxic substances produced under anaerobic conditions in water facultative ponds, sulfate and sulphide hydrogen at higher concentrations influence algal photosynthesis.[7,87] In this case, oxygen depletion lead to Thiobacteria proliferation and sulphide hydrogen production. This phenomenon, which sharply inhibits algal photosynthesis,[88] appears to be reversible.[89] According to a study by Pearson et al[7] on microalgae, acute toxicity on sulfate and sulphide hydrogen, we can arrange the species sensitivity sequence as follows: *Euglena, Scenedesmus, Chlorella* and *Chlamydomonas* (Table 10.3).

Dynamics of Picoplankton and Microplankton Flora in the Experimental Wastewater Stabilization Ponds in Marrakech, Morocco

This experiment was carried out at the wastewater spreading zone of Marrakesh, Morocco (31° 36' N, 8° 02' W). Annual precipitation and periods of drought are around 240 mm and 7 months, respectively. Mean minimum and maximum air temperatures are 5°C and 38°C, respectively. Average solar energy is 17 MJ/m²/day

Table 10.2. Ammonia concentrations causing inhibition of photosynthesis in four waste stabilization pond algae[7]

Alga	Ammonia causing 50% inhibition at pH 8.5 mg/l (NH_4-N)	mM	Free NH_3 causing 50% inhibition % at any pH mM
Chlorella	356	25.4	3.84
Euglena	87	6.2	0.94
Scenedesmus	150	10.7	1.62
Chlamydomonas	81	5.8	0.88

Table 10.3. Sulphide concentration causing inhibition of photosynthesis in the algae of four waste stabilization ponds[7]

Algae	Total sulfate concentrations at pH 7.25 μM	Free H_2S causing 50% inhibition at any pH μM
Chlorella	27.5	10.45
Euglena	58.3	22.15
Scenedesmus	80.0	30.40
Chlamydomonas	118.3	44.95

and monthly average insolation is about 216 hours, equivalent to 7-8 h/day. The experimental treatment plant studied receives sewage of essentially domestic origin. This form of stabilization pond treatment consists of two ponds linked in series, each of which is 0.25 hectares (ha) in area.

The experiment started in August 1985, and the first experimental phase lasted until December 1987. Each pond did not exceed 1.60 m in depth and theoretical retention time was 50 days while the installation was operating in a facultative mode. However, during the second phase from February 1988 to May 1989, the first pond's depth became 2.30 m while the second pond was unchanged. With these new dimensions, the system was able to receive 380 m³/day of raw flow, and the theoretical retention time was reduced to 22 days. This new situation changed the nature of the treatment process, which turned into a mixed "anaerobic-facultative" mode. In this chapter, we present only the results of the second experimental phase. In order to characterize wastewater pond inflow and outflow, some physicochemical analyses were done simultaneously with biological sampling using the methodology described by Rodier.[90]

Both Nanomicroplankton (2-200 μm) and Picoplankton (0.2 - 2 μm) samples from each pond were taken, respectively, with a 1-liter Van-Dorn bottle and a 2-m long PVC tube with a 60-mm internal diameter. Additional details on the sampling methodology may be found in references 38-39. The nano-microplankton algal biomass was estimated in two ways: 1) Direct count of cell number with an inverted microscope and using the Utermöhl technique;[40] and 2) Chlorophyll a concentration (Chl. a) determined by Lorenzen's[41] spectrophotometric equations, using 90% acetone as solvent extraction. The picoplanktonic biomass was followed only by determination of Chl. a concentration. Inverted microscope cell counting is not accurate, however, due to smaller cell size (0.2-2 μm). Nevertheless, the Whatman GF/C glass fiber filter (1.2 μm) for Chl. a analysis presented problems in collecting total phytoplankton as the cells, particularly the picoplanktonic cells, were able to pass through the Whatman GF/C filter.[42] This is one of the reasons many authors use epifluorescence microscopy for picoplanktonic cell counting.[43-44]

In order to follow the evolution of this picophytoplanktonic fraction, we used a method which required two different filters. After thoroughly mixing a vertical pond global sample taken by PVC tube (to avoid heterogeneous vertical distribution), one sub-sample (40 ml-100 ml) was filtered through a Whatman GF/C glass fiber filter (1.2 μm), which removed the Nanomicroplanktonic algae; the remaining filtrate was filtered again through Millipore filter type BA (0.45 μm), which presumably removed the picoplanktonic cells. Afterwards, both filters were destined for Chl.a extraction and analysis. This approach allowed us to evaluate the percentage contribution of picoplankton to total algal biomass.

Effect of Microalgae Growing on Wastewater Batch Culture on *Escherichia coli* and *Vibrio cholerae* Survival

The physiological state of *Vibrio cholerae* may be affected by various conditions during their lifetime in an aquatic environment.[57-59] Phytoplankton action is considered to be one of the most important biological factors controlling bacterial growth in aquatic habitats.[60-61] Various studies on biological interactions between bacteria and algae have confirmed their positive or negative mutual influence.[62-63]

Islam et al[64] observed that *Vibrio cholerae* can extend their survival time in an artificial aquatic environment in association with a green algae, *Rhizoclonium fontanum*.

This association is considered as a principal factor in sewage treatment systems, leading to biological waste purification efficiency.[21] Domestic sewages carry not only an organic load, but also a significant pathogenic bacteria load as *V. cholerae*. In wastewater, temporal evolution of *V. cholerae*, which is negatively correlated with fecal contamination, shows low levels during the cold season and high levels in the warm season.[65-67] Understanding the die-off and aftergrowth characteristics of *V. cholerae* and indicator bacteria in the presence of various algal species is important in system design. The available data do not yet clarify the effect of algal species on *V. cholerae* growth in wastewater treatment systems. In wastewater stabilization ponds of Marrakesh, high levels of *V. cholerae* were noted during the summer;[16] these levels coincided with picocyanobacteria blooms observed in the same study site.[68] In this investigation, we studied algae isolated from a wastewater stabilization pond, and the effects of algal growth. The effects were followed for a short period in order to appreciate how these algae can affect the behavior and survival of *E. coli* and *V. cholerae* in such wastewater treatment systems.

In this study, the bacterial strains tested were *E. coli* isolated from stabilization pond influent and *V. cholerae* E1 *tor ogawa* serotype (clinical strain). The experimental procedure used followed that described by Mezrioui et al.[91]

As for algal strains, the culture used was uni-algal, but not axenic. *Chlorella sorokiniana* (SOROKIN & KRAUSS) (green algae) was isolated from wastewater stabilization ponds of Marrakech, Morocco and maintained in culture on a solidified mineral medium suggested by Dauta.[69] The cyanobacteria strains *Synechococcus elongatus* NAGELI (Sync) and *Synechocystis parvula* PERIFILIEW (Synst) were also isolated from the same study site and were cultivated on BG 13 medium.[70] These microalgae cells were maintained in exponential phase by systematic replication (each 3 days) in a specific growth medium. In order to have a synchronized inoculum, the algal cells were suspended at a few ml of filtered wastewater and stored in darkness for 24 hours.

Algal growth (increase in cell number) was estimated by direct counting using an Hemacytometer-Malassez, 0.2 mm deep.[71] Growth rates were determined by the expression $\mu = Ln\ N_1/N_0$, in which N_0 was the concentration of starter cells and N_1 was the number of cells after 24 h of incubation, and the doubling time (dt) was calculated from $dt = 0.69/\mu$.[72] Enumeration on bacteria was done by indirect count of colony forming units (CFU) on a selective media: TCBS agar (Difco) incubated at 37°C for 24 hours (*V. cholerae*) and TTC-Tergitol 7 agar incubated at 44.5°C for 24 h (*E. coli*). Bacterial and algal growth counts were done twice daily after each clear and dark period. The pH was simultaneously measured using an ORION Research (model 601A) pH meter.

Results and Discussion—Sewage Physicochemical Characteristics

The pond systems discussed here are used mainly for domestic sewage treatment. The average delivery of sewage inflow is 4.4 l/s, corresponding to an average load equal to 109 Kg of BOD per day and 213 Kg of COD per day. The theoretical water retention time is 13 days and 9 days, respectively, for the first pond and the second pond. The physicochemical parameters of the pond influents and effluents are provided in Table 10.4.

Because of the important input of solid suspended matter and excessive microbial production, the euphotic zone did not exceed 50 cm. Therefore, light penetration was certainly a limiting factor for autotrophic activity, especially in the ponds' deep layers. Temperature fluctuations have shown similar trends in the surface layers of both ponds. The water temperature in Spring and early Autumn ranged from 15°-25°C, but went above 25°C (30-31°C, maximum) over the duration of the Summer. In general, higher pH were observed in the second pond, particularly in Summer. The pH decreased slightly during the cold period; this may be due to a decrease in algal activity. Although the first pond's depth (2.30 m), dissolved oxygen (DO) data indicated incomplete depletion, the average percentage oxygen saturation never exceeded 50%; however, it reached over 350% in the second pond. As for organic load, the system allowed for 53% of COD and 49% for BOD-load removal. The maximum efficiency in removal occured during hot periods. It appears that nutrient concentration (PO_4 and NH_4) was not significantly affected by biological activity, so perhaps other factors control the concentration of nutrients in this system. These last results agreed with those of Bouarab.[45]

Composition of Algal Flora

Nine genera and thirteen species of algae were identified in the sewage ponds (Table 10.5). The Euglenophyta represented the greatest number of species. The presence of Volvoccales, Chroococcales and Euglenales among phytoplankton flora of these ponds is in agreement with other reports in the literature,[3,6,28] although Chroococcales are rarely cited in sewage ponds.[46] It is interesting to note that

Table 10.4. Physicochemical characteristics for Marrakech sewage pond effluents

Parameters	Inflow	Pond 1	pond 2	outflow
Transparency, EZ (cm)	-	26	31	-
Temperature, °C	27.1	23.5	23.6	-
pH	7.6	8.05	8.25	-
Dissolved oxygen (DO mg/l)	0.95	8.5	10.8	5.7
% Oxygen saturation	17.9	107.3	140.6	66.5
Orthophosphate PO_4 (mgP/l)	8.25	5.62	6.27	9
Ammonia NH_4 (mg N/l)	38	37.6	25.8	35
Biochemical Oxygen Demand*	286	-	-	146
Chemical Oxygen Demand **	560	328	295	264
Total suspended matter (mg/l)	333	218	212	169

*BOD 5 days (mg O_2/l); ** COD (mg/l)

Oscillatoriales and Pennales species are never dominant among planktonic flora, while pennate flora and filamentous blue-green algae dominate the pond benthic flora (Oudra, unpublished). This result is confirmed by studies in other countries.[3,5]

The algal flora underwent notable qualitative and quantitative seasonal "turnovers." Some algae genera, such as *Euglena*, *Chlamydomonas* and *Chlorella*, were the most persistent and dominant species in the phytoplankton, except in the second pond, especially during Summer. When chroococcoid Cyanobacteria (*Synechocystis* and *Synechococcus* genera) produce "blooms", they are associated with Chlorococcales (*Micractinium pussillum*, *Coelastrum* sp and *Chlorella sorokiniana*.). It is important to note that during the cold period, as a result of dissolved oxygen decreases essentially in the primary pond, thiobacteria proliferation occured and generated production of sulphide hydrogen (H_2S) (Khamam, unpublished). These phenomena induce a sharp drop in qualitative and quantitative algae, but it appears that *Chlamydomonas* genus can persist and dominate, which reflects its tolerance to high sulphide concentration.[7] In contrast, the occurrence of seasonal-specific algae groups for each pond reflects the fact that algal periodicity in wastewater stabilization ponds is controlled by the physicochemical nature of effluent, as reported by others.[6,26,27]

Seasonal Variation of Nanomicroplankton Algal Species and Biomass in Sewage Ponds

In Marrakech primary sewage ponds, the total microalgae density ranged from 2.10^5 to 10^9 cell/l. The bimonthly data variation of microphytoplankton density indicate a marked reduction in quantitative content during the cold period. This

Table 10.5. Algal inventory in Marrakech wastewater stabilization ponds

Division and order	Species
Euglenophyta	
Euglenales	*Euglena clavata* SKUJA
	Euglena pseudoviridis CHADEF
	Euglena viridis EHRB.
	Euglena pisciformis KLEBS
Chlorophyta	
Volvoccales	*Chlamydomonas* sp
Chlorococcales	*Chlorella sorokiniana* SHIH. et KRAUS
	Micractinium pusillum FRES
	Coelastrum sp
Chrysophyta	
Pennales	*Nitzschia umbonata* (EHR.) LANGE-BERTALOT
Cyanophyta	
Oscillatoriales	*Oscillatoria irrigua* KUTZING GOMT.
	Oscillatoria sp
Chroococcales	*Synechococcus elongatus* NAGELI
	Synechocystis parvula PERFILIEW

decrease in cell number is mainly caused by a red water phenomenon which occurs periodically at this study site.[38,39,45,47] This period coincides with high sulphide concentration, which may generate photosynthesis algal and growth inhibition.[7]

In the secondary pond, the algal biomass variation ranged from 2.10^3 to 8.10^8 cells/l. The lowest density was observed paradoxically during Summer and Winter periods. First, it is interesting to note that the red water phenomenon, which occurred in the primary pond, also had an indirect influence on the second pond's micro-algae. Second, during the hot period, the secondary pond experienced a picophytoplanktonic bloom, which was not considered in the microflora content.

In the first pond, the seasonal and annual specific abundance diagrams show clearly the Euglenales and Volvoccales codominance in Spring (Fig. 10.1), with *Euglena clavata* and *Chlamydomonas* sp.; while during Summer the Chlorococcales (*Chlorella sorokiniana*) completely took over *Chlamydomonas*. It seems that *Chlamydomonas* was not as thermophilic as *Chlorella sorokiniana*.[48] This observation was confirmed in Autumn and Winter when the temperature decreased. The Volvoccales and Euglenales again became codominant in this toxic red water, however, *Chlamydomonas* was the most common genus.

In the secondary pond, seasonal and annual specific algae diagrammatic analysis indicated that algal groups Euglenales, Volvoccales and Chlorococcales were relatively codominant (Fig. 10.2). Except during Summer, the microplankton was represented by 95% Chlorococcales (*Chlorella, Ceolastrum* and *Micractinium* genera) and 5% Euglenales. The presence of *Micrac-tinium* and *Coelastrum* was ephemeral. Their density did not exceed 3.10^7 cells/l and their contribution to micro-algae biomass was not then strong. During this period, a qualitative sample examination revealed the existence of Chroococcoid picocyanobacteria, identified as *Synechocystis parvula* and *Synechococcus elangatus*. This picoplanktonic fraction was not taken into consideration in cell enumeration because of its smaller size. This is a specific feature of Marrakech's stabilization ponds, however.

Seasonal Variation of Picoplankton Biomass and its Contribution to Total Algal Biomass in Sewage Ponds

In various hydrosystems, picoplanktonic flora is made up generally of quite small Chroococcoid cyanobacteria species.[44-49] According to Pick and Caron,[43] they range from 0.2-2 µm in cell diameter. In Marrakech wastewater stabilization ponds, picoplanktonic flora is essentially represented by Chroococcoid cyanobacteria of *Synechocystis* and *Synechococcus* genera. These thermophilic species[46-50] have rarely been reported in sewage ponds.[38,39,46] Consequently, in order to know all about the native pond algae in such treatment systems, biomass should be very useful. The spatio-temporal evolution of this phytoplankton fraction in sewage ponds is given in Figure 10.5A-B.

The comparison between Nano-microplanktonic biomass and the Picoplanktonic chlorophyll a values (Chl. a) in the first pond (Fig. 10.3A) shows a general Chl. a concentration less than 50 µg/l. In the second pond (Fig. 10.3B), the picoplankton chl. a concentration underwent stronger fluctuations than in the primary pond. The chl a values reached 280 µg/l and 360 µg/l, respectively in June and July; whereas the low values were observed during the cold period (December, January and February). Seasonal variation of picophytoplanktonic biomass showed

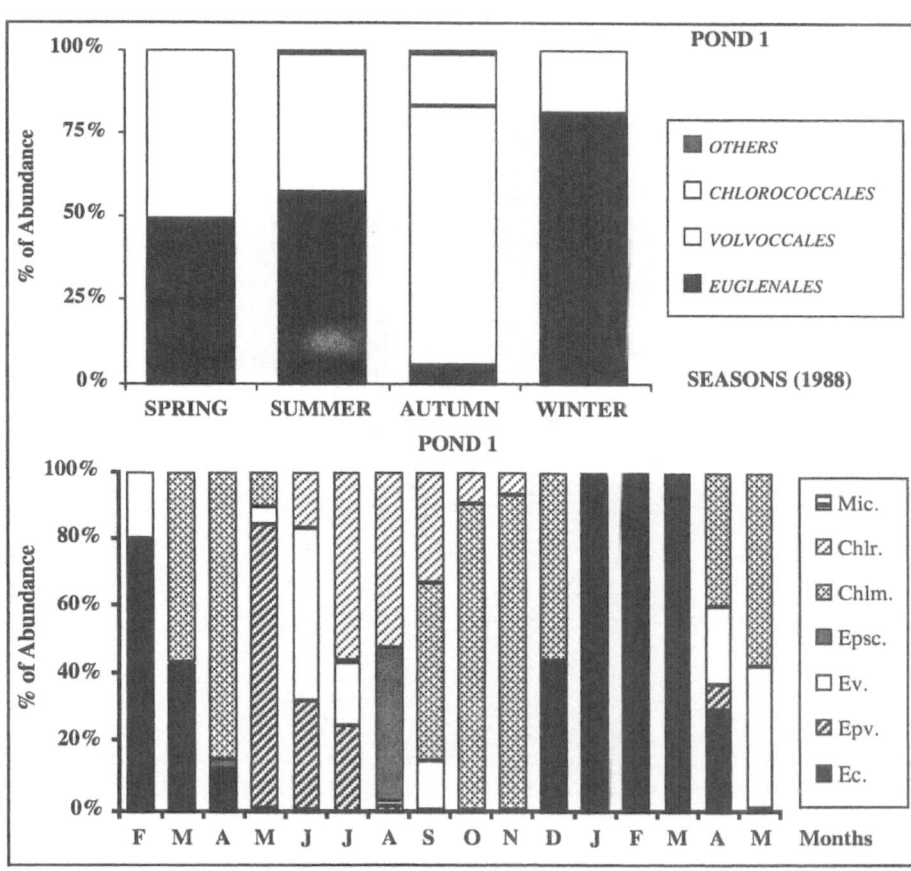

Fig. 10.1. Specific algae abundance diagram in primary pond.
Abbreviations: Ec: *Euglena clavata* ; Epv: *E. pseudoviridis* ; Ev: *E. viridis*; Epsc: *E. pisciformis* ;
Chlm: *Chlamydomonas*; Chlr: *Chlorella*; Mic: *Micractinium*; Coel: *Coelastrum*

maximum production during the hot period, essentially in the secondary pond. In
this pond physicochemical and biological conditions allowed its maintenance and
development over the Summer season (Table 10.6).

It is important to recall that total phytoplanktonic biomass in these water
samples was estimated by addition to the microplanktonic and picoplanktonic Chl.
a concentrations, estimated by double filtration method. This methodology allowed us to evaluate the seasonal percentage contribution of each fraction to total
phytoplankton biomass. The results (Table 10.6) showed clearly that in these
Marrakesh sewage ponds, picoplanktonic flora may contribute a maximum of 56.7%
of total biomass, essentially in the secondary pond; while this contribution was
less than 8.6% year-round in the primary pond. However, the Whatman GF/C (1.2
μ) cannot collect more than 43% of total phytoplanktonic biomass for photosynthetic pigment analysis, particularly during picoplanktonic blooms.

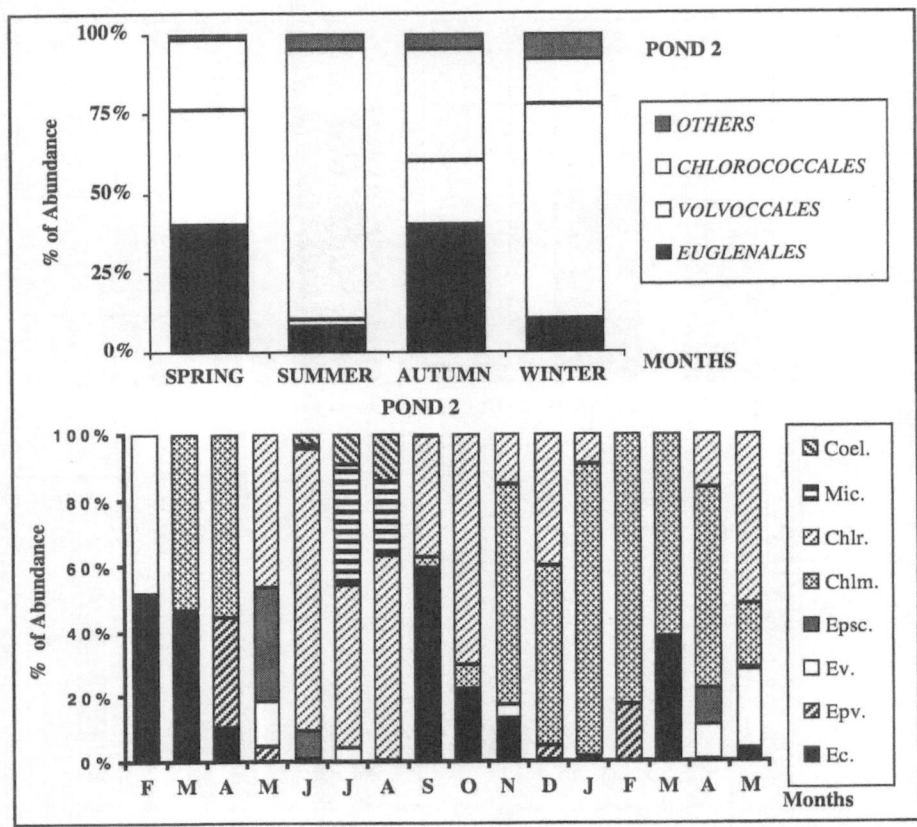

Fig. 10.2. Specific algae abundance diagram in secondary pond.
Abbreviations: Ec: *Euglena clavata*; Epv: *E. pseudoviridis*; Ev: *E. viridis*; Epsc: *E. pisciformis*;
Chlm: *Chlamydomonas*; Chlr: *Chlorella*; Mic: *Micractinium*; Coel: *Coelastrum*.

The increase in picocyanobacteria during the hot period reflects the thermophilic nature of these species. Lincoln and Hill[46] reported high rates of *Synechocystis*
proliferation in algal ponds in Israel; this too was correlated with temperature increase. There exists a scarcity of published literature on this topic in sewage ponds;
we therefore present here some observations concerning similar phenomenon in
the Great Lakes. A significant positive correlation between temperature increase
and picoplanktonic proliferation has been noted in various hydrosystems of different trophic status.[43,49,51] Weisse[44] did not deny the effects of temperature on
picoplankton production, but he suspected other intervention factors. In the
Marrakech sewage ponds, an increase in temperature seemed to induce development of these cyanobacteria, but the persistence of this fraction and its production—essentially in the second pond—may be attributed to other factors, such as
suitable effluent quality. Furthermore, zooplankton biomass in this study site, particularly rotifers species, reached its maximum density during the hot season in

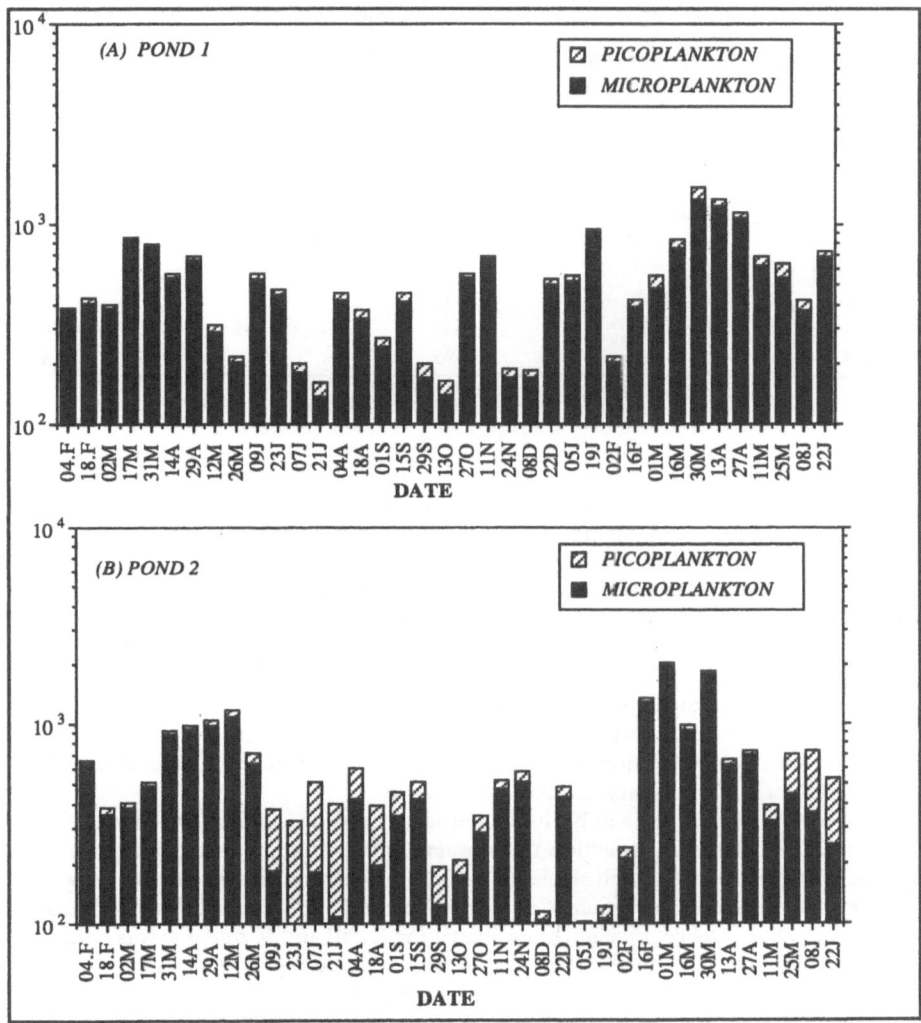

Fig. 10.3. Spatio-temporal variation of micro-and picoplankton Chl. a concentration during studied period (February 1988 - June 1989). A: pond 1; B: pond B.

the secondary pond.[52] Because of its herbivorous feeding, it causes a significant decrease in microalgae, particularly microplankton density.[39] In studies of cyanobacteria interactions, Barband et al[53] reported that members of microzooplankton, including rotifers, were often codominant with cyanbacterial species. Contrary to Cladocera, which is significantly affected in Marrakech sewage ponds particularly during summer, rotifers appear to show little sensitivity to cyanobacteria bloom species.[54] So, in agreement with many authors, algal periodicity in wastewater stabilization pond is usually attributed not only to physical or

Table 10.6. Average seasonal picoplankton Chl. a concentration (µg/L) and its percentage of contribution of total biomass, during 1988 in Marrakech sewage ponds

Season	Pond 1		Pond 2	
	Chl. a	% of contribution	Chl. a	% of contribution
Winter	31.8	7.32	33	11.12
Spring	17.28	6.33	50	8.08
Summer	28.16	8.45	236.5	56.70
Autumn	23.7	8.65	58	18.74

chemical control but also to zooplankton grazing. Furthermore, in Marrakech sewage pond statistical global analysis results confirmed the significant role played by the zooplanktonic component in overall pond operation.[56]

Relationship Between Cyanobacteria and Bacteria

Bacterial abundances of *E. coli* and *V. cholerae* in filtered and autoclaved wastewater stabilization pond effluent, with or without with algae and exposed to light or kept in darkness, are illustrated in Figure 10.4.

In order to compare the behavior of both bacteria studied in this microcosm study, the bacterial die-off coefficients (k) were calculated from data obtained under experimental conditions.[73,74] Table 10.7 gives bacterial survival, estimated by k calculation in the presence and absence of algae.

We noted an increase in bacterial abundance during the first two days of the experiment. Bacterial reduction was observed after 62 hours, particularly when bacteria was cultivated with algal cells. In fact, the bacterial die-off rate (k) in light is well expressed in the presence of algae (Table 10.7). In a control flask (bacteria without algae) the coefficient (k) evaluated for *E. coli* and *V. cholerae* was 0.0144 and 0.0375 h^{-1}, respectively. When the same reactional medium was mixed with *Chlorella* or Cyanobacteria and exposed to light, survival of both bacteria was significantly reduced compared with the control or with microcosm-containing bacteria and algae incubated in darkness. The die-off rate of *E. coli* in light is 0.0656 and 0.0731 h^{-1} in the presence of *Chlorella* and Cyanobacteria, respectively. The mortality of *V. cholerae* is also more expressed at light when it is associated with algae. The coefficient k is 0.0876 h^{-1} with *Chlorella* and 0.0536 h^{-1} with Cyanobacteria.

In absence of algal activity (incubation at darkness), survival of *E.coli* and *V. cholerae* is increased compared to that calculated in the same microcosm exposed to light (Table 10.7). Die-off rates of *E.coli* associated with *Chlorella*, which was 0.0656 h^{-1} in light, became 0.0191 h^{-1} in darkness; the same concentrations are observed with *V. cholerae*. In the presence of *Chlorella*, *V. cholerae* survives better in darkness (k = 0.0388 h^{-1}) than in light (k = 0.0876 h^{-1}). Therefore, the die-off of *E.*

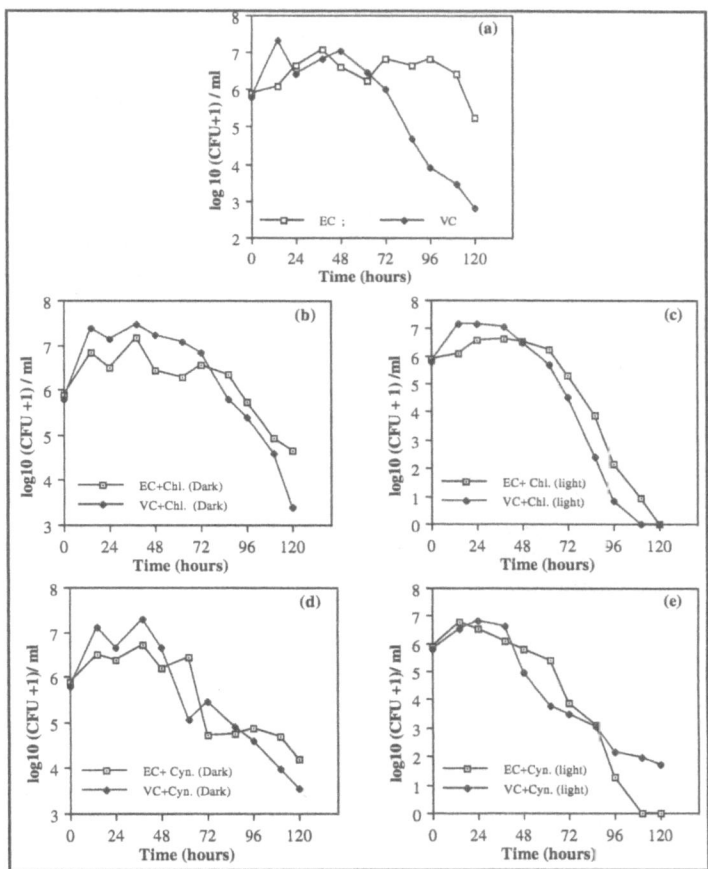

Fig. 10.4. Comparison of temporal evolution of *E. coli* and *V. cholerae* incubated in light and dark mixed or not with *Chlorella* (Chl) and Cyanobacteria. (Cyn).

Table 10.7. Bacteria survival in terms of die-off coefficient (k expressed by h⁻¹, n = 3 repetitions) in presence or absence of algae in light and dark microcosms

Bacterial strains	Bacteria only	Bacteria + *Chlorella*	Bacteria + *Chlorella*	Bacteria + Cyn.	Bacteria + Cyn.
	(light)*	(light)*	(dark)	(light)*	(dark)
E. coli	0.0144	0.0656	0.0191	0.0731	0.0226
	± 0.0014	± 0.0011	± 0.0006	± 0.0005	± 0.0002
V. cholerae	0.0375	0.0876	0.0388	0.0536	0.0321
	± 0.0012	± 0.0008	± 0.0002	± 0.0009	± 0.0004

Cyn. : Cyanobacteria = *Synechocystis* (Synst.) + *Synechococcus* (Sync.)
* 14 : 10 light / dark cycle.

coli and *V. cholerae* was faster when exposed with algae to light, compared to the control microcosm (bacteria cultivated without algae) or to microcosm-containing algae and incubated at darkness.

This increase in bacterial die-off in the presence of algae in light conditions can be explained in part by the increase in pH values, which reached 9.9 (Fig. 10.6). On the other hand, the bacterial reduction can be explained by algal activity, which is different in light and dark microcosms. In fact, the bacterial reduction correlates with the increased exponential phase of algal growth (Fig. 10.5). In light batch cultures, high growth of algae is observed compared with the dark microcosms. The global growth rate for *Chlorella* is of 4.44 day^{-1} in light and 0.48 day^{-1} in darkness. For Cyanobacteria, the two species *Synechococcus* and *Synechocystis* equally show a high growth in light. The global growth rate is 1.47 and 0.99 day^{-1} in light and darkness, respectively; we also noted a significant reduction in the number of cells (mortality) (Table 10.8). In a batch algal culture kept in the dark, the pH remained stable during incubation (Fig. 10.6) and the algal activity was less intense, therefore the bacteria die-off is attenuated.

However, if we compare the survival of pathogenic bacteria (*V. cholerae*) and indicator pollution bacteria (*E. coli)*, under these experimental conditions, we note that the behavior of these two species is different. In the presence of *Chlorella* incubated in light, *E. coli* survives better (k = 0.0656 h^{-1}) than *V. cholerae* (k = 0.0876 h^{-1}). An inversion of bacterial survival is noted with Cyanobacteria. The die-off coefficient is about 0.0536 h^{-1} for *V. cholerae* against 0.0731 h^{-1} for *E. coli.*

The same concentrations were also noted when we compared bacterial survival in light and dark microcosms. In the presence of *Chlorella*, it appears that *V. cholerae* was slightly more subject to a reduction in survival than *E. coli.* We noted a difference in mortality in light and dark cultures of 0.0488 h^{-1} for *V. cholerae* versus 0.0456 h^{-1} for *E. coli* ; the same difference, evaluated in the presence of Cyanobacteria, becomes reversible. *V. cholerae* showed only a difference in mortality (light/dark) of 0.0215 h^{-1} related to *E.coli* which presented a great difference of 0.0363 h^{-1}. Furthermore, although *E. coli* usually survived longer than *V. cholerae* in our experimental study, the survival of *V. cholerae* (k= 0.0536 h^{-1}) became higher than that of *E. coli* (k= 0.0731 h^{-1}) in the presence of Cyanobacteria.

It is apparent that Cyanobacteria promoted the survival of *V. cholerae* (k = 0.0536 h^{-1}) compared with *Chlorella* (k = 0.0876 h^{-1}). *E. coli*, however, survived better with *Chlorella* (k = 0.0656 h^{-1}) than when associated with Cyanobacteria (k = 0.0731 h^{-1}). This higher bacterial reduction of *V. cholerae* observed with *Chlorella* can be attributed principally to substances released by this green algae which have an important effect on *V. cholerae* than on *E. coli*. The difference in survival between *E. coli* and *V. cholerae* in the presence of Cyanobacteria in light microcosms can be attributed to several factors. The pH, which is influenced positively by algal activity, seems to be an important factor stimulating the growth of *V. cholerae*.[59-58] Nair et al[65] who studied the ecology of *V. cholerae* in aquatic environments in India, showed that the Summer peak of *V. cholerae* abundances are correlated with alkaline pH. Islam[75] also reported that algal blooms and alkaline pH are associated with cholera proliferation periods in Bangladesh.

In contrast with *E. coli*, it has largely been described in the literature that alkaline pH have a bactericidal effect on this species.[7,76] This difference in survival can be attributed to toxic substances released by this alga in aquatic environments. It

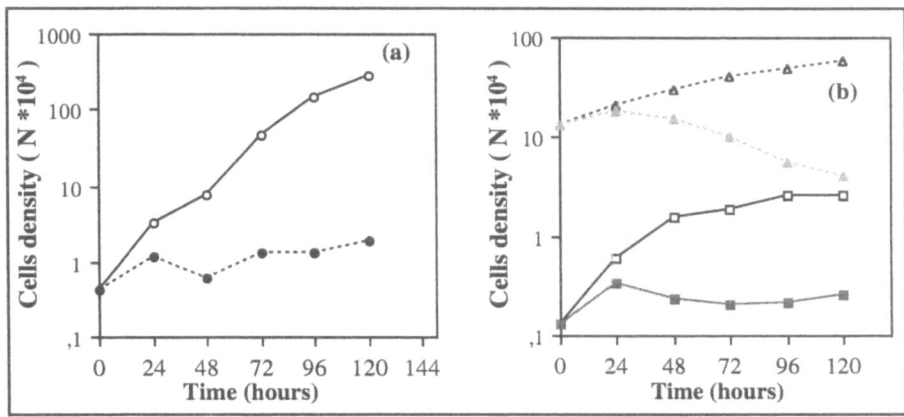

Fig. 10.5. Growth behavior of Chlorella and Cyanobacteria (Sync. and Synst.) incubated with bacteria in a light / dark regime and in complete dark (logarithmic scale).

———o——— : *Chlorella* (Light); ---●--- : *Chlorella* (Dark); ---▲--- : *Synst* (Dark)
---△--- : *Synst* (Light); ———■——— : *Sync* (Dark); ———□——— : *Sync* (Light)

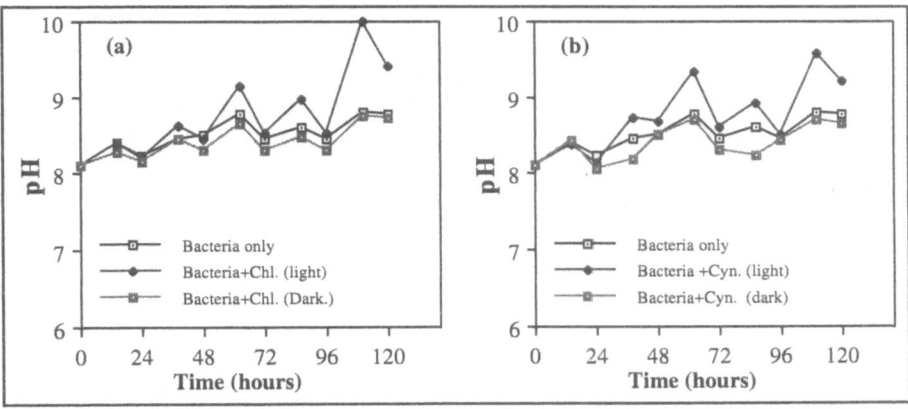

Fig. 10.6. Temporal evolution of pH in batch cultures containing bacteria and/or algae incubated in light and darkness.

is known that these algal populations (Cyanobacteria) are able to synthesize toxic substances which have antibacterial properties.[77-79] These substances seem to have a greater effect on *E. coli* than on *V. cholerae*.

This phenomenon was also observed when bacterial survival was compared in the microcosm incubated without algae and the microcosm containing algae incubated at dark. In this last medium, we observed that *V. cholerae* survived longer in darkness, particularly with Cyanobacteria ($k = 0.0321$ h^{-1}), compared with the control ($k = 0.0375$ h^{-1}). *E. coli* survived longer in the control microcosm ($k = 0.0144$ h^{-1}) than in batch cultures kept in the dark ($k = 0.0191$ h^{-1} with *Chlorella* and $k = 0.0226$ h^{-1} with Cyanobacteria). The relatively better behavior of *V. cholerae* observed

Table 10.8. Computation of microalgae global growth rate (m (d^{-1})) and global doubling time (dt), associated with bacteria in light and dark microcosms, during the experimental time (5 days).

Interval, day	Light* (1-5)		Dark (1-5)	
	m (d-1)	dt (h)	m (d-1)	dt (h)
Chlorella	4.44	0.16	0.48	1.46
Synechcoccus	1.47	0.48	- **	- **
Synechocystis	0.99	0.70	- **	- **

* 14: 10 light/dark cycle.
** negative values: reduction of cells number.

with Cyanobacteria in darkness than with *E. coli* can be attributed to the availability of *V. cholerae* to nutrients. Organics and other nutrients contained in algal protoplasm may serve as the nutrient source.[80-81] It is also probable that darkness-incubated algae which are stressed can release antibiotic substances in the medium. These substances seem to have a more toxic effect on *E. coli* than on *V. cholerae*.

These results are in concordance with observations of stabilization ponds, particularly those of Marrakech. The phytoplanktonic (picocyanobacteria) blooms observed during Summer[68] may be one of the factors explaining the abundant evolution of *V. cholerae*, which are negatively correlated with the low levels of fecal coliforms in Summer[82-83] against V. cholerae, which presents a high concentration.[16]

Conclusions

The evolution of the physicochemical conditions of sewage running through the ponds is changing due to the growth of bacteria, algae and other hydrobionts capable of using organic matter and inorganic nutrients. As a result, Marrakech sewage ponds, operating in an arid climate, show a reduction in organic load. Their average percent COD and BOD removal were 53% and 49%, respectively. Even better organic load removals may be achieved during hot periods.

Algal periodicity, which is attributed to climatic conditions and effluent quality, probably affect pond operation and system performance. The algal population in the primary pond was usually dominated by Euglenales except during the cold period (late Autumn and Winter) when volvocales (*Chlamydomonas* sp) completely took over *Euglena* genus. The drop in phytoplankton qualitative and quantitative content is mainly caused by red water phenomenon, due to thiobacteria proliferation and increases in sulphide concentration. In light of this toxicity, only suitably tolerant species will predominate as *Chlamydomonas* genus. In this pond, the contribution of picoplanktonic flora was less than 8.6% of total phytoplanktonic biomass.

In the secondary pond, where water quality was more suitable for better specific algal diversity, algal population was represented by: Euglenales, volvoccales, chlorococcales and chroococcoid species. These groups underwent seasonal turnovers according to their physiological ability and resistance to climatic conditions. In contrast, these groups coexisted during Spring, Autumn and Winter periods; however, the paradox occured in Summer when the microphytoplanktonic biomass indicated the lowest cell content (2.6 x 10^3 cells/l). A direct qualitative sample examination revealed the existence of chroococcoid cyanobacteria (*Synechocystis* and *Synechococcus* genera), associated with some chlorococcales green-algae (*Microctinium, Coelastrum* and *Chlorella* genera). It is interesting to note that this algal composition resulted in a bright green pond color with good water transparency. The estimation of picoplanktonic fraction biomass by Chlorophyll a analysis indicated that this last fraction contributed 56.7% to total algal biomass. Too much production was not considered neither in Chl. a concentration determined using the Whatman GF/C (1.2 μ).

Picocyanobacteria "blooms" occur in specific Marrakech sewage ponds. High temperatures appear to favor these phenomenon, however, their persistence over the Summer in the secondary pond were controlled by other factors, such as zooplanktonic grazing and effluent quality. Biomass and composition due to algal activity are the major parameters controlling waste stabilization pond operation. We conclude that Picoplanktonic biomass is an important component in these treatment plants, particularly in arid climates. Picocyanobacteria physiological abilities should therefore be exploited in sewage biological treatment.

As for the relationship between cyanobacteria and bacteria, these experimental studies show that microalgae growing in wastewater batch laboratory cultures have a net influence on bacterial behavior and establishment. It appears that *Chlorella* reduces *V. cholerae* (pathogenic bacteria) abundance more so than *E. coli* (fecal contamination bacteria). This green algae seems to secrete products which have a great toxic effect on *V. cholerae*. Moreover, better survival of this pathogenic bacteria was observed in the presence of Cyanobacteria. These algae induce an alkaline pH which promotes the survival of *V. cholerae* and reduce *E. coli* abundance; these algae also secrete substances which seem to lead to a greater die-off of *E. coli* than *V. cholerae*. The phytoplanktonic (Cyanobacteria) blooms during Summer should therefore be considered among the most important factors regarding temporal distributions of these bacteria, with a strong abundance of *V. cholerae* and weak concentrations of coliform bacteria during stabilization treatment of wastewaters of Marrakech.

In this work, we have also shown that the survival of the pathogenic bacteria (*V. cholerae*) does not usually correlate with the indicator pollution bacteria (*E. coli*). Consequently, measurement of fecal coliform abundance is not a reliable way to obtain information for predicting the abundance of pathogens like *V. cholerae* in wastewater stabilization ponds, which contain appreciable concentrations of algae as Cyanobacteria, especially during the Summer.

References

1. Gloyna EF. Waste stabilization ponds. World Health Organization 1972; 1-185.
2. Sauze F. Interaction des algues et des microorganismes dans les milieux pollués. Industrie Alimentaire et Agricole 1978; 95:1234-1243.
3. Raschke RL. Algal periodicity and waste reclamation in stabilization ponds system. J Wat Pol Cont Fed 1970; 42:518-530.

4. Palmer CM. Algae and Water Pollution. US EPA Ohio, 1977:1-124.

5. Ergashev AE, Tajiev SH. Seasonal variation of phytoplankton in a series of waste treatment lagoons (Chimkent, Central Asia). 1-Artificial inoculation and role of algae in sewage purification. Int Res Der Ges Hydrobiol 1986; 17:545-555.

6. Barbe J, Steiner B. Méthodologie de la caractérisation des installations de lagunage naturel par le phytoplancton. Revue Sciences de l'eau 1987; 6:137-150.

7. Pearson HW, Mara DD, Mills SW et al. Factors determining algal populations in waste stabilization ponds and influence of algae on pond performance. Wat Sci Technol 1987; 19:131-140.

8. Becker EW. Elimination of heavy metal removal from wastewater by means of algae. Water Research; 17:459-466.

9. Abelovich A, Weisman D. Role of heterotrophic nutrition in growth of alga *Scenedesmus obliquus* in high rate oxidation ponds. Appl Environ Microbiol 1978; 35:32-37.

10. Kalisz L. Role of algae in sewage purification. I-oxygen production. Pol Arch Hydrol 1973; 20:389-412.

11. Pouliot Y, de la Noüe J. Mise au point d'une usine-pilote d'épuration des eaux usées par production de micro-algues. Revue Sciences de l'Eau 1985; 4:207-222.

12. Sérodes JB, Walsh E, Goulet O et al. Tertiary treatment of municipal wastewater using bioflocculating micro-algae. Can J Civ Eng 1991; 18:940-944.

13. De la Noüe J, Laliberté G, Proulx D. Algae and waste water. J Appl Phycol 1992; 4:247-254.

14. Talbot P, de la Noüe J. Tertiary treatment of wastewater with *Phormidium bonheri* (Schmidle) under various light and temperature conditions. Water Res 1993; 27:153-159.

15. Mara DD, Pearson HW. Artificial freshwater enviroments : waste stabilzation ponds. In: Rehm H, Reed G, eds. Biotechnology - A Comprehensive Treatise,VCH, Weinheim, 1986: 177-206.

16. Mezrioui N, Oufdou KH, Baleux B. Dynamics of non-O1 *Vibrio cholerae* and fecal coliforms in experimental stabilization ponds in the arid region of Marrakesh, Morocco, and the effect of pH, temperature and sunlight on their experimental survival. J Can Microbiol 1995; 41:489-498.

17. Neel KJ, Mc-Dermott JH, Monday CA. Experimental lagooning of raw sewage at Fayette Missouri. J Wat Pol Cont Fed 1961; 33:603-641.

18. Vuillot M, Pujol R. Traitement des eaux usées domestiques par le lagunage naturel. Performances et suivi d'installation. Rapport: traitement des E.U. des petites collectivités. 35 Journée internationale du CEBEDEAU 1982. Liège (France).

19. Yang LB, Wing LK, McGarry, MG et al. Overview of wastewater treatment and resource recovery. Report of a workshop on light-rate algae ponds 1980 Singapore, February 47p.

20. Culminance MJ, Shafer JRA. Algae and phosphorus removal from lagoon effluents by in situ chemical addition. Tech Report EL/82/6. US army Enninier waterways experiment station CE. Vieksburg, Miss.

21. Oswald WJ. Microalgae and wastewater treatment. In Borowitzka, MA, Borowitzka, LJ eds. "Microalgal Biotechnology." New York, 1988: 305-328.

22. King DL. The role of carbon in eutrophication. J Water Pollution Control Federation 1970; 42:2035-2051.

23. Humenik FJ, Hanna GP. Algal bacterial symbiosis for removal and conservation of wastewater nutrients. J Water Pollution Control Federation 1971; 43:580-594.

24. Harris GP. Photosynthesis, production and growth: the physiological ecology of phytoplankton. Arch Hydrobiol Beich Ergebm Limnol 1978; 10:1-171.

25. Kalisz L. Role of algae in sewage purification. II Nutrient removal. Pol Arch Hydrobiol 1973; 20:413-434.
26. Van der post DC, Toerien DF. The retardment of algal growth in maturation ponds. Water Res 1974; 8:593-600.
27. Patil HS, Dodakundi GB, Rodgi SS. Succession in zoo- and phytoplankton in sewage stabilization ponds. Hydrobiologia 1975; 7:253-264.
28. Shillanglow SN, Pieterse AJH. Observation on algal populations in experimental maturation pond system. Water SA 1977; 3:183-192.
29. Steiner B. Evolution des peuplements planctoniques d'une installation de lagunage naturel (chaucenne, Doubs). Actes Colloque. Montpellier 1 - 5 juin/lagunage 1982:13.
30. Roques H. Fondements théoriques du traitement biologique des eaux I et II. Techniques et Documents: Paris. 1979.
31. Kankalaa P. The relative importance of algae and bacteria as food for *Daphnia longispina* (Cladocera) in a polyhumic lake. Fresh Water Biology 1988; 19: 285-296.
32. Raschke RL. Algal periodicity and waste reclamation in a stabilization ponds system. J Water Pollution Control Federation 1970; 42:518-530.
33. Oswald WJ. Productivity of algae in sewage disposal. Solar Energy 1973; 15: 107-117.
34. Delannoy M. Etude du phytoplancton dans les bassins d'autoépuration de la station de Wavrin. Rapport du Labo d'Algologie, UER de Biologie, Université de Lille I 1973:83.
35. Martin JA. Phytoplankton - zooplankton relationship in Narragausett Bay 4. Limnol Oceanogr 1970; 15:413-418.
36. Bailey Watts AE. Seasonal variation in size spectra of phytoplankton assemblages in Loch leven Scotland. Hydrobiologia 1986; 138:25-42.
37. Pourriot R, Champ P. Consommateurs et production secondaire. In Pourriot, R, Masson eds. "Ecologie du plancton des eaux continentales." Paris 1982:49-112.
38. Chifaa A. Etude de la dynamique des peuplements phytoplanctoniques et interactions avec la qualité de l'eau des bassins expérimentaux de lagunage sous climat aride. Thèse de Doctorat de 3ème Cycle, Univ Cadi Ayyad Fac Sci. Marrakech 1987;1-173.
39. Oudra B. Bassins de stabilisation anaérobie et aérobie facultatif pour le traitement des eaux usées à Marrakech: Dynamique du phytoplancton (Microplancton et Picoplancton) et évaluation de la biomasse primaire. Thèse de Doctorat de 3ème Cycle, Univ. Cadi Ayyad, Fac, Sci. Marrakech 1990; 1-144.
40. Komarek J. Taxonomic review of genera Synechocystis 1892, Synechococcales 1849 and Cyanophyceae. Arch Protistark Ed 1976; 118:119-179
41. Lorenzen CJ. Determination of Chlorophyll and phaeopigments spectrophotometric equations. Limnol Oceanogr 1967; 12:343-346.
42. Prepas EE, Dunnigan ME, Trimbee AM. Comparisons *in situ* estimates of chlorophyll a obtained with Whatman GF/F and GF/C glass fiber filters in mesotrophic to hypertrophic lakes. J Can Fish Aquat Sci 1988; 45:910-914.
43. Pick FR, Caron DA. Picoplankton and nannoplankton biomass in lake Ontario: relative contribution of phototrophic and heterotrophic communities. J Can Fish Aquat Sci 1987; 44: 2164-2172.
44. Weisse T. Dynamics of autotrophic picoplankton in lake Constance. J Plankton Res 1988; 10: 1179-1188.
45. Bouarab L. Contribution à l'étude des differentes formes du phosphore dans le lagunage naturel : station expérimentale de Marrakech. Thèse de Doctorat de 3ème Cycle, Univ Cadi Ayyad Fac Sci Marrakech 1988:1-125.
46. Lincoln EP, Hill DT. An integrated microalgae system. In: Shelef G, Soeder CJ, eds. "Algal Biomass" North-Holland, Biomedical Press 1980:224-244.

47. Ouazzani N. Lagunage expérimental sous climat aride: variation des paramètres physico-chimiques à Marrakech. Thèse de Doctorat de 3ème Cycle, Univ Cadi Ayyad Fac Sci Marrakech 1987:1-84.

48. Belkoura M, Dauta A. Interaction lumière-température et influence de la photopériode sur le taux de croissance de *Chlorella Sorokiniana* Shih et Kraus. Annales de Limnologie 1992; 28:101-107.

49. Craig SR. The distribution and contribution of picoplankton to deep photosynthetic layers in some monomictic lakes. Acta Acad Aboenis 1987; 47:55-81.

50. Yamaoka T, Satah K, Kotak S. Photosynthetic activities of thermophilic Blue-green alga. Plant Cell Physiol 1978; 19:943-954.

51. Caron DA, Pick FR, Lean DRS. Chroococcoid cyanobacteria in lake Ontario: seasonal and vertical distribution during 1982. J Phycol 1985; 21: 171-175.

52. Tifnouti A. Zooplancton des bassins de lagunage de Marrakech: structure du peuplement et dynamique des principales populations. Thèse de Doctorat de 3ème Cycle, Univ. Cadi Ayyad, Fac, Sci. Marrakech 1987; 1-198.

53. Braband A, Faafeng BA, Kallqvist T et al. Biological control of undesirable cyanobacteria in culturally eutrophic lake. Oecologia 1983; 60:1-5.

54. Paerl HW. Growth and reproductive strategies of freshwater blue-green algae (Cyanobacteria). In: Craig DS, ed. Growth and Reproductive Strategies of Freshwater Phytoplancton. New York, 1988:261-315.

55. Tifnouti A, Pourriot R. Dynamique d'une population de *Moina micrura* (Crustacea, Cladocera) dans un bassin de lagunage à Marrakech (Maroc). Rev Hydrobiol Trop 1989; 22:239-250.

56. Casellas C. Dynamique des ecosystèmes de lagunes d'épuration d'eaux usées, application au laguange experimental de Marrakeck (Maroc) et comparaison avec le laguange de Meze (Herault). Doctorat d'Etat es-Sciences pharmaceutiques 1990; 1-122.

57. Singleton FL, Atwell RW, Jangi MS et al. Influence of salinity and organic nutrient concentration on survival and growth of *V. cholerae* in aquatic microcosms. Appl Environ Microbiol 1982; 43:1080-1085.

58. Huq A, Small EB, West PA et al. The role of planktonic copepods in the survival and multiplication of *V. cholerae* in the aquatic environment. In Colwell, RR eds. "Vibrios in the Environment." John Wiley and Sons. New York, 1984: 521-534.

59. Colwell RR. *Vibrio cholerae* and related Vibrios in the aquatic environment - an ecological paradigm. Proc IV ISME 1986:426-434.

60. Rhee GY. Competition between an algae and an aquatic bacterium for phosphate. Limnol and Oceanogr 1972; 17:505-513.

62. Dor I, Svi B. Effect of heterotrophic bacteria on the green algae growing in wastewater. In: Shelef G, Soeder CG, eds. "Algal Biomass." Elsevier/North-Holland. Biomedicall Press, 1980:425-429.

63. Dor I, Svi B. Effect of the green algae isolated from wastewater on the activity of sewage bacteria. In Shelef, G, Soeder, CG eds. "Algal Biomass." Elsevier/North-Holland. Biomedicall Press, 1980 (presented as a poster communication).

64. Islam MS, Drakar BS, Bradley DJ. Attachment of toxigenic *V. cholerae* O1 to various freshwater plants and survival with a green algae, *Rhizoclonium fontanum.* J Trop Med Hyg 1989; 92:396-401.

65. Nair GB, Sarkar BL, De SP et al. Ecology of *V. cholerae* in the fresh water environs of Calcutta, India. Microbiol Ecol 1988; 15:203-215.

66. Hofer E, Ernandez D. Incidencia de *Vibrio cholerae* nào O1 em affluentes de estaçoês de ratamento de esgotos da cidade do Rio de Janeiro, RJ Rev Microbiol sào Paulo 1990; 21:31-40.

67. Martin YP, Bonnefont JL. Variations annuelles et identification des vibrions cultivant à 37°C dans un effluent urbain dans des moules et dans l'eau de mer en rade de Toulon (Méditerranée, France). J Can Microbiol 1990; 36:47-52.
68. Oudra B, Lâakari A, Loudiki M et al. Blooms à picocyanobacteria dans un bassin expérimental de stabilisation des eaux usées de Marrakech: Causes et conséquences. Proc of fourth C.I.L.E.F. 25-28 Avril 1994, Marrakesh.
69. Dauta A. Conditions de développement du phytoplancton. Etude comparative du comportement de huits espèces en culture. 1. Détermination des paramètres de croissance en fonction de la lumière et de la température. Annales Limnologie 1982; 18:3,217-262.
70. Ferris JM, Hirsch CF. Method for isolation and purification of Cyanobacteria. Appl Environ Microbiol 1991; 57:1448-1452.
71. Sournia A. Phtoplanktonic Manual. UNESCO eds. Paris, 1978: 1-337.
72. Guillard RRL. Divison rates. In: Stein JR, eds. "Hand Book of Physiological Methods- Cultures Methodes & Growth Measurements." Cambridge University Press, 1973:289-311.
73. Chamberlin CE, Mitchell R. A decay model for enteric bacteria in naturel waters. In: Mitchell R, ed. Water Pollution Microbiology. John Wiley and Sons, New York: 1978; 2:325-348.
74. Burdyl P, Post FJ. Survival of *Escherichia coli* in great salt lake water. Water Air Soil Pollut. 1979; 12:237-246.
75. Islam MS. Increased toxin production by *V. cholerae* O1 survival with a green algae, *Rhizoclonium fontanum*, in an artificial aquatic environment. Microbiol Immunol 1990; 34:557-563.
76. Mezrioui N, Baleux B. Effets de la température, du pH et du rayonnement solaire sur la survie de différentes bactéries d'intérêt sanitaire dans une eau usée épurée par lagunage. Revue Sciences l'eau 1992; 5:573-591.
77. Himber K, Keijola AM, Husvirta L et al. The effect of water treatment processes on the removal of hepatotoxines from *Microcystis* and *Oscillatoria* Cyanobacteria: Laboratory study. Wat Res 1989; 23:979-984.
78. Repavich WM, Svonzogni WC, Standridge JH et al. Cyanobacteria (blue green algae) in Wisconsin waters: Acute and chronic toxicity. Wat Res 1990; 24:225-231.
79. Yasumo M, Sugaya Y. Toxicities of *Microcystis viridis* and the isolated hepatotoxic polypeptides on cladocerans. Verh Intern Verein Limnol 1991; 24:2622-2626.
80. Augier, H. Contribution à l'étude biochimique et physicochimique des substances de croissance chez les algues. Thèse Doc. Univ. Marseille 1972; 1-323.
81. Soeder CJ, Fingerhut U, Groneweg J. Microalgae and bacteria in wastewater: cooperation and interaction. In: Megunsar, F, Gantar, M eds. Perspective in Microbial Ecology. Proc of Fourth Intenational Symposium on Microbiol Ecology. Ljubljana, 1986: 80-85.
82. Imziln B. Traitement des eaux usées par lagunage anaérobie et aérobie facultatif à Marrakech: Etude bactériologique quantitative et qualitative. Antibiorésistance des bactéries d'intérêt sanitaire. Thèse de 3ème cycle. Univ Cadi Ayyad Fac Sci Semlalia Marrakech 1990:1-122.
83. Hassani L. Traitement des eaux usées par lagunage expérimental à Marrakech: Evolution spatio-temporelle et antibiorésistance des coliformes fécaux et d'*Aeromonas* spp, Influence du transfert conjucatif et des rayons solaires sur la résistance d'*E. coli* aux antibiotiques. Doctorat d'Etat Es-Sciences, Univ Cadi Ayyad Fac Sci Semlalia Marrakech 1993:1-141.
84. Abeliovich A, Weisman D. Role of heterotrophic nutrition in growth of algae *Scenedesmus obliqus* in high rate oxidation ponds. Appl Environ Microbiol 1978; 35:32-37.

85. Azov Y, Goldman JC. Free ammonia inhibition of algal photosynthesis in intensive culture. Appl Environ Microbiol 1982; 43:735-739.
86. Konig A, Pearson HW, Silva SA. Ammonia toxicity to algal growth in wastewater stabilization ponds. Wat Sci Technol 1987; 19:115-122.
87. Pinhero HM, Reis MT, Novais LM. A study of the performance of a high-rate photosynthesis pond system. Wat Sci Technol 1987; 19:237-241.
88. Edline F. L'épuration biologique des eaux usées résiduaires. Théorie et technologie. CEBEDEAU:Thiége 1980:299.
89. Mara DD, Pearson HW. Design manual for waste. World Health Organization 1986:104.
90. Rodier J. L'analyse de l'eau: Eaux naturelles, eaux résiduaires et eau de mer. Chimie, physico-chimie, bactériologie et biologie. 7ème eds, DUNOD, Paris 1984: 1365.
91. Mezrioui N, Oudra B, Oufdou KH et al. Effect of microalgae growing on wastewater batch culture on *Escherichia coli* and *Vibrio cholerae* survival. Wat Sci Technol 1994; 30:295-302.

CHAPTER 11

A Material Transformation Model for Biological Stabilization Ponds

Xianghua Wen

Introduction

B iological stabilization ponds have been accepted as an effective and economical means of waste water treatment and have been applied for many years throughout the world. This is one kind of semi-natural ecosystem in which various material transformation processes take place. Among these processes, carbon, nitrogen and phosphorus transformations are the most important ones. From an ecological point of view, material transformation and cycling processes are crucial for keeping the system stable.[1] Some researchers[2-10] have discovered the importance of material transformation in biological stabilization ponds. However, systematic and quantitative studies of nutrient transformation processes by applying the principles of ecology are few and insufficient. The objective of this chapter is to establish a good model of material transformation in biological stabilization pond systems based upon principles of ecology. This model shall completely describe the transformation processes of carbon, nitrogen and phosphorus in a pond system. In an application of the model, it was successfully used in making material transformation graphics for visual and quantitative analysis of material transformation processes in biological stabilization pond systems. As a result, a better and more comprehensive understanding of pond mechanisms can be achieved. This model is also very reliable for pond design and operational control. Furthermore, these study procedures are uniquely complete in terms of systematization of the designs for the experiments, as well as model calibration and verification.

Model Development

Modeling techniques are a very effective means to analyze a complex system, because modeling can clearly demonstrate the properties of the system and remedy deficiencies in our knowledge. Mass conservation is the most basic law of

Wastewater Treatment with Algae, edited by Yuk-Shan Wong and Nora F.Y. Tam.
© Springer - Verlag and Landes Bioscience 1998.

ecology. Our model is based on this law and the hypothesis of a completely mixed flow in the pond. The mathematical description of the above mentioned items is Equation 1.

$$\frac{dc}{dt} = QC_i - Q_e C + V \sum_{j=1}^{n} (v_c)_j \qquad\qquad 1)$$

Where:

C_i, C influent and effluent concentration of substance, respectively (mg/l);

dc/dt change rate of substance in pond (mg/l.d)

V pond volume (l);

Q, Q_e influent and effluent flow rate, respectively (l/d);

n number of reactions that involve the substance

Material transformation processes in biological stabilization pond systems are very complicated. It is impossible to simulate all the details of their transformation processes, even only for carbon, nitrogen and phosphorus. We have made some assumptions in order to develop a proper model. For carbon, as shown in Figure 11.1 we assume that it exists in four forms: 1) Dissolved inorganic carbon(C_I); 2) Carbon content of bacterial cells (C_B); 3) Dissolved organic carbon (C_O); and 4) Carbon content of algal cells (C_A). Carbon experiences seven routes of transformation in the pond: 1) Conversion of organic carbon to bacterial cells; 2) Bacterial decay producing inorganic carbon; 3) Decomposition of organic carbon to inorganic carbon; 4) Algal cell formation through photosynthesis using inorganic carbon as carbon source; 5) Algal decay producing inorganic carbon; 6) Lysis of algal cells; and 7) Sedimentation of algae.

Additionally, the following considerations about the carbon cycle were introduced to simplify the model: 1) Inorganic carbon dioxide interfacial transfer can be neglected; this has been proven by many previous researchers and was also supported by the authors' experimental results;[11] 2) The settling and decay of benthal deposits are considered as one item only the settling rate is used to describe these two processes; this is also applied to nitrogen and phosphorus. This is due to the fact that balancing between sedimentation and decomposition of sediment is such a complicated process that no equations can describe it precisely. As a matter of fact, because the experimental processing was not very long and the pond was not very deep, very little sedimentation was produced. Decomposition is ignored in this study.

Figure 11.2 shows the assumptions for the nitrogen transformation routes. Nitrogen is assumed to exist in four forms: 1)Dissolved inorganic nitrogen (N_I); 2) Nitrogen content of bacterial cells (N_B); 3) Dissolved organic nitrogen(N_O); and 4) Nitrogen content of algal cells (N_A). There are seven transformations of nitrogen: 1) From organic nitrogen to inorganic nitrogen resulting from bacterial action; 2) Inorganic nitrogen becoming part of a bacterial cell; 3) Bacterial decay producing inorganic nitrogen; 4) Inorganic nitrogen becoming an element of algae; 5) Algal decay producing inorganic nitrogen; 6) From lysis of algae to organic nitrogen; 7) Sedimentation of algae.

We simplified nitrogen reactions in order to clearly express the main process of nitrogen cycling in pond systems: 1) The ammonia stripping process cannot be involved in the modeling, because NH_3-N present is less than 10% of NH_3-N plus

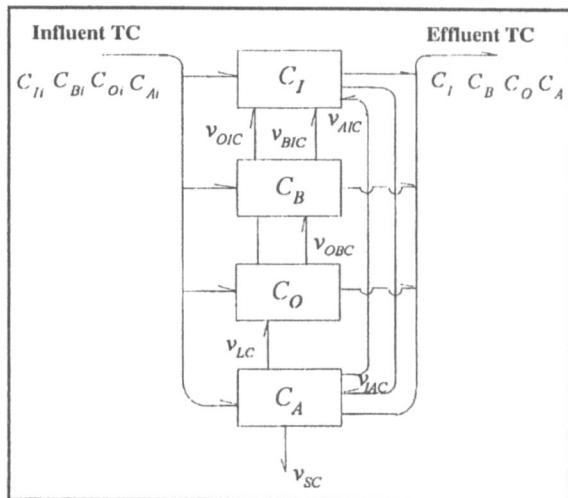

Fig. 11.1. Carbon transformation routes.

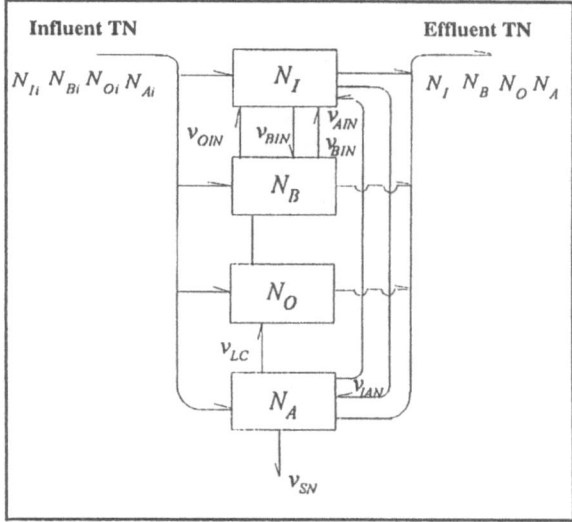

Fig. 11.2. Nitrogen transformation routes.

NH_4^+-N in the experiments of this study; 2) The nitrification-denitrification process is not the main nitrogen reaction in a pond system; this has been supported by previous studies.[8-14]

As shown in Figure 11.3, phosphorus in the pond is assumed in three forms: 1) Total dissolved phosphorus (mostly inorganic phosphorus) (P_T); 2) phosphorus content of bacterial cells (P_B); and 3) phosphorus content of algal cells (P_A). There are six transformations of phosphorus in the pond system: 1) Formation of bacterial cell by using phosphorus; 2) Bacterial decay producing dissolved phosphorus; 3) Formation of algae cell by using phosphorus; 4) Algal decay producing dissolved phosphorus; 5) From lysis of algae to phosphorus; and 6) Sedimentation of algae.

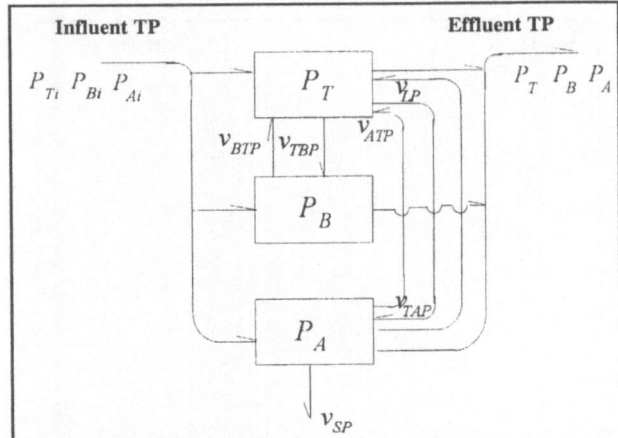

Fig. 11.3. Phosphorus
transformation routes.

Based on Equation 1 and earlier discussion, the mass balance equation for carbon, nitrogen and phosphorus can be drawn. After organizing each of the mass balance equations, the following steady-state ecological models from Eq.2 to Eq.13 for biological stabilization ponds can be obtained. It consists of twelve non-linear equations, mainly with two types of sub-model: One is the Monod equation describing the growth of microbes, and the other is the first-order expression describing all the other biochemical reactions such as bio-decay, biolysis and organism settling. In the equations, subscript i, I, B, O and A represent influent, inorganic, bacterial, organic and algal, respectively. μ, K, k, Y represent the corresponding biochemical rate constants. The variables of the model are listed in Table 11.1.

Carbon:

$$\frac{C_{Ii}}{\theta} - \frac{C_I}{\theta} + (1-Y)\mu_{mBC}\frac{C_o}{K_{BC}+C_o} \cdot C_B\Big/Y + k_{dBC}\cdot C_B + k_{dAC}\cdot C_A$$
$$-\mu_{mAC}\frac{C_I}{K_{AC}+C_I}\cdot C_A \cdot f(I) = 0 \qquad\qquad 2)$$

$$\frac{C_{Bi}}{\theta} - \frac{C_B}{\theta} + \mu_{mBC}\frac{C_o}{K_{BC}+C_o}\cdot C_B - k_{dBC}\cdot C_B = 0 \qquad\qquad 3)$$

$$\frac{C_{oi}}{\theta} - \frac{C_o}{\theta} - \mu_{mBC}\frac{C_o}{K_{BC}+C_o}\cdot C_B\Big/Y + k_{LC}\cdot C_A = 0 \qquad\qquad 4)$$

$$\frac{C_{Ai}}{\theta} - \frac{C_A}{\theta} + \mu_{mAC}\cdot\frac{C_I}{K_{AC}+C_I}\cdot C_A \cdot f(I) - k_{dAC}\cdot C_A$$
$$-k_{LC}\cdot C_A - k_{SC}\cdot C_A = 0 \qquad\qquad 5)$$

Nitrogen:

$$\frac{N_{Ii}}{\theta} - \frac{N_I}{\theta} + \alpha_N \cdot N_o + k_{dBN} \cdot N_B + k_{dAN} \cdot N_A$$

$$-\mu_{mBN} \frac{N_I}{K_{BN} + N_I} \cdot N_B$$

$$-\mu_{mAN} \frac{N_I}{K_{AN} + N_I} \cdot N_A \cdot f(I) = 0 \qquad\qquad 6)$$

$$\frac{N_{Bi}}{\theta} - \frac{N_B}{\theta} + \mu_{mBN} \frac{N_I}{K_{BN} + N_I} \cdot N_B - k_{dBN} \cdot N_B = 0 \qquad\qquad 7)$$

$$\frac{N_{oi}}{\theta} - \frac{N_o}{\theta} - \alpha_N \cdot N_o + k_{LN} \cdot N_A = 0 \qquad\qquad 8)$$

$$\frac{N_{Ai}}{\theta} - \frac{N_A}{\theta} + \mu_{mAN} \cdot \frac{N_I}{K_{AN} + N_I} \cdot N_A \cdot f(I)$$

$$-k_{dAN} \cdot N_A - k_{LN} \cdot N_A - k_{SN} \cdot N_A = 0 \qquad\qquad 9)$$

Phosphorus:

$$\frac{P_{Ti}}{\theta} - \frac{P_T}{\theta} + k_{dBP} \cdot P_B + k_{dAP} \cdot P_A - \mu_{mBP} \frac{P_T}{K_{BP} + P_T}$$

$$\cdot P_B - \mu_{mAP} \frac{P_T}{K_{AP} + P_T} \cdot P_A \cdot f(I) + k_{LP} \cdot P_A = 0 \qquad\qquad 10)$$

$$\frac{P_{Bi}}{\theta} - \frac{P_B}{\theta} + \mu_{mBP} \frac{P_T}{K_{BP} + P_T} \cdot P_B - k_{dBP} \cdot P_B = 0 \qquad\qquad 11)$$

$$\frac{P_{Ai}}{\theta} - \frac{P_A}{\theta} + \mu_{mAP} \cdot \frac{P_T}{K_{AP} + N_P} \cdot P_A \cdot f(I)$$

$$-k_{dAP} \cdot P_A - k_{LP} \cdot P_A - k_{SP} \cdot P_A = 0 \qquad\qquad 12)$$

Function of light:

$$f(I) = \frac{I_{av}}{I_m} \exp(1 - \frac{I_{av}}{I_m}) \qquad\qquad 13)$$

Model Calibration and Parameter Sensitivity Analysis

Experimental Procedure

Lab-scale experiments were conducted to obtain the data needed for determining the parameters in the model. A three-stage pond was designed with the dimensions of 78.5 x 40.0 x 49.5 cm for the first pond and 78.5 x 22.5 x 49.5 cm for both the second and third ponds. The pond was placed in a room with constant temperature at 20°C. Fluorescent tubes were the light source, which operated 12 hours a day. The light intensity of the water surface in the pond was about 6000 Lx. Synthetic sewage composed of urea, KH_2PO_4 and industrial glucose, etc. was the influent. The reactor was continuously operated at different conditions, as shown in Table 11.2.

Total carbon (TC), inorganic carbon (IC), nitrogen (NO_3-N, NO_2-N, NH_3-N, TN), total phosphorus (TP), pH, and alkalinity in the influent and in the pond were measured. The total biomass settled and attached to the wall of the pond,

Table 11.1. Variables in the model

Name	Symbol	Unit
Dissolved inorganic carbon	C_I	mg/l
Carbon content of bacterial cells	C_B	mg/l
Dissolved organic carbon	C_O	mg/l
Carbon content of algal cells	C_A	mg/l
Dissolved inorganic nitrogen	N_I	mg/l
Nitrogen content of bacterial cells	N_B	mg/l
Dissolved organic nitrogen	N_O	mg/l
Nitrogen content of algal cells	N_A	mg/l
Dissolved total phosphorus	P_T	mg/l
Phosphorus content of bacterial cells	P_B	mg/l
Phosphorus content of algal cells	P_A	mg/l

Table 11.2. Operation conditions of the experiments

No.	Temp. (°C)	F.R.[1] (l/d)	HRT (d)	COD (mg/l)	TC (mg/l)	TN (mg/l)	TP (mg/l)
1	20	38.3	8.6	109.72	83.89	14.55	1.63
2	20	34.8	9.4	187.21	110.84	21.01	2.21
3	20	30.0	11.0	246.95	126.58	26.56	2.65
4	20	30.0	11.0	302.81	147.04	33.07	3.36

1. Flow rate

dissolved oxygen content and other variables were also determined. The nutrient contents in microbial cells were determined by the concentration difference of filtrate going through different size membranes (0.2 μm and 0.45 μm).

Some traditional analysis was also conducted in order to validate the results of this laboratory experiment. It shows that the results well represent the general characteristics of a biological stabilization pond. Therefore, the experimental data were used in determining the parameters of the model.

Parameter Determination
Parameter determination was conducted mathematically by the constrained change step way. The construction process of the objective function for carbon is illustrated as follows. Taking the equation for carbon as an example, including reformulating Eq.2-Eq.5 and introducing the function symbol of "F".

$$C_{1i} = F_i(\theta, C_1, C_2, C_3, C_4, \mu_{mBC}, \mu_{mAC},$$
$$K_{BC}, K_{AC}, k_{dBC}, k_{dAC}, Y, f(I)) \tag{14}$$

$$C_{2i} = F_2(\theta, C_2, C_3, \mu_{mBC}, K_{BC}, k_{dBC}) \tag{15}$$

$$C_{3i} = F_3(\theta, C_2, C_3, C_4, \mu_{mBC}, K_{BC}, k_{dBC}, k_{LC}, Y) \tag{16}$$

$$C_{4i} = F_4(\theta, C_1, C_4, \mu_{mAC}, K_{AC}, k_{dAC},$$
$$k_{LC}, k_{SC}, f(I)) \tag{17}$$

Introducing symbol "J" into the objective function, we have:

$$J(\mu, K, k, Y) = \sum_{j=1}^{N} \sum_{K=1}^{M} (F_K - C_{Kim})_j^2 \Rightarrow 0 \tag{18}$$

Where:

$C_{1i}, C_{2i}, C_{3i}, C_{4i} = C_{Ii}, C_{Bi}, C_{Oi}, C_{Ai}$, respectively;
N = number of the experiments;
M = number of the variables;
C_{Kim} = influent values of the variables.

Constrains: $\mu, K, k, Y > 0$ (19)

The objective functions for the nitrogen and phosphorus model are similar to the above equations. The microorganism species, nutrient levels and biochemical reactions are different for each stage of a multistage pond, therefore the dynamic parameters for each pond should be different. In the study, the parameters of carbon, nitrogen and phosphorus model for each pond have been determined based on our experimental data and the above mentioned method. The parameters of the model are shown in Tables 11.3-11.5.

Sensitivity Analysis
Since other factors in the experiments may influence the reliability of the parameter determination, sensitivity analysis was also conducted. The result of the analysis indicates that most of the determined parameters for the first pond are good and reliable. Some parameters for ponds 2 and 3 have shown either high sensitivity or low sensitivity, which are considered reasonable.

Model Verification

Model Solution Technique
This model is a set of non-linear equations. The Marquardt method by computer was used for model verification. The construction step for the expressions is illustrated below, again, by taking the carbon model as an example. First, we must reformulate Eq.2-Eq.5 to the form of Eq.14-Eq.17, and then set the objective function as:

$$Z(C_1, C_2, C_3, C_4) = \sum_{k=1}^{M} (F_k - C_{km})^2 \Rightarrow 0 \tag{20}$$

where: $C_1, C_2, C_3, C_4 = C_I, C_B, C_O, C_A$, respectively;

Constrains: $C_1, C_2, C_3, C_4 > 0$ (21)

Table 11.3. Parameters of carbon model

Parameters	Pond 1	Pond 2	Pond 3
Y	0.323	0.195	0.185
μmBC	1.003	0.126	0.090
μmAC	12.975	3.407	7.456
K_{BC}	0.001	0.001	2.689
K_{AC}	0.129	0.123	0.051
k_{dBC}	0.327	0.120	0.068
k_{dAC}	0.221	0.001	0.159
k_{LC}	0.001	0.001	0.071
k_{SC}	0.200	0.203	0.297

Table 11.4. Parameters of nitrogen model

Parameters	Pond 1	Pond 2	Pond 3
αN	0.547	0.080	0.001
μmBN	0.680	0.697	0.195
μmAN	12.026	13.649	2.529
K_{BN}	0.001	10.314	2.529
K_{AN}	0.001	9.953	8.811
k_{dBN}	0.981	0.313	0.001
k_{dAN}	0.412	0.001	0.011
k_{LN}	0.001	0.001	0.001
k_{SN}	0.001	0.346	0.104

Table 11.5. Parameters of phosphorus model

Parameters	Pond 1	Pond 2	Pond 3
μ_{mBP}	0.369	0.316	0.052
μ_{mAP}	8.686	0.167	0.011
K_{BP}	6.962	0.001	0.009
K_{AP}	0.001	0.001	0.001
k_{dBP}	0.001	0.001	0.001
k_{dAP}	0.028	0.001	0.001
k_{LC}	0.028	0.001	0.001
k_{SC}	0.161	0.191	0.144

Model Verification

The purpose of establishing a model is to further understand the properties of the biological stabilization pond and to provide useful information for pond design. It is obvious that the lab-scale pond is different from the full-scale pond in terms of light intensity and temperature. To ensure that the model established by the lab-scale pond operation data can properly describe the characteristics of a full-scale pond, information of two full-scale pond with 400 records has been used. The ponds are in Chengdu and Wuhan in China. In the process of verification, model parameters related to temperature were modified by Arrhenius equation. These parameters include K, k and α. Light limitation function, as shown in Equation 13 was used to modify light differences.

Fairly good consistence between the model outputs and the measurements from the full scale-ponds were obtained. Figure 11.4 provides a good example of this and demonstrates that the model structure is correct. The parameters determined are reliable and the model solution technique is successful. It also conveys that the law of material conservation is well represented by this model and that this static model can reflect the dynamic characteristics of a pond system and correctly describe pond operation behavior in different seasons.

Material Transformation Graphics

One application of the model is to simulate the general operation behavior of a three-stage pond system with hydraulic detention time of 5 days for pond 1 and 2.5 days for the others, in order to further understand the mechanism of biological stabilization pond systems. The simulation conditions of two cases with different concentration of materials are listed in Table 11.6. In addition, the original and transformation quantities of each variable in all three ponds for both cases are calculated by using the model. The material transformation graphics are depicted in accordance with the calculation results, Figures 11.5 and 11.6 show examples for carbon, where the numbers in rectangles are the original quantities, the arrow lines show the transformation directions and the widths of the lines vary to reflect

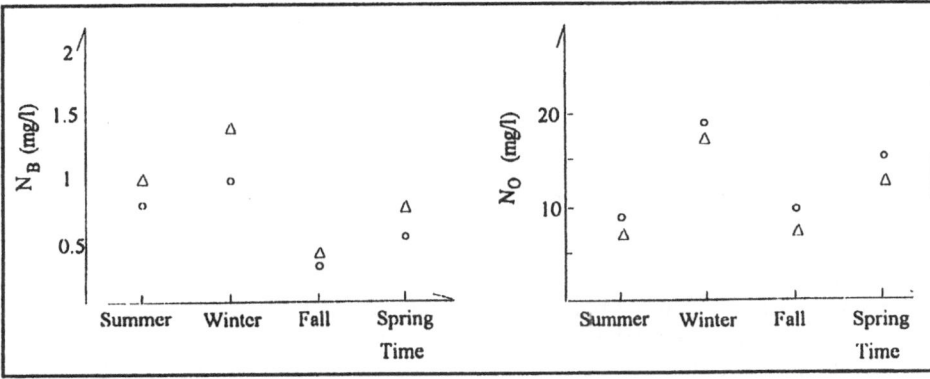

Fig. 11.4. Model verification examples. △ represents the model calculation result. O represents the real pond record.

Table 11.6. Simulation conditions (in mg/d)

Element (mg/l)	case 1	case 2
C_{Oi}	24.00	72.00
T_{Ni}	14.20	23.80
N_{Ii}	2.13	3.57
N_{Oi}	12.00	20.23
P_{Ti}	1.47	2.46

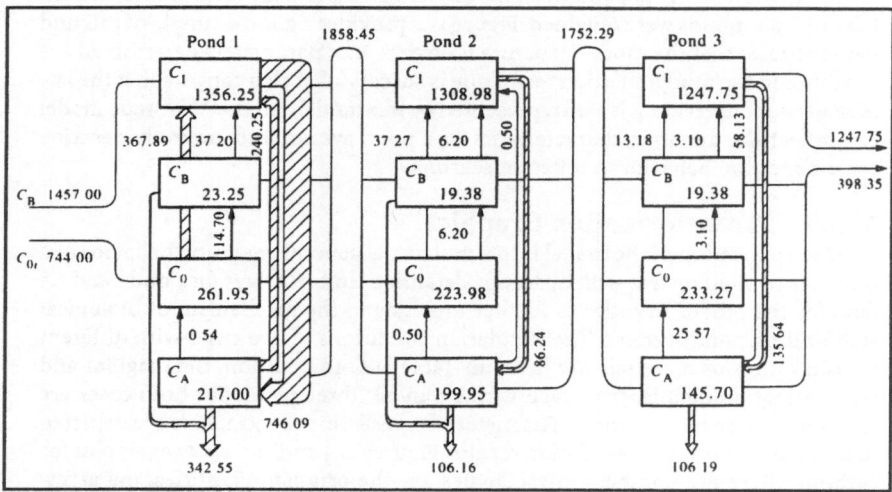

Fig. 11.5. Carbon transformation graphic (case 1). Reprinted from Water Res 1994; 28(7):1659-1665, with kind permission from Elsevier Science Ltd.

different quantities in the transformation. The analysis of the transformation of individual elements in the experiments follows. The graphs for nitrogen and phosphorus have been analyzed in a separate report.[15]

The organic carbon removal rate in the case 1 simulation is 68.65% in the entire pond system, but most of the carbon (64.74%) was removed in pond 1. In case 2, total organic carbon removal rate is 86.96% while pond 1 removed 80.51% of the total carbon in circulation. As we can see from Figures 11.5 and 11.6, the transformation flux of each kind of carbon element reaches its highest value in pond 1. This indicates that there is a close relation between carbon transformation and carbon removal. The more active the carbon transformation process, the higher the removal of organic carbon. This is also supported by the different results of case 1 and case 2. Comparing Figures 11.5 and 11.6, we find that the arrow line representing transformation flux of a pond in case 2 is much wider than that of the corresponding pond in case 1, and as mentioned earlier, the organic carbon removal rate is higher in case 2 than in case 1.

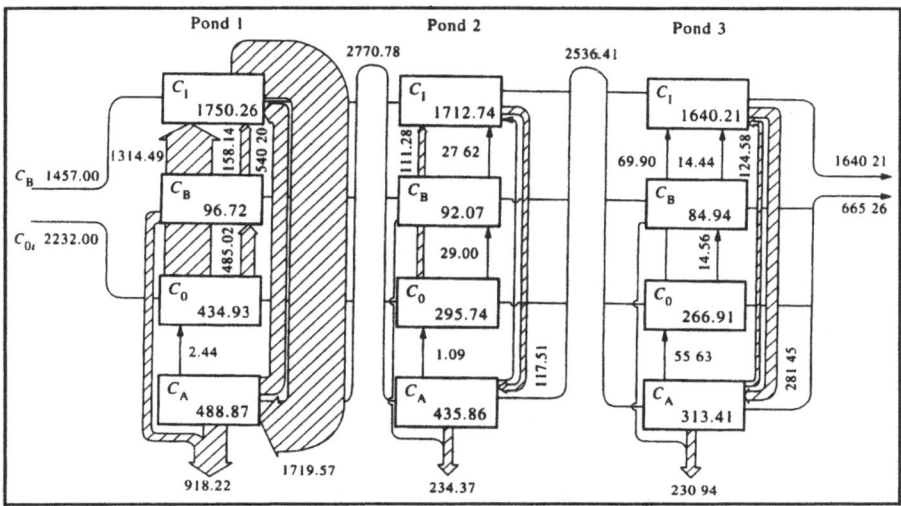

Fig. 11.6. Carbon transformation graphic (case 2). Reprinted from Water Res 1994; 28(7):1659-1665, with kind permission from Elsevier Science Ltd.

In view of the active carbon transformation in pond 1, let us look at the fate, or the ultimate materials after transformation in this pond. The proportion of the fate of different carbon in transformation is listed in Table 11.7. It shows that in case 1 the fate of carbon after transformation is mainly in three forms in the sequence of total weight: Sediment > Algae > Bacteria. The total weight of inorganic carbon did not increase but instead declined, although organic carbon was constantly transforming. The inorganic carbon resulting from bacterial decomposition of organic carbon was not enough to support the growth of algae. Therefore, the inorganic carbon reservoir had to make some supplemental contributions. In case 2, the fate of carbon is mainly in four forms in the sequence of total weight: Sedimentation > Algae > Inorganic Carbon > Bacteria. In this case, the inorganic carbon reservoir enlarged slightly because the algae needed less than the bacterial decomposition of organic carbon produced. Figures 11.5 and 11.6 also show that the majority of the sedimentation is algae which caused 84.16% and 74.93% of the sedimentation in case 1 and case 2, respectively. Also, 58.30% of the removed organic carbon transformed to algae in case 1, while 52.74% of that transformed to algae in case 2. In other words, algae play a very significant role in the process of organic carbon removal, and are the largest source of energy and nutrition in a biological stabilization pond.

Once algae discharges into a receiving water body, it creates turbidity and releases a lot of nutrition from lysis and decay of algae. It also accelerates the eutrophication of natural water. Therefore, close attention must be given to algae disposal systems in practice.

The above analysis indicates that the symbiotic relationship of algae and bacteria is a fundamental mechanism in biological stabilization ponds. This is very different from the mechanism in traditional biological treatment systems. From a carbon transformation point of view, organic carbon mainly transforms to bacteria

Table 11.7. Transformation fates of organic carbon in Pond 1

C_{oi} 100	Fate	case 1	case 2	Notice
	C_I	-13.53	13.14	1. unit: %
	C_B	3.12	4.32	2. The minus value of
	C_A	23.36	21.91	C_I in case 1 shows
	Sediment	38.31	41.14	its decrease
	Removal	64.79	80.51	

and CO_2 in an activated sludge process. In this process, oxygen demanded by bacteria should be supplied by mechanical aeration, and the produced CO_2 is stripped into the air. As a result, organic carbon is eliminated substantially in the waste water. It is obvious that energy and carbon sources are wasted in this way. Furthermore, extra facilities must be put in place to dispose surplus sludge and to stabilize the proliferated bacteria. In a stabilization pond system, the oxygen needed for propagation of bacteria is provided by algal photosynthesis and in turn, algae absorbs the CO_2 produced by bacterial decomposition of organic carbon. It becomes possible to retrieve the carbon source. Therefore, waste water stabilization ponds are more economical than activated sludge systems in terms of oxygen provision and carbon utilization. Additionally, since the food chain in a pond system is longer than that in a traditional biological waste water treatment system, the pond system has better operational stability. However, it should be pointed out that to ensure the stabilization of waste water, disposal of algae from the effluent needs to be considered. In addition, to maximize the economic value of a stabilization pond system, it is important to consider using algae, or growing aquatic plants and breeding aquatic animals.

As per the transformation graphics, the inorganic carbon reservoir in a pond is relatively stable in both cases 1 and 2. This means that the carbon in transformation is active and that the carbon dioxide absorbed by algae comes primarily from bacterial decomposition of organic carbon. The release and absorption of carbon dioxide is just the critical regime between bacteria and algae.

Prospect of the Model Application

Design

If the influent quality is known and the effluent quality is defined, the model can calculate out a proper pond retention time. Take the carbon model as an example. Make an objective function S(0):

$$S(\theta) = \sum_{k=1}^{M} (F_k - C_{ik})^2 \Rightarrow 0 \qquad\qquad 22)$$

Where: C_{ik} = the influent quality;
M = Number of the water criteria.

With Equation 22, an optimized θ can be obtained.

Predication of the Pond Operation Situations
The pond operation situations can be calculated with the model. This is helpful for pond control and management.

Conclusions and Recommendations
The following conclusions can be reached based on this study:

1) A material transformation model which can systematically describe the carbon, nitrogen and phosphorus transformation in biological stabilization pond has been established based on ecological principles and experimental analyses. All parameter values in the model were determined by model calibration on experimental research data.

2) Parameter sensitivity analysis and model verification have indicated that: the model structure is correct; the determined parameters are valid; the model solution technique is successful and the model has performed well in representing the law of material conservation. This model will therefore have unique advantages in real applications.

3) Graphic analysis shows that there is a close relationship between the removal and transformation of carbon. When carbon transformation is active, the organic carbon removal rate is high. Through analysis of the transformation fates of organic carbon, the mechanisms of a waste water stabilization pond and its differences from traditional biological treatment systems may be recognized and further interpreted.

References
1. Odum EP. Basis of Ecology (Chinese version). Education Publisher, China 1981.
2. McKinney RF. Functional characteristics unique to ponds, In: Ponds as a Waste Water Treatment Alternative. University of Texas, Austin 1976.
3. Hong HJ, Gloyna EF. Phosphorus models for waste stabilization ponds. J Envir Eng Div ASCE 1984; 110:550-561.
4. Buhr HO, Miller SB. A dynamic model of the high-rate algae-bacteria waste water treatment pond. Water Res 1979; 7:29-37.
5. Fritz JJ. Dynamic process modeling of wastewater stabilization ponds. JWPCF. 1979; 51:2724-2743.
6. Ferarra RA, Harleman DRF. Dynamic nutrient cycle model for waste stabilization ponds. J Environ Eng Div ASCE 1980; 106:37-54.
7. Isao Somiya, Shigeo Fujii. Material balance of organic nutrients in an oxidation pond. Wat Res 1984; 106:325-333.
8. Gu R, Stefan HG. Stratification dynamics in wastewater stabilization ponds. Wat Res 1995; 28:1909-1923.
9. Azov Y, Tregubova I. Nitrification processes in stabilization reservoirs. Water Sci Technol 1995; 31:313-323.
10. Wang B, Dong W, Zhang J, Cao X. Experimental study of high-rate pond system treating piggery wastewater. Water Sci Technol 1996; 34:125-132.
11. Wen XH. A study on the transformation of carbon, nitrogen and phosphorus in biological stabilization pond system, Dissertation of Ph D. Tsinghua University, China 1991.

12. Ferarra RA, Avci CB. Nitrogen dynamics in waste stabilization ponds. JWPCF 1982; 54:361-369.
13. Pano A, Middlebrooks EJ. Ammonia nitrogen removal in facultative wastewater stabilization ponds. JWPCF 1982; 54:352-360.
14. Shelef G. High-rate algae ponds for wastewater treatment and protein production. Water Sci Tech 1982; 4:439-452.
15. Wen XH, Qian Y, Gu XS. Graphical presentation of the transformation of some nutrients in a wastewater stabilization pond system. Water Res 1994; 28: 1659-1665.

CHAPTER 12

Limits to Growth

Michael A. Borowitzka

Introduction

Urban, industrial and agricultural wastewaters contain up to three magnitudes higher concentrations of total nitrogen and phosphorous, compared with natural water bodies.[1] Normal primary and secondary treatment of these wastewaters eliminates the easily settled materials and oxidizes the organic material present, but does not remove the nutrients which will cause eutrophication of the rivers or lakes into which these wastewaters may be discharged. Tertiary treatment of the effluent is therefore required, and both chemical and physical methods which are used are very expensive. Oswald[2] estimates that the relative cost of tertiary treatment to remove PO_4^{3-}, NH_4^+ and NO_3^- is about 4 times the cost of primary treatment. Higher orders of treatment, such as quaternary treatment required to remove refractory organics and organic and inorganic toxicants and quinary treatment to remove inorganic salts and heavy metals, are 8 to 16 times as expensive as primary treatment. Algae can be used as a biological alternative tertiary treatment and also for the removal of heavy metals and possibly other toxic substances.[3,4] The possibility exists that the algae produced in these systems can be used as animal feed supplements,[5,6] or be composted. The use of waste-grown algae may ultimately also have application in closed cycle life-support systems,[7,8] or may be used in conjunction with power stations, not only to treat wastewaters, but also to act as a CO_2 sink for the amelioration of the impact of greenhouse gases.[9-13]

The biotreatment of wastewaters with algae to remove nutrients such as nitrogen and phosphorous, and to provide oxygen for aerobic bacteria was proposed over 40 years ago by Oswald and Gotaas.[14] Since then there have been numerous laboratory and pilot studies of this process and several treatment plants using various algal treatment systems have been constructed.[2,15] Microalgal systems for the treatment of other wastewaters such as piggery effluent,[16-18] the effluent from food processing factories,[19,20] highly saline tannery wastewaters[21] and agricultural wastes[22,23] have also been studied. More recently, algae-based systems for the removal of toxic minerals such as Pb, Cd, Hg, Se, Sn, Ni, As and Br are also being developed.[24-28]

Wastewater Treatment with Algae, edited by Yuk-Shan Wong and Nora F.Y. Tam.
© Springer - Verlag and Landes Bioscience 1998.

Because of the large volumes to be treated, most algal wastewater treatment systems use large outdoor ponds. These may either be waste stabilization ponds (WSP) or high rate oxidation ponds (HROP). The WSPs usually consist of facultative and maturation ponds in series and rely to some degree on algae for their successful operation. A non-algal anaerobic pretreatment pond also may be included. The HROP system, on the other hand relies wholly on the interaction between algae and bacteria for complete treatment. Hybrid systems, consisting either of anaerobic pretreatment ponds followed by HROPs, or of WSPs followed by HROPs have also been considered.[29] Some smaller closed systems are also under consideration for more specialized applications.[30]

In high rate oxidation ponds the algae species of particular interest are: 1) for freshwater systems—the green algae *Chlorella, Chlamydomonas, Scenedesmus, Micractinium* and *Oocystis,* the euglenophyte *Euglena,* and the blue-green algae (cyanobacteria) *Oscillatoria* and *Phormidium;* 2) for marine systems—the diatoms *Phaeodactylum, Nitzschia* and *Skeletonema* and the green alga *Tetraselmis.*

Over the last 25 years there have been major advances in the large-scale culture of microalgae with the primary purpose being to produce algal biomass for food or feed, and to use the algae as a source of valuable biochemicals; commercial operations exist in Australia, USA, Israel, Japan, China, India, Thailand and Taiwan.[31,32] The development of these operations, in particular, has led to significant advances in our understanding of the biology and ecology of commercial-scale algal culture, and in the engineering of large-scale culture systems and harvesting methods. The experiences gained with these systems are equally relevant to wastewater treatment systems as to 'clean' systems. This chapter will focus on the limiting factors on large-scale algal culture, with particular emphasis on high rate wastewater oxidation ponds.

Limiting Factors to Algal Growth

Algal growth in HROPs is influenced by a number of physical and biotic factors and these are summarized in Table 12.1. Furthermore, effective operation of an HROP requires a stable algal community, and the first prerequisite for this is to establish an equilibrium between algal growth rate and the rate of algal dilution by the wastewater input stream. If this is achieved, good algal growth rates can be maintained for a long time. The maintenance of good algal growth rates, and therefore reduced residence time, requires an optimization of light and nutrient utilization, the avoidance of toxic levels of certain compounds in the feed stream, a control of potential predators and algal pathogens and the maintenance of an optimal algal population size by regular harvesting of the algal biomass. These are considered in more detail below.

High algal productivity is essential for successful economical operation of high-rate oxidation ponds. The problems of optimizing growth rate and photosynthetic efficiency in HROPs are the same as those which affect the microalgal industry in general. These problems are the maximization of photosynthetic efficiency, nutrient supply and assimilation efficiency (especially the supply of CO_2) and the control of potential predators and competitors. HROPs have additional problems resulting from the variable nutrient load, the presence of light absorbing substances in the water and the possible presence of inhibitors and/or toxins in the waste stream.[33-35]

Table 12.1. Factors influencing algal growth

Abiotic factors	
	- Light (quality, quantity)
	- Temperature
	- Nutrient concentration (esp. N, P and organic C)
	- O_2, CO_2
	- pH
	- Salinity
	- Toxic chemicals
Biotic factors	
	- Pathogens (bacteria, fungi, viruses)
	- Predation by zooplankton
	- Competition
Operational Factors	
	- Mixing
	- Dilution rate
	- Depth
	- Addition of bicarbonate
	- Harvest frequency

Light and Temperature

Although some algal species can grow heterotrophically in the dark, most algae are phototrophs and may grow either photoautotrophically or photoheterotrophically (mixotrophically). Light is therefore of fundamental importance and, in algal cultures, light rapidly becomes limiting due to absorption by the algal cells in the water column.[36] In HROPs this is exacerbated by the high content of particulate matter and the presence of colored substances in the wastewater. For this reason, almost all open algal culture systems are shallow, with depths ranging from about 15 to 30 cm.[37,38] The generally observed relationship between pond depth and productivity observed in many raceway systems is shown in Figure 12.1.

In outdoor cultures, algal yields are broadly correlated with incident radiation,[39] however, the relationship between growth, biomass yield and irradiance is complex.[37,40,41] The relationship between photosynthetic rate and irradiance in dilute cultures, at constant temperature and in a constant light field, can be approximated by several mathematical models such as the rectangular hyperbola or by a logistical equation. At high irradiances photoinhibition may occur and this has been incorporated into some models (Table 12.2).

In large-scale cultures the relationship between light and actual productivity is further complicated as the algal population density may vary, and in outdoor cultures the amount of light also varies over the day as well as over the seasons.

The amount of light I at depth d can be described by Beer Lambert's Law:[42]

$$I = I_o e^{(-\eta_r d)} \qquad \text{1)}$$

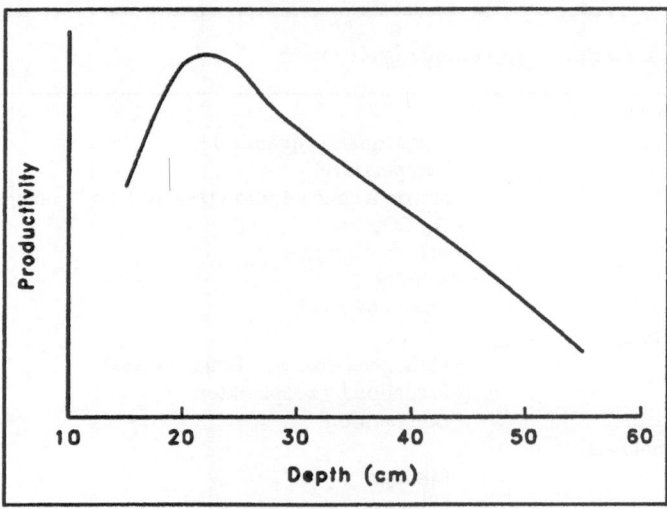

Fig. 12.1. The general relationship between pond depth and productivity (g.m⁻².d⁻¹) observed in many raceway systems.[37,153]

---

*Table 12.2. Mathematical models of the relationship between irradiance and
photosynthesis in algal cultures*

---

(1) Hyperbolic tangent model[151]

$$P_i = P_{max}\, tanh(\frac{\alpha I_i}{P_{max}})$$

(2) Exponential model (without photoinhibition)

$$P_i = P_{max}(1 - e^{(-\alpha I_i/P_{max})})$$

(3) Exponential model (including photoinhibition term)[152]

$$P_i = P_{max}(1 - e^{(-\alpha I_i/P_{max})})e^{(-\beta I_i/P_{max})}$$

Where P_i is the gross photosynthetic rate at irradiance I_i, P_{max} is the maximum gross
photosynthetic rate, I = the irradiance, α = the initial slope of the P/I curve, and β = a factor
representing the degree of photoinhibition.

---

where I_o is the surface irradiance and ν_e is the overall extinction coefficient. In
shallow ponds there is no significant effect of water on ν, and in 'clean' algal cultures ν_e can be formulated as

$$\eta_e = \eta_s + \eta_c\, [Chl\, a] \qquad\qquad 2)$$

where ν_s is the extinction coefficient of any suspended solids (~ 15 m⁻¹), ν_c is
the specific extinction coefficient of chlorophyll a (0.11 m⁻².mg⁻¹ Chl a) and [Chl
a] is the concentration of chlorophyll a (mg.m⁻³). In wastewaters a third term, ν_d,
the extinction coefficient of any dissolved colored material present, needs to be
added to Equation 2.

These equations show that increasing cell densities (increasing chlorophyll) rapidly reduce the amount of light reaching the algae deeper in the pond, leading to light limitation of the cells. What these equations do not show is that there is also a narrowing of the light field due to differential absorption by the algal cells and any colored material as the light passes through the water. This has been modelled by Kroon et al.[43] The outcome of this is shown in Figure 12.2 which illustrates the relationship between algal cell density, cell yield (productivity) and specific growth rate over a range of algal population densities for outdoor cultures of *Spirulina* under conditions where nutrients and temperature are not growth-limiting. However, the maximum yield outdoors does not coincide with the highest specific growth rate; the highest specific growth rate is achieved at the lowest cell density when mutual shading and light limitation are minimal, and, with increasing population density, the specific growth rate decreases as the culture becomes more light-limited. The highest yield, however, is achieved where the specific growth rate is not at maximum, but where the net photosynthetic efficiency is at its highest; thus the optimum irradiance is not the maximum irradiance. At low cell densities the photosynthetic efficiency is actually reduced due to photoinhibition,[44,45] however, this is unlikely to be important in most wastewater treatment ponds due to the relatively high algal and bacterial biomass, the presence of light absorbing dissolved matter in the water and the depth of the ponds.

In dense cultures, not all the cells receive the maximum amount of light all the time, rather the cells are exposed to a variable and intermittent light regime depending upon the degree of mixing (turbulence) and the length of the pond. If there is little mixing, as is the case in WSPs, the algae at the surface receive high, and possibly injurious, levels of irradiance whilst deeper in the water column the algae are light-limited. The higher productivities of HROPs are due not only to the shallower depths, but most importantly due to mixing which ensures that all cells are kept in suspension in the water column and are exposed regularly to light. There are many studies which have shown that increased flow rates in raceway ponds result in increased productivities.[41,46,47] Increased mixing also increases the optimum cell density due to the improved light regime and the positive effect of increased turbulence (mixing) is potentially greater at high irradiances.[48,49] The relationship between yield, mixing rate and population density is shown diagrammatically in Figure 12.3.

The exposure of the cells alternately to high-light/ low-light conditions also appears to enhance growth in some cases. This has been shown practically by Laws et al[50] who placed airplane-type wings in the flow of their cultures in order to increase turbulence and thus the frequency of the light/dark cycle; by appropriate spacing of the 'wings', they could optimize the high-light/low-light regime and thus enhance the yield. The exact light/dark cycle required to enhance productivity remains a matter of controversy, but Terry[51] has shown that in order to achieve any significant effect from alternating light/dark conditions the frequency must be ≥ 1 Hz, with high irradiances and a light/dark ratio of <1. However, in raceway ponds with depths of 20-30 cm and flow rates of 10-30 cm.sec^{-1} the normal frequency of the light/dark cycles is in the range of seconds to minutes and it appears unlikely that this phenomenon can be exploited practically other than in shallow systems such as the sloping reactors used at Trebon in the Czech Republic for the culture of *Chlorella*.[52,53]

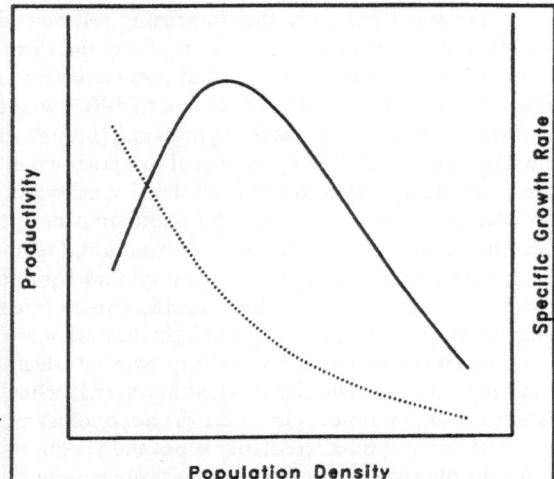

Fig. 12.2. Effect of population density (g.l⁻¹) on the specific growth rate (dotted line; day⁻¹) and the productivity (solid line, g.m⁻².d⁻¹) in outdoor raceway cultures of *Spirulina platensis*. Based on data in references 40,41,47.

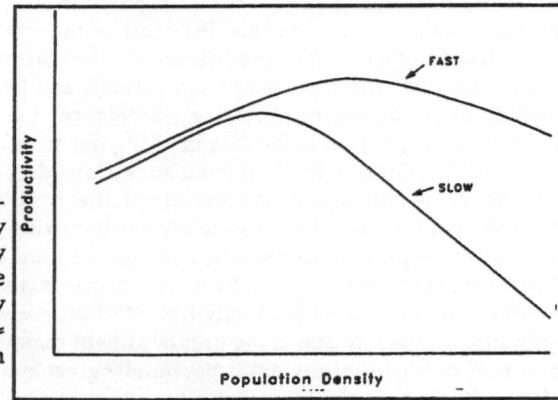

Fig. 12.3. Relationship between population density (g.l⁻¹) and productivity (g.m⁻².d⁻¹) as affected by the rate of mixing in a raceway pond. Slow ≈ 15 cm.s⁻¹, Fast ≈ 50 cm.s⁻¹ (based on data from *Spirulina platensis*).[40,41,47]

Irradiance also interacts strongly with temperature. The growth rate of algae increases with increasing temperature until the optimum temperature is reached. Further increases in temperature, beyond the optimum, usually lead to a rapid decline in the growth rate.[54,55] At temperatures close to the optimum for growth, the algae are also better able to tolerate much higher irradiances before photoinhibition sets in.[39,55,56] Figure 12.4 shows the interaction between temperature and irradiance for three species of algae. *Oscillatoria agardhii* is a planktonic filamentous cyanobacterium typical of cold temperate eutrophic waters, *Ankistrodesmus falcatus* is a unicellular chlorophyte typically found in oxidation ponds, and *Phormidium bohneri* is a benthic cyanobacterium typically found lining sewerage pipes in warm to tropical climates. The temperature response of these algae reflects their adaptation to the temperature conditions found in their natural environment.

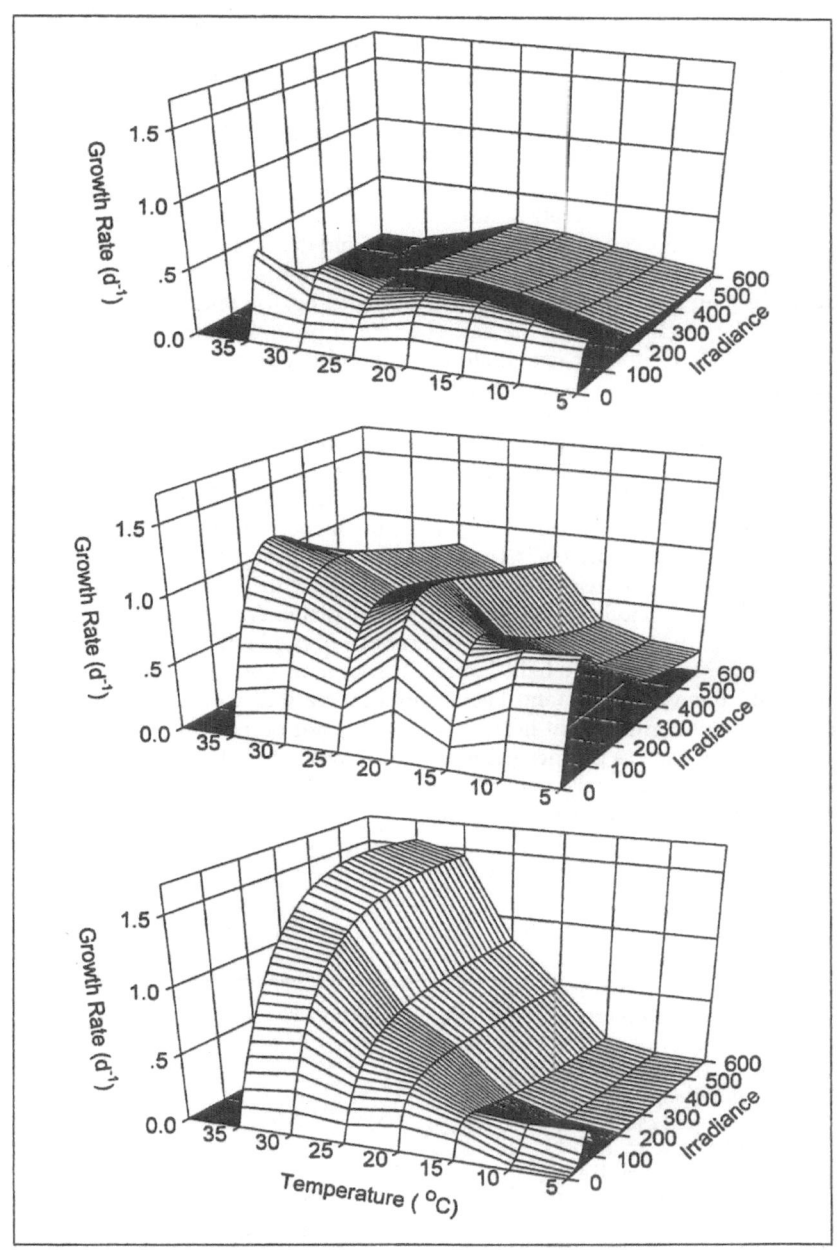

Fig. 12.4. Effect of irradiance and temperature on the growth rate of three species of
micro–algae. Top = *Oscillatoria agardhii*, Middle = *Ankistrodesmus falcatus*, Bottom
= *Phormidium bohneri*. Irradiance = μmol photons.m-2.s-1. (Recalculated from data
in ref. 56).

This interaction between light and temperature and its effects on productivity also has been observed in outdoor cultures. For example, Vonshak et al[41] found that the specific growth rate of *Spirulina* declined from summer to winter. This decline was most pronounced at low algal densities compared with higher densities (Fig 12.5). Similar results have also been obtained by Castillo et al[57] in Peru with *Scenedesmus*. Not only are there seasonal changes in temperature and total irradiance, but there are also diurnal variations, and maximum pond productivity will only occur when both pond temperature and irradiance are near their optimum, something which rarely occurs. Ponds cool down at night and in the morning the pond temperature is often sub-optimal and the algae therefore cannot utilize all the available light; i.e., they are temperature-limited for part of the day. Depending on how long the pond takes to warm up, the optimum productivity will only be reached later in the day. The rate of pond warming will depend, of course, mainly on the air temperature and pond depth; shallower ponds warm faster but, they also cool faster at night. Therefore, consideration might be given to altering the pond depth between night and day.

In cold climates algal ponds will have to be heated if adequate algal growth is to occur.[58,59] Temperature limitation in colder climates also can be overcome, in part, by using algae with lower temperature optima.[56] An alternative is to use greenhouses and this has been tested in studies in Quebec, Canada,[60] New England, USA and in Stensund, Sweden.[61] The improvement in productivity with increasing temperature has been well demonstrated by Lee et al[62] in cultures of *Chlorella*. On the other hand, Svoboda and Fallowfield[18] found that intermittent heating of algal raceways treating piggery waste had little effect on productivity and nutrient removal as the cultures were light-limited. However, in non-light-limited cultures temperature is very important. The economical feasibility of heating ponds remains to be determined, but heating is probably not economical unless a cheap source of heat such as geothermal hot water is available.[63] The fact remains that low temperatures generally coincide with low irradiances so that any benefit of heating is counteracted by a lack of light.

Fig. 12.5. Specific growth rate of *Spirulina platensis* grown in a raceway pond at Sede Boqer, Israel, as affected by population density (absorbance at 450 nm) and time of the year. Top curve = summer; midle curve = autumn and spring; bottom curve = winter (redrawn from ref. 41).

Not only do photosynthetic rates and productivity increase with increasing temperature, so does the respiration rate of the algae. This is of particular relevance at night where there is a net loss of up to 35% of the algal biomass for 'clean' *Spirulina* cultures[64,65] and even greater for algae in an HROP.[66] This net loss, due to night-time respiration, may reduce the productivity on the next day. Dark respiration rates increase with growth rates, and the ratio between respiration and growth increases even further under non-optimal conditions.[67] Since high algal growth rates are generally correlated with a high efficiency of nutrient removal in algal wastewater treatment systems, the losses due to dark respiration are therefore a cost which cannot be avoided. However, the losses can be minimized by ensuring the algae are growing under near-optimal conditions. Changes in temperature may also lead to shifts in the species present in the pond and this will also affect performance.[58]

The relationship between irradiance and temperature has been modelled by several workers, and such models can be used to determine the optimum conditions for biomass production or to estimate the potential yield of an algal system.[68] The following describes the model developed by Grobbelaar et al.[69]

A basic equation for productivity can be written as:

PROD(mg dry wt. $m^{-2}.h^{-1}$)=

PRD-RES-INB   3)

where:

PROD = 'productivity'
PRD = 'gross productivity'
RES = 'respiration'
INB = 'photoinhibition'.

PRD can be calculated from inputs of biomass concentration present in the culture, culture temperature, and light impinging on the surface of the culture:

$$PRD = A_1 X_1 (A_2)^T \frac{I_z I_s (A_3)^T}{I_z + I_s (A_3)^T}$$ 4)

where A_1 - A_3 are constants, X_1 is the biomass concentration in mg dry wt.l^{-1}, I_s is the light half saturation constant in mol photons.$m^{-2}.h^{-1}$, and T is a temperature factor. The constant A_1 can be interpreted as the efficiency of light utilization (photosynthetic efficiency), A_2 as the Q_{10} of photosynthesis, A_3 as the Q_{10} for light half saturation, and the factor $I_s(A_3)^T$ as the temperature dependence of I_s (light half saturation constant) in the photosynthesis versus irradiance curve of photosynthesis.[70]

The component RES is a total loss factor, i.e., losses due to respiration and exudation of organic compounds from the cells. RES increases with an increase in temperature[71] and, on the basis of experience with outdoor cultures,[72] can be approximated by

$$RES = X_1 \frac{1.5^T - 0.54}{100}$$ 5)

Photoinhibition is also temperature-dependent,[55] and an approximation based on experience with outdoor cultures[69] is given by

$$INB + PRD \frac{2.5^T}{7.5} I_z$$ 6)

when $I_z > 1$ mol photon.$m^{-2}.h^{-1}$.

The above model gives good agreement with the data collected over a 16-month period for cultures of *Spirulina* and *Coelastrum* grown in several different raceway ponds at Dortmund in Germany. However, the model could not fully account for the afternoon reduction in photosynthesis, which occurred sometimes.[73] Other models for microalgal production in HROPs have been developed[43,68,74] and Guterman et al[75] have developed a macromodel for outdoor algal mass production. These models show reasonable agreement with actual HROP pond performance.[74] They can be used as an aid in pond design and management, and focus primarily on the effects of light and temperature, however, they cannot yet account satisfactorily for other factors such as zooplankton grazing, micronutrient deficiency, ammonia toxicity etc.[76]

Nutrients

Carbon

The algae in HROPs utilize both inorganic carbon (CO_2 and HCO_3^-) and organic carbon. Unlike 'clean' algal culture systems, HROPs have a high organic load and this organic carbon may be utilized by the algae either photoheterotrophically near the surface, or heterotrophically in the deeper parts of the pond. The organic load is probably responsible for the usual pattern of species dominance observed in HROPs; i.e., at high loads mainly *Euglena* and *Chlamydomonas*, at intermediate load mainly *Chlorella, Scenedesmus* and *Micractinium* and, at low loads, other algae.[77,78]

Inorganic carbon in HROPs is derived from both atmospheric CO_2 and CO_2 produced by bacterial respiration. Data presented by Abeliovich[35] and Abeliovich and Weismann[79] indicate that the amount of bacterially-generated CO_2 must be quite small, and that about 25 to 50% of the algal carbon is derived from heterotrophic utilization of organic carbon. Algae which are able to grow mixotrophically (photo-heterotrophically) grow faster in the light when both inorganic and organic carbon sources are available.[80-82] Although little work has been done to study the (photo) heterotrophic potential of microalgae found in HROPs, it is likely that these algae are very efficient at organic carbon utilization.[83,84] For example, Martínez et al[85] have isolated a strain of *Chlorella vulgaris* from the wastewater ponds of a sugar refinery which requires glucose to reach maximal growth rate, even under saturating light conditions. In fact, the growth of this alga with glucose in light was equal to or greater than the sum of photoautotrophic (light without glucose) and heterotrophic (dark with glucose) growth. This has also been observed by Ogawa and Aiba[86] with *Chlorella* and *Scenedesmus*. It is therefore very likely that the dominant algae in HROPs are efficient (photo)heterotrophs.

Nitrogen

The nature of the available organic-N sources may also influence the performance of the algal wastewater treatment system. Lau et al[87] have shown that the degree of N and P uptake and their reduction in the wastewater by *Chlorella pyrenoidosa* depended on the types of organic-N sources in the water. This was due to both direct effects on algal metabolism as well as their effects on bacterial metabolism.

Algae in HROPs are generally not nutrient limited, however, the ammonia concentration in the pond is an important factor in the successful management of the HROP. 2.0 mM total ammonia (NH_4^+ + NH_3) at pH 8.1 has been shown to inhibit photosynthetic O_2 evolution by about 50% in many algal species,[88,89] and high concentrations of ammonia can lead to death of the algae.[90,91] This effect is pH-dependent and is intensified at higher temperatures since a higher proportion of the total ammonia occurs as free ammonia.[92] Therefore, one of the main concerns in managing HROPs is to determine the dilution time of water in HROPs so as to avoid this critical concentration of total ammonia since it would reduce the algal growth rate, causing washout of the algae; total ammonia concentration generally should not exceed 1.5 mM. High nutrient containing wastewaters, such as aerobic or anaerobic secondary effluents of piggery wastewaters, therefore may have to be diluted before tertiary treatment with microalgae.[93,94] An alternative strategy is to use algae with a higher ammonia tolerance such as the strain of *Phormidium* used by Cañizares-Villanueva et al.[17]

Phosphorous

Phosphorous is rarely limiting in sewage or wastewaters derived from animal husbandry, however, nitrogen may become limiting. The optimum N:P ratio for phytoplankton growth generally is about 15:1, and high rations (i.e., about 30:1) suggest P-limitation, whereas low ratios of about 5:1 suggest N-limitation.[95] Individual species may, however, deviate from these ratios.[96] This conclusion is supported by the study of Tam and Wong[97] who compared growth of *Chlorella pyrenoidosa* on primary-settled sewage and secondarily-treated effluent and show that the alga grew best on the primary-settled sewage which has a N:P ratio of 5.7 compared to a ratio of <2 for the secondarily-treated sewage.

Other Nutrients

Although the supply of N and C is critical, other nutrients also affect the growth and metabolism of the algae. If diatoms are grown in the wastewater system then silicon can rapidly become limiting. For example, in outdoor ponds used to treat marine fish-farm effluent a silicon:phosphorous ratio of about 4:1 was found to be optimal for biomass production and nutrient removal.[98] The main alga in this system was the diatom *Skeletonema costatum*. Langis and coworkers[99] have also postulated that the growth inhibition of *Selenastrum capricornicum* which resulted from the addition of dissolved organic matter from treated secondary wastewater effluent was caused by binding of some trace elements, possibly iron.

pH

Photosynthetic CO_2-uptake will raise the pH in the pond and, if there is a high concentration of total ammonia, i.e., >2.0 mM, inhibition of photosynthesis will occur when the pH reaches about 8.1. The only effective method of lowering the pH is by respiration and, therefore, a balance must be maintained between photosynthesis which occurs mainly near the surface of the pond, and respiration which occurs throughout the water column. Thus, when ammonia concentration is high, a careful balance should be maintained between its concentration and BOD, to enable the continued lowering of pH via respiration; i.e., if ammonia concentration is high then the BOD should also be high.[100] Furthermore, if algal ponds are fed by diluted wastewater or by secondary effluents, carbon limitation may develop[78,101] and this will, in turn, limit the rate of nutrient uptake, especially nitrogen uptake.

pH also affects the availability of inorganic carbon; at a pH greater than 9, most of the inorganic carbon is in the form of carbonate and cannot be taken up by the algae for photosynthesis. High pH may also lead to flocculation of the algae[102,103] and this may in turn lead to reduced nutrient uptake and reduced growth. Such high pH values may be reached in the cultures during the day when photosynthetic rates are high and where the buffering capacity of the medium is low.[104,105] High pH can also lead to phosphate precipitation, however, this phosphate will be resolubilized when the pH drops at night.

Photorespiration and Photoinhibition

Photorespiration results from the competition between O_2 and CO_2 at the active site of ribulose bis-phosphate carboxylase (RuBisco) and may often be greater than conventional dark respiration. Although photorespiration has often been demonstrated in laboratory cultures, it is more difficult to demonstrate in large outdoor cultures and its effect on pond productivity is unclear. In wastewater ponds the rate of photosynthetically produced O_2 build-up is reduced compared to 'clean' laboratory cultures because of the presence of bacteria and zooplankton, however, it is possible that the afternoon inhibition in productivity observed in some HROPs may be caused by photorespiration.

Photoinhibition has been suggested as a factor influencing pond productivity,[106] however, it unlikely to be important in well mixed HROPs as there is intensive self-shading and the cells are only exposed to damaging irradiances for short periods. However, in very long channel lengths turbulence is reduced and the system may approach laminar flow in the sections away from the paddle wheel, resulting in the surface cells being exposed to high light for longer periods. Ultraviolet radiation may also induce photoinhibition,[107] however, the UV light is rapidly absorbed in HROPs.[43]

Turbulence

Turbulence affects the algae in several ways. Some degree of turbulence is necessary to ensure that the algae remain suspended in the water column and can reduce the possibility of some parts of the ponds becoming anoxic, something which is not desirable in HROPs. Turbulence also exposes all the cells to light, thus enhancing productivity in high density cultures (see above). Turbulence also reduces the unstirred boundary layer around the algal cell, enhancing the mass transfer rate of nutrients and metabolites. An increase in algal productivity with increasing turbulence (mixing) has been observed by several workers,[46,50] however, the actual causes of the increased productivity remain obscure. Grobbelaar[108] has been able to separate at least part of the effects of the light/dark fluctuations and of the variations in mass transfer rates caused by changes in turbulence in laboratory studies. High turbulences may, however, be damaging to the algae.[109,110]

The mode of mixing may also affect the algal species composition in the ponds. In the study of Azov et al,[111] it was found that mixing by pumps and jet aerators favored *Chlorella* sp., while mixing with cage aerators favored *Scenedesmus* and *Microactinium* as well. Where as the growth of *Euglena* was favored in the absence of mixing. In another study with a HROP receiving piggery waste, the cyanophyte

Synechocystis sp. grew best in summer and autumn in the absence of mixing or with intermittent mixing. Mixing at 20 to 30 cm.sec^{-1} suppressed the growth of this alga.[112]

Inhibitory Substances

Clearly, not all wastewaters will be suitable for algal treatment. If the wastewater contains too high concentrations of ammonia then this will affect the performance of the plant (see above) and the wastewater must be either diluted or the feed rate reduced. Extreme concentrations of heavy metals, herbicides or pesticides may also prove toxic, and in these cases alternative treatment strategies must be used. If the wastewater is mainly domestic sewage, then it may also contain surfactants from detergents and household cleaning and personal care products etc., and these may be toxic to the algae. The toxicity of these surfactants varies with algal species and type of surfactant. Reported toxic levels vary from 0.003 to 17,784 mg.l^{-1} and it has been shown that laboratory populations are more susceptible to surfactants than natural algal populations under natural conditions.[113-115] Furthermore, cationic detergents have been shown to be more toxic than nonionic or anionic detergents.[33] Unfortunately, the amount of information on the effects of detergents on the algae in HROPs is extremely limited and their impact on HROP performance cannot be estimated at this time.

Heavy metals in the effluent may also affect the performance of the system.[116] Wong and Chang[117] have shown that Cu^{2+}, Cr^{6+} and Ni^{2+} inhibit growth, nitrate and phosphate uptake in a *Chlorella* sp. isolated from activated sludge. 2 nM Cu^{2+} was sufficient to cause a significant inhibition, while Cr^{6+} and Ni^{2+} were less inhibitory. Such concentrations of heavy metals have been observed in sewage effluent in Hong Kong. The presence of organic material such as glucose and glutamate in the wastewater, however, may reduce the toxicity of some of the heavy metals to the algae.[118] The addition of glutathione or cysteine also reduced the inhibitory effect of Cd^{2+} on *Chlorella*.[119] It has also been shown that the concentration of phosphate affects the toxicity of mercury to *Selenastrum*[120] and the toxicity of arsenic to *Scenedesmus*.[121]

The occurrence of autoinhibitory substances produced by some algae when grown at high densities has been postulated for some time.[122,123] A recent study, however, has failed to find any evidence for these in cultures of *Chlorella vulgaris*,[124] and the growth inhibition has been attributed to high bicarbonate concentrations and nutrient limitation. Re-examination of earlier studies and further studies on other strains and species are necessary to confirm whether autoinhibitors exist and how important they may be in large-scale cultures.

Predators, Parasites and Pathogens

Both HROPs and WSP contain a range of zooplankton species as well as algae and bacteria.[125] In order to maintain a stable steady state in the HROP, one must avoid the development of complex food chains. The development of rotifers, copepods, mosquito larvae, fly larvae, and other invertebrates will result in an unstable algal population at the mercy of the population dynamics of herbivorous grazing zooplankters and their predators.[76,126] This can lead to sudden collapses of the algal population. The development of these animal populations can be prevented by establishing a pond regime which leads to diurnal anaerobic conditions for a short period. Grobbelaar[127] has also suggested that acidifying the pond to pH 3.5 for a few hours is adequate to kill most rotifers and protozoa, however, this is probably not possible in large ponds.

Another potential problem is infection of the algae by parasitic fungi, mainly of the genus *Chytridium.*[128,129] These fungi also cannot complete their life cycle if anaerobic conditions prevail for several hours. Algal viruses may also be a problem although, as yet, these have not been reported in HROPs.

The bacteria in HROPs are not only CO_2-generators, but there is also evidence that heterotrophic bacteria may reduce algal growth.[130,131] However, algae produce substances inhibiting the growth of bacteria such as *Escherichia coli* and *Vibrio chlolerae.*[132-135] The main role of bacteria in HROPs is probably to break down complex organic molecules to make them available to the algae, however, the interaction between algae and bacteria and the exact role of the bacteria in these systems requires more detailed study.

Operation of High Rate Oxidation Ponds

The practical outcomes of considering some of the main limiting factors discussed above can be seen when we consider the operation of high rate oxidation ponds:

Algal growth can be described by the first order equation:

$$\frac{dX}{dt} = \mu X \qquad\qquad 7)$$

where X is the algal biomass (g.m^{-3}) and μ is the specific growth rate (day^{-1}). Equation 7 can be expressed in terms of the pond productivity:

$$P = \mu X d \qquad\qquad 8)$$

where P is the productivity (g.m^{-2}.day^{-1}) and d is the pond depth (m). Accumulation of algae in the pond operating as a completely mixed reactor can be described as:

$$\frac{dX}{dt} = \mu(X - D) \qquad\qquad 9)$$

where D is the dilution rate (sewage inflow rate per unit volume per day). At steady state $dX/dt = o$ and $\mu = D$. Thus, at steady state, Equation 8 becomes:

$$P = DXd \qquad\qquad 10)$$

Thus, the two controllable variables which determine pond productivity are dilution rate (or its reciprocal, retention time) and depth. Because of seasonal variations in temperature and light, retention time and depth may need to be varied over the various seasons if maximum productivity is to be maintained. There are four alternative strategies for doing this:[34]

1) Constant Retention Time, Area and Depth

This is the simplest alternative, however, it means that in winter the algal concentration may become too low to supply all the oxygen required for waste biodegradation through photosynthesis. This means that supplemental oxygen has to be supplied.

2) *Variable Retention Time*

A variable retention time takes into account the effects of temperature and light on algal growth. For example, at the experimental HROP operated at Haifa in Israel[136] a 2 day retention time was required in summer, and a 6 day retention time in winter. The variable retention time can be achieved by either (a) varying both area and depth, (b) varying area at constant depth, or (c) varying depth at constant area.

(a) Varying Both Area and Depth

This method was first suggested by Oswald.[137] For Californian conditions he suggested an optimum depth of 20 cm in winter and 40 cm in summer for maximal areal yields, and 75 cm in winter and 120 cm in summer for optimal sewage treatment. The problem with this method is that a pond system operating at a constant sewage inflow at 20 cm depth and 6 days retention time in winter and 40 cm depth and 2 days retention time in summer would require six times as much area in winter as in summer. The cost of constructing the extra pond area would probably make the whole operation uneconomical.[138]

(b) Varying Area at Constant Depth

This method also requires extra pond area, and has been tested at Haifa.[34,136] The pond area not required during the warmer months was used to grow fish. Although this sort of polyculture may be economical under some circumstances, it complicates operations and requires a workforce skilled both in algal culture and fish culture.

(c) Varying Depth at Constant Area

This option appears to be the most effective, both in terms of the capital cost of pond construction and in the ease of operation.

The above strategies assume that the sewage stream composition remains fairly constant over the year. Where this is not the case, a strategy must be developed to vary the retention time and/or pond depth in response to changing nutrient inputs. The above strategies also do not take account of diurnal variations in photosynthesis and pH in the pond. Nutrient removal is greatest during the day when the algae actively take up N and P, as well as by volatilization of NH_3 due to the increased pond pH in the day. Picot et al[104] have shown that a greater efficiency in nutrient removal can be achieved if the input of fresh wastewater is controlled according to the photosynthetic activity in the pond.

Effective and reliable operation of any large-scale algal culture, including HROPs, requires effective monitoring of the culture and clearly defined strategies for pond management. As previously mentioned, the concentration of ammonium is a critical, and easily monitored, factor for maintaining an effective culture. Other easily monitored variables are pH and O_2 concentration. pH in HROPs is usually between 7.5 and 10. High pH (>8.0), coupled with high concentrations of ammonia and low BOD load mean long detention times, whereas low pH (<7.5) and low ammonia concentrations mean short detention times. pH values less than about 6.0 usually lead to algal death and should be avoided. If such low pH values do occur, then the pH can be adjusted by the addition of lime.[139] With optimal photosynthesis, HROP pH values may reach > pH 10 at the pond surface without having any adverse effects on the stability of the algal community or on biomass production.

High oxygen concentrations (> 30 mg.l^{-1}) should also be avoided since these may lead to photooxidative damage to algal cells. This can be avoided with proper mixing which will increase the transfer of the supersaturated oxygen from the water to the atmosphere and will also transfer oxygen for algal and bacterial respiration below the photic zone.[78] Table 12.3 gives a general outline of suggested operation times for HROPs on domestic sewage.

Ultimately, the type of algal wastewater system and the operation system will be determined by economic factors. It is therefore always desirable to keep the effects on the cost of wastewater treatment in mind when studying algal wastewater treatment systems and in research on reducing the effects of limiting factors.

Limits in immobilized cell systems

One approach to overcoming the expense of harvesting the algal biomass as well as improving light utilization efficiency is to immobilize the algae either in carrageenan beads, alginate beads, polyvinyl foam, chitosan flakes[140-143] or on screens and surfaces.[144] As in the systems using free-living cells, the algae in these systems are light limited[143] thus requiring optically thin bioreactors. Furthermore, Garbisu et al[141] have shown that there are also diffusional limitations in foam immobilized cells.

Conclusions

Algal systems have several general limitations:

1) *They require light*. The cultures therefore have to be relatively shallow and well mixed, thus necessitating a large land area.

2) *Generation times are long*. This results in the need of long residence times and large ponds.

3) *The active biomass concentration is generally low*. This also results in long residence times and large ponds. This is further exacerbated if the prevailing temperatures are sub-optimal.

Novel solutions to the above limitations which have been proposed are the use of very concentrated cultures or a two-stage process. In hyperconcentrated cultures the algae are first concentrated (to >2.5 g dry wt.l^{-1}) before being used to

---

Table 12.3. Parameters for a stable operation of a high rate oxidation pond (domestic sewage only) based on data compiled by Abeliovich[78]

RAW SEWAGE		DETENTION TIME (DAYS)		EFFLUENT		
Ammonia (mg.l^{-1})	Total BOD (mg.l^{-1})	8-14°C	22-25°C	Algal conc. (g dry wt.l^{-1})	Ammonia (% of initial conc.)	BOD (% removed; less removed algal biomass)
80 - 90	400 - 500	10 - 12		0.25 - 0.5	70 - 80	90 - 99
80 - 90	400 - 500		4	0.5 - 1.0	80 - 90	90 - 99
40 - 50	300 - 400	6 - 7		0.1 - 0.2	40 - 50	90 - 95
40 - 50	300 - 400		2 - 3	0.2 - 0.3	40 - 50	90 - 95

remove the nutrients from the wastewater. Hyperconcentrated cultures of *Scenedesmus obliquus* were found to be very efficient at nutrient removal.[145,146] In the two-stage process the algal biomass is first grown at low nutrient levels so the cells become nutrient depleted, after which they are transferred to the high nutrient waste stream where they rapidly take up the nutrients.[147] Although both of these systems are very efficient in the laboratory, it is difficult to see how they can be scaled-up and still be cost effective.

Another alternative is to use algal culture systems with better light penetration characteristics, more efficient temperature control and better turbulence than open ponds or raceways; i.e., closed photobioreactors. The most promising of these are tubular photobioreactors, especially the helical photobioreactor or 'Biocoil'[30] and plate reactors.[148] In these reactors the light path is usually less than 30 mm and the cultures are circulated through the reactor by pumps or air. The short light path and high turbulence means that they can sustain a high biomass and productivity[149,150] and this means that the residence time needed to treat the wastewater is considerably shortened. Whether the improved productivity and shorter residence time can offset the significantly higher capital costs remains to be determined.

4) *Harvesting is difficult and costly.* If the algal biomass is not removed from the treated wastewater then the nutrients taken up by the algae will be discharged with the wastewater stream, largely nullifying the nutrient reduction effects of the algae. The need for harvesting is a major limiting factor for some algal wastewater treatment systems. Controlled autoflocculation is possible in tubular photobioreactors (unpublished results) and this may be of economic benefit.

Application of algae for wastewater treatment requires careful consideration of the type of wastewater to be treated and the prevailing climatic conditions. Sufficient information is available to allow the design of an effective and reliable plant for the treatment of wastewaters with algae. In the last three decades major advances have been made in our understanding of algal physiology and ecology and in the area of algal mass culture to allow the rational assessment of the suitability of an algal treatment process compared to alternative processes for a given waste stream and site. Algal wastewater treatment has proven to be cost-effective in many situations, but it must be recognized that these systems are not a panacea for all forms of wastewater treatment in all parts of the world. However, increased experience with algal wastewater systems, and improved system design is overcoming some of the limitations listed above and this is providing the opportunity for wider applications of algal wastewater systems.

References

1. de la Noüe J, Laliberté G, Proulx D. Algae and wastewater. J Appl Phycol 1992; 4:247-54.
2. Oswald WJ. Micro algae and waste-water treatment. In: Borowitzka MA, Borowitzka LJ, eds. Micro-algal Biotechnology. Cambridge: Cambridge University Press, 1988:305-28.
3. Gerhardt MB, Green FB, Newman RD et al. Removal of selenium using a novel algal bacterial process. Res J Water Pollut Cont Fed 1991; 63:799-805.
4. Wu XF, Kosaric N. Removal of organochlorine compounds in an upflow flocculated algae photobioreactor. Wat Sci Technol 1991; 24:221-32.
5. Lipstein B, Hurwitz S. The nutritional value of algae for poultry. Dried *Chlorella* in broiler diets. Brit Poult Sci 1980; 21:9-21.

6. Sandbank E, Hepher B. Microalgae grown in wastewater as an ingredient in the diet of warmwater fish. In: Shelef G, Soeder CJ eds. Algal Biomass. Amsterdam: Elsevier/North Holland Biomedical Press, 1980:697-706.

7. Wharton RA, Smernoff DT, Averner MM. Algae in space. In: Lembi CA, Waaland JR eds. Algae and Human Affairs. Cambridge: Cambridge University Press, 1988:485-509.

8. Oguchi M, Otsubo K, Nitta K, et al. Closed and continuous algae cultivation system for food production and gas exchange in CELSS. Adv Space Res 1989; 9:169-77.

9. Laws EA, Berning JL. A study of the energetics and economics of microalgal mass culture with the marine chlorophyte *Tetraselmis suecica*-Implications for use of power plant stack gases. Biotechnol Bioeng 1991; 37:936-47.

10. Anon. New micro-alga for fixing carbon dioxide. Japan Patent 8009963. 1996.

11. Kishimoto M, Okakura T, Nagashima H et al. CO_2 fixation and oil production using micro-algae. J Ferment Bioeng 1994; 78:479-82.

12. Negoro M, Shioji N, Miyamoto K et al. Growth of microalgae in high CO_2 gas and effects of SOx and NOx. Appl Biochem Biotechnol 1991; 28-9:877-86.

13. Takano H, Takeyama H, Nakamura N et al. CO_2 removal by high-density culture of a marine cyanobacterium *Synechococcus* sp using an improved photobioreactor employing light-diffusing optical fibers. Appl Biochem Biotechnol 1992; 34-5:449-58.

14. Oswald WJ, Gotaas HB. Photosynthesis in sewage treatment. Trans Am Soc Civ Eng 1957; 122:73-105.

15. Oswald WJ. Large-scale algal culture systems (engineering aspects). In: Borowitzka MA, Borowitzka LJ eds. Micro-Algal Biotechnology. Cambridge: Cambridge University Press, 1988:357-94.

16. Martin C, de la Noüe J, Picard G. Intensive culture of freshwater microalgae on aerated pig manure. Biomass 1985; 7:245-59.

17. Cañizares-Villanueva RO, Ramos A, Lemus R et al. Growth of *Phormidium* sp in aerobic secondary piggery waste-water. Appl Microbiol Biotech 1994; 42:487-91.

18. Svoboda IF, Fallowfield HJ. An aerobic piggery slurry treatment system with integrated heat recovery and high rate algal ponds. Wat Sci Technol 1989; 21: 277-87.

19. Tanticharoen M, Bunnag B, Vonshak A. Cultivation of *Spirulina* using secondary treated starch wastewater. Australas Biotechnol 1993; 3:223-6.

20. Rodrigues AM, Oliviera JFS. Treatment of wastewaters from the tomato concentrate industry in high rate algal ponds. Wat Sci Technol 1987; 19:43-9.

21. Rose PD, Maart BA, Dunn KM et al. High rate algal oxidation ponding for the treatment of tannery effluents. Wat Sci Technol 1996; 33:219-27.

22. Phang SM, Ong KC. Algal biomass production on digested palm oil mill effluent. Biological Wastes 1988; 25:177-91.

23. Geeta PK, Phang SM, Hashim MA et al. Rubber effluent treatment in a high-rate algal pond system. In: Phang SM, Lee YK, Borowitzka MA et al, eds. Algal Biotechnology in the Asia-Pacific Region. Kuala Lumpur: Institute of Advanced Studies, University of Malaya, 1994: 306-12.

24. Aksu Z, Kutsal T. The usage of *Chlorella vulgaris* in waste water treatment containing heavy metal ions. Proc 4th Eur Cong Biotech 1987; 2:80-3.

25. Mahan CA, Holcombe JA. Immobilization of algae cells on silica gel and their characterization for trace metal preconcentration. Analyt Chem 1992; 64:1933-9.

26. Mallick N, Rai LC. Influence of culture density, pH, organic acids and divalent cations on the removal of nutrients and metals by immobilized *Anabaena doliolum* and *Chlorella vulgaris*. World J Microbiol Biotechnol 1993; 9:196-201.

27. Wilde EW, Benemann JR. Bioremoval of heavy metals by the use of microalgae. Biotechnol Adv 1993; 11:781-812.

28. Corder SL, Reeves M. Biosorption of nickel in complex aqueous waste streams by cyanobacteria. Appl Biochem Biotechnol 1994; 45-6:847-59.
29. Oswald WJ. Design basis for facultative and high rate ponds. In: Borowitzka MA, Mathew K eds. Waste Treatment by Algal Cultivation. Perth: Murdoch University, 1991:1-10.
30. Borowitzka MA. Closed algal photobioreactors: design considerations for large-scale systems. J Mar Biotechnol 1996; 4:185-91.
31. Borowitzka MA. Products from Algae. In: Phang SM, Lee K, Borowitzka MA, et al. eds. Algal Biotechnology in the Asia-Pacific Region. Kuala Lumpur: Institute of Advanced Studies, University of Malaya, 1994:5-15.
32. Borowitzka LJ. Commercial *Dunaliella* production: history of development. In: Villa TG, Abalde J, eds. Profiles on Biotechnology. Santiago de Compostela: Universidade de Compostela, 1992: 233-45.
33. Lewis MA. Chronic toxicities of surfactants and detergent builders to algae - A review and risk assessment. Ecotoxicol Env Safety 1990; 20:123-40.
34. Azov Y, Shelef G, Moraine R, et al. Alternative operating strategies for high-rate sewage oxidation ponds. In: Shelef G, Soeder CJ, eds. Algal Biomass. Amsterdam: Elsevier/North Holland Biomedical Press, 1980:523-9.
35. Abeliovich A. Factors limiting algal growth in high-rate oxidation ponds. In: Shelef G, Soeder CJ, eds. Algal Biomass. Amsterdam: Elsevier/North Holland Biomedical Press, 1980:205-15.
36. Richmond A. The challenge confronting industrial microalgaculture: High photosynthetic efficiency in large-scale reactors. Hydrobiologia 1987; 151/152: 117-21.
37. Fontes AG, Vargas MA, Moreno J et al. Factors affecting the production of biomass by a nitrogen-fixing blue-green alga in outdoor culture. Biomass 1987; 13: 33-43.
38. Richmond A. Large scale microalgal culture and applications. Prog Phycol Res 1990; 7:269-330.
39. Tamiya H, Hase E, Shibata K et al. Kinetics of growth of *Chlorella*, with special reference to its dependance on quantity of available light and on temperature. In: Burlew JS ed. Algal Culture. From Laboratory to Pilot Plant. Washington DC: Carnegie Institution of Washington, 1953:204-32.
40. Richmond A. Open systems for the mass production of photoautotrophic microalgae outdoors - Physiological principles. J Appl Phycol 1992; 4:281-6.
41. Vonshak A, Abeliovich A, Boussiba S et al. Production of *Spirulina* biomass: effects of environmental factors and population density. Biomass 1982; 2:175-85.
42. Kirk JTO, Light and Photosynthesis in Aquatic Ecosystems. Cambridge: Cambridge University Press, 1983.
43. Kroon BMA, Ketelaars HAM, Fallowfield HJ et al. Modelling microalgal productivity in a high rate algal pond based on wavelength dependent optical properties. J Appl Phycol 1989; 1:247-56.
44. Falkowski PG, Greene R, Kolber Z. Light utilization and photoinhibition of photosynthesis in marine phytoplankton. In: Baker NR, Bowyer JR, eds. The Photo-Inhibition of Photosynthesis: From Molecular Mechanisms to the Field. Oxford: Bios Scientific Publishers, 1994: 407-32.
45. Grande KD, Bender ML, Irwin B et al. A comparison of net and gross rates of oxygen production as a function of light intensity in some natural plankton populations and in a *Synechococcus* culture. J Plankt Res 1991; 13:1-16.
46. Richmond A, Vonshak A. *Spirulina* culture in Israel. Arch Hydrobiol 1978; 11:274-80.

47. Richmond A, Grobbelaar JU. Factors affecting the output rate of *Spirulina platensis* with reference to mass cultivation. Biomass 1986; 10:253-64.
48. Hu Q, Richmond A. Optimising the population density in *Isochrysis galbana* grown outdoors in a glass column photobioreactor. J Appl Phycol 1994; 6:391-6.
49. Richmond A. Efficient utilization of high irradiance for production of photoautotrophic cell mass: a survey. J Appl Phycol 1996; 8:381-7.
50. Laws EA, Terry KL, Wickman J et al. A simple algal production system designed to utilize the flashing light effect. Biotechnol Bioeng 1983; 25:2319-35.
51. Terry KL. Photosynthesis in modulated light: quantitative dependence of photosynthetic enhancement on flashing rate. Biotechnol Bioeng 1986; 28:988-95.
52. Grobbelaar JU, Nedbal L, Tichy V. Influence of high frequency light/dark fluctuations on photosynthetic characteristics of microalgae photoacclimated to different light intensities and implications for mass algal cultivation. J Appl Phycol 1996; 8:335-43.
53. Doucha J, Livansky K. Novel outdoor thin-layer high density microalgal culture system: Productivity and operational parameters. Algol Stud 1995; 76:129-47.
54. Jitts HR, McAllister CD, Stephens K et al. The cell division rates of some marine phytoplankters as a function of light and temperature. J Fish Res Bd Canada 1964; 21:139-57.
55. Dauta A, Devaux J, Piquemal F et al. Growth rate of four freshwater algae in relation to light and temperature. Hydrobiologia 1990; 207:221-6.
56. Talbot P, Thébault JM, Dauta A et al. A comparative study and mathematical modeling of temperature, light and growth of three microalgae potentially useful for wastewater treatment. Wat Res 1991; 25:465-72.
57. Castillo J, Merino F, Heussler P. Production and ecological implications of algae mass culture under Peruvian conditions. In: Shelef G, Soeder CJ, eds. Algal Biomass. Amsterdam: Elsevier/North-Holland Biomedical Press, 1980:123-34.
58. De Pauw N, Verlet H, De Leenheer L. Heated and unheated outdoor cultures of marine algae with animal manure. In: Shelef G, Soeder CJ eds. Algal Biomass. Amsterdam: Elsevier/North Holland Biomedical Press, 1980:315-41.
59. Richmond A, Vonshak A, Arad S. Environmental limitations in outdoor production of algal biomass. In: Shelef G, Soeder CJ, eds. Algal Biomass. Amsterdam: Elsevier/North Holland Biomedical Press, 1980:65-72.
60. Pouliot Y, de la Noüe J. Development of a pilot-scale facility for wastewater treatment and microalgae production (In French). Rev Franc Sci de L'eau 1985; 4:207-22.
61. Guterstam B, Todd J. Ecological engineering for wastewater treatment and its application in New England and Sweden. Ambio 1990; 19:173-5.
62. Lee Y-K, Tan H-M, Hew C-S. The effect of growth temperature on the bioenergetics of photosynthetic algal cultures. Biotechnol Bioeng 1985; 27:555-61.
63. Bedell GW. Stimulation of commercial algal biomass production by the use of geothermal water for temperature control. Biotechnol Bioeng 1985; 27:1063-6.
64. Torzillo G, Sacchi A, Materassi R et al. Effect of temperature on yield and night biomass loss in *Spirulina platensis* grown outdoors in tubular photobioreactors. J Appl Phycol 1991; 3:103-9.
65. Torzillo G, Sacchi A, Materassi R. Temperature as an important factor affecting productivity and night biomass loss in *Spirulina platensis* grown outdoors in tubular photobioreactors. Bioresource Technol 1991; 38:95-100.
66. Cromar NJ, Fallowfield HJ. Separation of components of the biomass from high rate algal ponds using percoll density gradient centrifugation. J Appl Phycol 1992; 4:157-63.

67. Geider RJ, Osborne BA. Respiration and microalgal growth: a review of the quantitative relationship between dark respiration and growth. New Phytol 1989; 112:327-41.
68. Martin NJ, Fallowfield HJ. Computer modelling of algal waste treatment systems. Wat Sci Technol 1989; 21:1657-60.
69. Grobbelaar JU, Soeder CJ, Stengel E. Modeling algal productivity in large outdoor cultures and waste treatment systems. Biomass 1990; 21:297-314.
70. Harris GP. Photosynthesis, productivity and growth: The physiological ecology of phytoplankton. Arch Hydrobiol Beih, Ergebn Limnol 1978; 10:1-171.
71. Grobbelaar JU, Soeder CJ. Respiration losses in planktonic green algae cultivated in raceway ponds. J Plankt Res 1985; 7:497-506.
72. Soeder CJ. Massive cultivation of microalgae: Results and prospects. Hydrobiologia 1980; 72:197-209.
73. Berner T, Dubinsky Z, Schanz F et al. The measurement of primary productivity in a high-rate oxidation pond (HROP). J Plankt Res 1986; 8:659-72.
74. Fallowfield HJ, Mesple F, Martin NJ et al. Validation of computer models for high rate algal pond operation for wastewater treatment using data from Mediterranean and Scottish pilot scale systems-Implications for management in coastal regions. Wat Sci Technol 1992; 25:215-24.
75. Guterman H, Vonshak A, Ben-Yaakov S. A macromodel for outdoor algal mass production. Biotechnol Bioeng 1990; 35:809-19.
76. Mesple F, Casellas C, Troussellier M et al. Some difficulties in modelling chlorophyll a evolution in a high rate algal pond ecosystem. Ecol Mod 1995; 78: 25-36.
77. Palmer CM. A composite rating of algae tolerating organic loading. J Phycol 1969; 5:78-81.
78. Abeliovich A. Algae in wastewater oxidation ponds. In: Richmond A, ed. CRC Handbook of Microalgal Mass Culture. Boca Raton: CRC Press, 1986:331-8.
79. Abeliovich A, Weisman D. Role of heterotrophic nutrition in growth of the alga *Scenedesmus obliquus* in high rate oxidation ponds. Appl Env Microbiol 1978; 35:32-7.
80. Marquez FJ, Nishio N, Nagai S et al. Enhancement of biomass and pigment production during growth of *Spirulina platensis* in mixotrophic culture. J Chem Tech Biotechnol 1995; 62:159-64.
81. Neilson AH, Lewin RA. The uptake and utilization of organic carbon by algae: an essay in comparative biochemistry. Phycologia 1974; 13:227-64.
82. Cid A, Abalde J, Herrero C. High yield mixotrophic cultures of the marine microalga *Tetraselmis suecica* (Kylin) Butcher (Prasinophyceae). J Appl Phycol 1992; 4:31-7.
83. Burrell RE, Mayfield CI, Inniss WE. Biomass production from the green algae *Chlorella vulgaris* and *Ankistrodesmus braunii* cultured heterotrophically. Biotech Lett 1984; 6:507-10.
84. Martinez F, Avendaño MC, Marco E et al. Algal population and auxotrophic adaptation in a sugar refinery wastewater environment. J Gen Appl Microbiol Tokyo 1987; 33:331-41.
85. Martinez F, Orús MI. Interactions between glucose and inorganic carbon metabolism in *Chlorella vulgaris* strain UAM-101. Plant Physiol 1991; 95:1150-5.
86. Ogawa T, Aiba S. Bioenergetic analysis of mixotrophic growth in *Chlorella vulgaris* and *Scenedesmus acutus*. Biotechnol Bioeng 1981; 23:1121-32.
87. Lau PS, Tam NFY, Wong YS. Influence of organic-N sources on an algal wastewater treatment system. Resources Conservation and Recycling 1994; 11:197-208.

88. Abeliovich A, Azov Y. Toxicity of ammonia to algae in sewage oxidation ponds. Appl Env Microbiol 1976; 31:801-6.

89. Azov Y, Goldman JC. Free ammonia inhibition of algal photosynthesis in intensive cultures. Appl Env Microbiol 1982; 43:735-9.

90. Thomas WH, Hastings J, Fujita M. Ammonium input to the sea via large sewage outfalls. 2. Effects of ammonia on growth and photosynthesis of southern California phytoplankton cultures. Mar Env Res 1980; 3:291-6.

91. Borowitzka MA, Borowitzka LJ. Limits to growth and carotenogenesis in laboratory and large-scale outdoor cultures of *Dunaliella salina*. In: Stadler T, Mollion J, Verdus MC et al, eds. Algal Biotechnology. Barking: Elsevier Applied Science, 1988:371-81.

92. Konig A, Pearson HW, Silva SA. Ammonia toxicity to algal growth in waste stabilisation ponds. Wat Sci Technol 1987; 19:115-22.

93. Pouliot Y, Buelna G, Racine C et al. Culture of cyanobacteria for tertiary wastewater treatment and biomass production. Biological Wastes 1989; 29:81-91.

94. Talbot P, de la Noüe J. Tertiary treatment of wastewater with *Phormidium bohneri* (Schmidle) under various light and temperature conditions. Wat Res 1993; 27:153-9.

95. Darley WM. Algal biology: A Physiological Approach. Basic Microbiology Vol. 9, Oxford: Blackwell Scientific Publications, 1982:168.

96. Kunikane S, Kaneko M. Growth and nutrient uptake of green alga *Scenedesmus dimorphus*, under a wide range of nitrogen/phosphorus ratio. II. Kinetic model. Wat Res 1984; 18:1313-26.

97. Tam NFY, Wong YS. Feasibility of using *Chlorella pyrenoidosa* in the removal of inorganic nutrients from primary settled sewage. In: Phang SM, Lee YK, Borowitzka MA et al, eds. Algal Biotechnology in the Asia-Pacific Region. Kuala Lumpur: Institute of Advanced Studies, University of Malaya, 1994:291-9.

98. Lefebvre S, Hussenot J, Brossard N. Water treatment of land-based fish farm efluents by outdoor culture of marine diatoms. J Appl Phycol 1996; 8:193-200.

99. Langis R, Couture P, de la Noüe J et al. Induced responses on algal growth and phosphate removal by three molecular weight DOM fractions from a secondary effluent. J Wat Pollut Contr Fed 1986; 58:1073-7.

100. Abeliovich A. The effect of unbalanced ammonia and BOD concentrations on oxidation ponds. Wat Res 1983; 17: 299-305.

101. de la Noüe J, Clouthier-Mantha L, Walsh P et al. Influence of agitation and aeration modes on biomass production by *Oocystis* sp. grown on wastewaters. Biomass 1984; 4:43-58.

102. Yahi H, Elmaleh S, Coma J. Algal flocculation-sedimentation by pH increase in a continuous reactor. Wat Sci Technol 1994; 30:259-67.

103. Sukenik A, Shelef G. Algal autoflocculation-Verfication and proposed mechanism. Biotechnol Bioeng 1984; 26:142-7.

104. Picot B, Moersidik S, Casellas C et al. Using diurnal variations in a high rate algal pond for management pattern. Wat Sci Technol 1993; 28:169-75.

105. Lessard P, Proulx D, Delanoue J. Nutrient removal using cyanobacteria (Phormidium bohneri): Experimental results with a batch reactor. Wat Sci Technol 1994; 30:365-8.

106. Sukenik A, Falkowski PG, Bennett J. Potential enhancement of photosynthetic energy conversion in algal mass culture. Biotechnol Bioeng 1987; 30:970-7.

107. Manabe E, Hirosawa T, Tsuzuki M et al. Effect of near ultraviolet on growth of *Chlorella* cells. Physiol Plant 1986; 67:598-603.

108. Grobbelaar JU. Turbulence in mass algal cultures and the role of light dark fluctuations. J Appl Phycol 1994; 6:331-5.

109. Mitsuhashi S, Hosaka K, Tomonaga E et al. Effects of shear flow on photosynthesis in a dilute suspension of microalgae. Appl Microbiol Biotech 1995; 42:744-9.
110. Gudin C, Chaumont D. Cell fragility-the key problem of microalgae mass production in closed photobioreactors. Bioresource Technol 1991; 38:145-51.
111. Azov Y, Shelef G, Moraine R et al. Controlling algal genera in high rate wastewater oxidation ponds. In: Shelef G, Soeder CJ, eds. Algal Biomass. Amsterdam: Elsevier/North Holland Biomedical Press, 1980:245-53.
112. Lincoln EP, Koopman B, Hall TS. Control of a unicellular blue-green alga, *Synechocystis* sp. in mass algal culture. Aquacult 1984; 42:349-60.
113. Ukeles R. Inhibition of unicellular algae by synthetic surface active agents. J Phycol 1965; 1:102-10.
114. Yamane A, Okada M, Sudo R. The growth inhibition of planktonic algae due to surfactants used in washing detergents. Wat Res 1984; 9:1101-5.
115. Wängberg S, Blanck H. Multivariate patterns of algal sensitivity to chemicals in relation to phylogeny. Ecotoxicol Env Safety 1988; 16:72-82.
116. Wong SL, Wainwright JF, Pimenta J. Quantification of total and metal toxicity in wastewater using algal bioassays. Aquat Toxicol 1995; 31:57-75.
117. Wong PK, Chang L. Effects of copper, chromium and nickel ions on inorganic nitrogen and phosphorus uptake in *Chlorella* species. Microbios 1991; 67:107-15.
118. Mohanty RC, Mohanty L, Mohapatra PK. Change in toxicity effect of mercury at static concentration to *Chlorella vulgaris* with addition of organic carbon Sources. Acta Biologica Hungarica 1993; 44:211-22.
119. Kaplan D, Heimer YM, Abeliovich A et al. Cadmium toxicity and resistance in *Chlorella* sp. Plant Sci 1995; 109:129-37.
120. Chen CY. Theoretical evaluation of the inhibitory effects of mercury on algal growth at various orthophosphate levels. Wat Res 1994; 28:931-7.
121. Chen FH, Chen WQ, Dai SG. Toxicities of four arsenic species to *Scenedesmus obliquus* and influence of phosphate on inorganic arsenic toxicities. Toxicological and Environmental Chemistry 1994; 41:1-7.
122. Scott J. Autoinhibitor production by *Chlorella vulgaris*. Amer J Bot 1964; 51:581-4.
123. Pratt R, Fong J. Studies on *Chlorella vulgaris*. II. Further evidence that *Chlorella* cells form a growth-inhibiting substance. Amer J Bot 1940; 27:431-6.
124. Mandalam RK, Palsson BO. *Chlorella vulgaris* (Chlorellaceae) does not secrete autoinhibitors at high cell densities. Amer J Bot 1995; 82:955-63.
125. Canovas S, Casellas C, Picot B et al. Evolution annuelle du peuplement zooplanktonique dans un lagunage à haut rendement et incidence du temps de séjour. Rev Sci l'Eau 1991; 4:263-83.
126. Groeneweg J, Klein B, Mohn FH et al. First results of outdoor treatment of pig manure with algal-bacterial systems. In: Shelef G, Soeder CJ, eds. Algal Biomass. Amsterdam: Elsevier/North Holland Biomedical Press, 1980:255-64.
127. Grobbelaar JU. Infections: Experiences in Miniponds. UOFS Publ 1981; 3:116-23.
128. Abeliovich A, Dickbuck S. Factors affecting infection of *Scenedesmus obliquus* by a *Chytridium* sp. in sewage oxidation ponds. Appl Env Microbiol 1977; 34:32-7.
129. Payer HD, Pithakpol B, Nguitragool M et al. Major results of the Thai-German microalgae project at Bangkok. Arch Hydrobiol Beih 1978; 11:41-55.
130. Berger PS, Rho J, Gunner HB. Bacterial suppression of *Chlorella* by hydroxylamine production. Wat Res 1979; 13:267-73.
131. Dor I, Svi B. Effect of heterotrophic bacteria on the green algae growing in wastewater. In: Shelef G, Soeder CJ eds. Algal Biomass. Amsterdam: Elsevier/North-Holland Biomedical Press, 1980: 421-9.
132. Dor I. Effect of the green algae isolated from wastewater on the activity of sewage bacteria. In: Shelef G, Soeder CJ eds. Algal Biomass. Amsterdam: Elsevier/North-Holland Biomedical Press, 1980: 431-5.

133. Kobbia IA, Zaki D. Biological evaluation of algal filtrates. Planta Med 1976; 30:90-2.
134. Mezrioui N, Oudra B, Oufdou K, et al. Effect of microalgae growing on wastewater batch culture on *Escherichia coli* and *Vibrio cholerae* survival. Wat Sci Technol 1994; 30:295-302.
135. Sebastian S, Nair KVK. Total removal of coliforms and *E. coli* from domestic sewage by high-rate pond mass culture of *Scenedesmus obliquus*. Environ Pollut 1984; 34:197-206.
136. Shelef G, Azov Y, Moraine R et al. Algal mass production as an integral part of a wastewater treatment and reclamation system. In: Shelef G, Soeder CJ, eds. Algal Biomass. Amsterdam: Elsevier/North Holland Biomedical Press, 1980: 163-89.
137. Oswald WJ. Growth characteristics of microalgae in domestic sewage: environmental effects on productivity: Proceedings of the IBP/PP Technical Meeting, 1969.
138. Borowitzka MA. Algal biotechnology products and processes: Matching science and economics. J Appl Phycol 1992; 4:267-79.
139. Carberry JB, Brunner CM. Predictions of diurnal fluctuations in an algal bacterial clay wastewater treatment system. Wat Sci Technol 1991; 23:1553-61.
140. de la Noüe J, Proulx D. Biological tertiary treatment of urban wastewater with chitosan-immobilized *Phormidium*. Appl Microbiol Biotech 1988; 29:292-7.
141. Garbisu C, Hall DO, Llama MJ, et al. Inorganic nitrogen and phosphate removal from water by free-living and polyvinyl-immobilized *Phormidium laminosum* in batch and continuous-flow bioreactors. Enzyme Microb Technol 1994; 16:395-401.
142. Tam NFY, Lau PS, Wong YS. Wastewater inorganic N and P removal by immobilized *Chlorella vulgaris*. Wat Sci Technol 1994; 30:369-74.
143. Travesio L, Benitez F, Dupeiron R. Sewage treatment using immobilized microalgae. Bioresource Technol 1992; 40:183-7.
144. Kaya VM, de la Noüe J, Picard G. A comparative study of four systems for tertiary wastewater treatment by *Scenedesmus bicellularis*: new technology for immobilization. J Appl Phycol 1995; 7:85-95.
145. Lavoie A, de la Noüe J. Hyperconcentrated culture of *Scenedesmus obliquus*: a new approach for wastewater tertiary treatment? Wat Res 1985; 19:1437-42.
146. Chevalier P, de la Noüe J. Efficiency of immobilized hyperconcentrated algae for ammonium and orthophosphate removal from wastewaters. Biotech Lett 1985; 7:395-400.
147. de la Noüe J, Picard GA, Piette JP et al. Utilisation de l'algue *Oocystis* sp. pour le traitement tertiare des eaux usees II. Effet du conditionnement prealable des cellules ex cyclostat sur leur vitesse de prise en charge de l'azote lors d'incubations de longue duree. Wat Res 1980; 14:1125-30.
148. Storandt R, Farber I, Pulz O et al. Procedure and pilot plant for the disposal of inorganic loads from circulating water in aquaculture by microalgal cultivation. In: Kretschmer P, Pulz O, Gudin C et al. eds. 2nd European Workshop Biotechnology of Microalgae. Bergholz/ Rehbrücke: Institut fur Getreideverarbeitung GmbH, 1995:101-4.
149. Hu Q, Guterman H, Richmond A. A flat inclined modular photobioreactor for outdoor mass cultivation of photoautotrophs. Biotechnol Bioeng 1996; 51:51-60.
150. Richmond A, Boussiba S, Vonshak A et al. A new tubular reactor for mass production of microalgae outdoors. J Appl Phycol 1993; 5:327-32.
151. Jassby AD, Platt T. Mathematical formulation of the relationship between photosynthesis and light for phytoplankton. Limnol Oceanogr 1976; 21:540-7.
152. Platt T, Gallegos CL. Modelling primary production. In: Falkowski PG, ed. Primary Production in the Sea. NY: Plenum Press, 1980:339-51.
153. Richmond A. Outdoor mass culture of microalgae. In: Richmond A ed. CRC Handbook of Microalgal Mass Culture. Boca Raton: CRC Press, 1986:285-329.

Index